I0047643

PHARMACOGNOSY

PHARMACOGNOSY

Associate Professor in Botany
Holy Cross College (Autonomous)
Trichy, Tamil Nadu

MJP Publishers

Honour Copyright
&
Exclude Piracy

This book is protected by copyright. Reproduction of any part in any form including photocopying shall not be done except with authorization from the publisher.

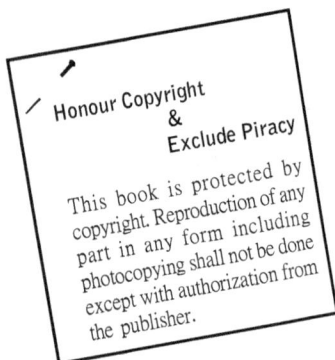

Cataloguing-in-Publication Data

Roseline, A (1956-   ).
    Pharmacognosy / by A. Roseline. –
Chennai : MJP Publishers, 2011
    xvi, 392 p.; 24 cm.
    Includes Glossary, References and Index.
    ISBN 978-81-8094-120-7 (hb.)
    1. Pharmacognosy   I. Title.
615.321 dc 22 ROS MJP 090

ISBN 978-81-8094-120-7
© Publishers, 2011
All rights reserved
Printed and bound in India

MJP PUBLISHERS
47, Nallathambi Street
Triplicane
Chennai 600 005

Publisher   : J.C. Pillai
Managing Editor   : C. Sajeesh Kumar
Marketing Manager   : S.Y. Sekar
Project Editor   : P. Parvath Radha
Acquisitions Editor : C. Janarthanan
Editorial Team   : B. Ramalakshmi, Lissy John,
                        M. Gnanasoundari, N. Thilagavathi
CIP Data   : Prof. K. Hariharan, Librarian
                        RKM Vivekananda College, Chennai.

This book has been published in good faith that the work of the author is original. All efforts have been taken to make the material error-free. However, the author and publisher disclaim responsibility for any inadvertent errors.

PREFACE

The urgent need for a standard text on pharmacognosy in simple language was greatly felt by the teachers and students of botany. This prompted me to write this book on pharmacognosy.

My teaching experience of more than thirty years has urged me to simplify the explanation of the concepts.

The book discusses the classification, preparation and storage of crude drugs, phytoconstituents of therapeutic values with suitable examples, drug adulteration and method of evaluation, systematic analysis of powdered drugs, different fibres, sutures and surgical dressings, and pharmaceutical products of plant and animal origin. Each chapter is provided with relevant questions of different types for the benefit of the students.

Special attention has been given to information on various histochemical tests for authentic identification of different phytoconstituents of natural origin and gross anatomical features of interest that are currently receiving so much attention in research laboratories.

In addition, a glossary of technical terms relevant to pharmacognosy is given at the end for easy reference.

Every effort has been made to provide detailed information, even if not comprehensive, on the plants used as drugs, in simple language, well-supported with suitable illustrations.

This book would adequately meet the needs of the students and will also serve as an excellent reference for faculty members, undergraduate and postgraduate students and research scholars of all Indian universities and other workers in areas of similar interest.

I wish to express my deep sense of gratitude to Dr. (Sr.) Sarguna, Principal, Holy Cross College (Autonomous), for her enthusiastic encouragement and unwavering support and also for providing me the facilities at all levels, which helped me to bring out this book in its present form.

It is my pleasant and heartfelt duty to thank my own beloved sisters Dr. (Sr.) John Marie, Retired Professor and Head of Tamil Department, Holy Cross College (Autonomous), Trichy, and Dr. (Sr.) Metti, Professor of Theology, whose constant prayerful support and blessings have always prompted me to pursue deep-rooted academic excellence.

I owe a great deal of gratitude to Dr. Miss. R. Parvathi, Head of the Department of Botany, Holy Cross College, for her constant motivation and timely help rendered during the preparation of the manuscript.

My sincere thanks are also due to my department colleagues, supportive staff and all my well wishers, who have freely accorded their cooperation in various capacities during the preparation of this book.

I whole-heartedly appreciate and acknowledge the painstaking sincere effort of my Department colleague Miss. E. Neelamathi in the correction of manuscript with great perseverance.

I also wish to record a special thanks to my friend Dr. Miss. D. Radha, Associate Professor of Economics of my college, whose support and untiring inspirations have played a major catalyst in the onward march of this book at every step.

I am very much indebted to my beloved husband Mr. S. Xavier and my loving son X. Infant Reni, for their fullest cooperation in many invisible ways and also for their endurance during the preparation of this text.

I thank and appreciate the all-round cooperation and support of the members of MJP Publishers for publishing the book in such a nice form with meticulous care.

Above all I thank God Almighty for his abundant blessings showered on me to complete this book successfully.

Suggestions for improvement of this book will be greatly accepted and acknowledged.

A. Roseline

CONTENTS

1

INTRODUCTION

DEFINITION

Pharmacognosy may be defined as a branch of bioscience (applied science), which deals with biological, biochemical and therapeutic features of crude drugs obtained from plants and animals and from mineral origin.

In short, it is an objective study of crude drugs from natural sources, which are treated scientifically, and it encompasses the knowledge of the history, distribution, cultivation, collection, processing for market and preservation. It is also the study of physical, chemical and structural characters and use of crude drugs.

Pharmacognosy also includes study of other materials used in pharmacy such as suspending, disintegrating and flavouring agents, filtering aids, etc. and substances like antibiotics, allergens, hallucinogenic and poisonous plants, immunizing agents, etc.

The word "Pharmacognosy" was coined by a German scientist, C.A. Seydler, at Halle, Germany, in 1815. (He coined the term pharmacognosy, while studying Sarsaparilla, in his work *Analecta Pharmacognostica*).

The word pharmacognosy is derived from two Greek words *Pharmacon* meaning "a drug" and *gnosis* meaning "to acquire knowledge".

HISTORY OF PHARMACOGNOSY

The origin of pharmacognosy is somewhat obscure, for no one knows when crude drugs of animal and plant origin had been in use. The use of plants as sources of medicinal agents lies deep in the roots of antiquity. No one will ever know, what actually led primitive humans to select certain plant materials like roots, leaves, barks, fruits and even plant exudates for the treatment of various ailments and diseases.

The three important basic needs of life—food, clothing and shelter—and a host of other useful products are supplied to humans by the plant kingdom. Nature has provided a complete storehouse of remedies to cure all ailments of humankind.

Human beings are affected by more diseases than any other animal species. Therefore, at the very dawn of civilization itself, humans would have been trying to alleviate their sufferings from injury and disease by using plants growing around them. Therefore in the past, almost all medicines used were mostly from the plants.

Today, there is a vast store of knowledge concerning therapeutic properties of different plants. Therefore, the history of herbal medicines is as old as human civilization. The documents, many of which are of great antiquity, revealed that plants were used medicinally in China, India, Egypt and Greece, long before the beginning of the Christian era. Most of the medicinally active substances identified in the 19th and the 20th centuries were used in the form of crude extract. In China, many medicinal plants had been in use since 5000 BC. The oldest known book on herbs is *Pen-t-Sao* written by Emperor Shen Nung in 3000 BC. It contains 365 drugs, one for each day of the year.

Our Indian population mostly depends on the Indian System of Medicine such as "Ayurveda", the ancient science of life and the Unani system. The well-known treatises in Ayurveda are the Charaka Samhita and the Sushruta Samhita. Galen was the first pharmacist, who had the number of pain-relieving materials including opium in his apothecary shop.

In the 19th century, the term *Materia Medica* was used for the subject now known as "Pharmacognosy". The knowledge about the medicinal use of the plants was passed onto others by word of mouth. Later, as different civilizations developed, humans were able to communicate their knowledge and ideas, first by carving into stones or clay and later by writing on parchment or paper with the result that their knowledge became known to much later generations.

Many of the crude drugs used during those civilizations are still used in the world today and to these the other plant materials are added continuously for the welfare of human beings.

CLASSICAL PHARMACOGNOSY

Religious beliefs have frequently influenced the use of plants, the most significant in Europe being the "Doctrine of Signatures" introduced by Paracelsus, a Swiss physician, who lived during the period 1493–1541. This "Doctrine of Signatures" had considerable influence on the use of plants in medicine throughout the world and its impact is still seen today.

REASON FOR STUDYING PHARMACOGNOSY

Whether the crude drugs were being used on a rational basis or not, there is no doubt that the development of medicine as a profession in the 12th century led to an increased trade in crude drugs which increased as the years went by. Trade and commerce have always provided opportunities for unscrupulous individuals to indulge in fraudulent activities; the trade in crude drugs was also not an exception to this. Thus came into existence, the practice of adulteration, for example, the drug opium was mixed (adulterated) with gum or latex of *Lactuca*.

It was, therefore, felt necessary to prepare descriptions of plants and animals used for medicinal purposes and of the crude drugs obtained from them. One of the most outstanding works in this direction was the publication of the four volumes of medicinal plants written by Bently, which remained the standard book of reference for many years.

DEVELOPMENT OF MODERN PHARMACOGNOSY

With the advent of European scientific methods, many of the reputed medicinal plants came under chemical scrutiny, leading to the isolation of active principles. Therefore many of the well-known medicinal plants were chemically analysed and their active principles were characterized. Soon after their isolation and characterization, these compounds, either in pure state or in the form of extracts, became part of Pharmacopoeias of several countries.

Starting from the year 1815, there was rapid growth in the subject of pharmacognosy. This was due to development of various methods of isolation and characterization of active principles of crude drugs. Morphine was isolated from opium poppy (*Papaver somniferum*) by Segium in 1804 and was introduced as a Medicine in 1818. Similarly quinine was isolated by Pelletier in 1820 from cinchona and it was introduced as an antimalarial drug in 1825. Penicillin was isolated from a fungus *Penicillium notatum* by Alexander Flemming in 1922. Thus pharmacists started paying attention to the chemical constituents of different drugs which are used in medicine. In recent years, importance is being given to the preparation of semi-synthetic derivatives and synthetic substitutes of naturally occurring drugs. Modern pharmacognosy, therefore, grows with the growth of other sciences like chemistry, biochemistry, pharmacology, phytochemistry, etc.

SCOPE AND IMPORTANCE OF PHARMACOGNOSY

The scope of pharmacognosy includes a detailed study of

1. crude drugs obtained from plants, animals and minerals, and
2. other substances used by doctors and pharmacists like synthetic sutures and surgical dressings.

Pharmacognosy is an applied science that has played a crucial role in the development of different disciplines of science. Its importance can be understood from the following.

  i. It is concerned with an enormous variety of substances that are synthesized and accumulated by plants, animals, minerals and other substances and elucidation of these substances.

 ii. The technology involving extraction, purification and characterization of pharmaceuticals from natural sources is a significant contribution to the advancement of natural and physical sciences.

iii. Pharmacognosy forms an important link between pharmacology and medicinal chemistry. As a result of rapid development of phytochemistry and pharmacological testing methods in recent years, new plant drugs are produced in medicine as purified phytochemicals rather than in the form of traditional or galenical preparations.

 iv. The knowledge of pharmacology through pharmacognosy is very essential for understanding the action of drugs on animals and the human system.

  v. Pharmacognosy forms the infrastructure, on which depends evolution of novel medicines, as it is seen that several crude drugs are utilized for preparation of galenicals or as sources of therapeutically significant substances that cannot be synthesized economically.

 vi. The crude drugs also provide essential intermediates for final synthesis of active compounds.

vii. It provides a system wherein active principles of crude drugs derived from natural sources could be dispensed, formulated and manufactured in dosage forms, acceptable to allopathic system of Medicine.

viii. Pharmacognosy forms a vital link between Ayurveda and Allopathic systems of medicine.

 ix. In a nutshell, pharmacognosy forms an important bridge between the pharmaceuticals and basic sciences.

SCHEME FOR PHARMACOGNOSTIC STUDIES OF CRUDE DRUGS

A systematic study of crude drugs involves a study of the following.

  i. *Origin of drugs*—the biological and geographical source, common names and history of drugs.

 ii. *Classification of drugs*—the taxonomical, morphological, pharmacological and chemical nature of drugs.

iii. *Cultivation and preparation*—the methods of cultivation, collection, drying, processing, packing and storage of drugs for marketing.

iv. *Characteristics of drugs*—the surface, physical, internal and sensory characters.

v. *Constituents of drugs*—the study of active and inert constituents, devising methods for their extraction, chemical tests, etc.

vi. *Evaluation of adulterants*—the methods of studying adulteration, substitutes and drug evaluation by means of morphological, microscopical, physical, chemical and biological methods.

vii. *Pharmacology*—the study of pharmacological actions, therapeutic and other pharmaceutical uses and pharmaceutical preparations or formulations.

viii. *Quality control of drugs*—the systematic identification of crude drugs and their quality assurance that forms an integral part of drug description.

Today, the herbs are staging a comeback and a herbal renaissance is blooming across the world. Therefore the need for proper dissemination of information on safety, efficacy and also the potential hazards involved in their use represent a challenge for the pharmaceutical profession and points out to the need for pharmacologically oriented pharmacognosy. Thus, the popularity of natural drugs all over the world in recent years is an indication of significant contributions of pharmacognosy in modern medicine.

REVIEW QUESTIONS

State whether the following sentences are true or false

1. The oldest known herbal is "*Pent-t-Sao*".
2. The word "pharmacon" refers to a tablet.
3. The drug quinine was first isolated by Pelletier.
4. *Penicillium notatum* belongs to the group algae.
5. The drug morphine was isolated from poppy plant.

Choose the correct answer

6. The treatise Charaka Samhita belongs to _____.
   a. Unani   b. Ayurveda   c. Siddha

7. Paracelsus introduced _____
   a. horticultural plants   b. medicinal plants   c. Doctrine of Signatures

8. Antimalarial drug is _____
   a. quinine　　　　b. vincristin　　　　c. benzoin
9. The antibiotic penicillin was isolated from a _____
   a. alga　　　　b. fungus　　　　c. bacterium
10. The study of crude drugs of natural origin is called _____
    a. pharmacognosy　　　b. pharmacology　　　c. pharmagenetics

Answers				
1. True	2. False	3. True	4. False	5. True
6. b	7. c	8. a	9. b	10. a

Answer the following

1. Who coined the term pharmacognosy?
2. Who isolated the antibiotic penicillin?
3. Define the term pharmacognosy.
4. Critically comment on the history of phamacognosy.
5. What are the scope and importance of pharmacognosy?
6. List out the significance of pharmacognosy?

INDIAN SYSTEMS OF MEDICINE AND AROMATHERAPY

2

The system of medicine which is native to a particular country is called indigenous system of medicine. Since it is practised traditionally in a particular country, it is also called traditional system of medicine. The indigenous or traditional systems of medicine practised in India are given below:

1. Ayurveda
2. Siddha
3. Unani
4. Homeopathy
5. Aromatherapy

In principle, all these four systems of medicine mostly use plant products, herbs and other natural substances. Many of the plant products are important therapeutic agents. These are represented by alkaloids, cardiac glycosides, anthraquinones, flavonoids, mucilages, enzymes, polyphenols, tannins, volatile oils, gums, resins, etc.

AYURVEDIC SYSTEM

It is the oldest and most ancient system of medicine practised in India, the origin of which has been taken from the Gods. Ayurveda was first perceived by Brahma, who taught this science to Daksa-Prajapati, who taught it to the Aswini-Kumaras, and later on was adopted in India and so on.

The name Ayurveda is derived from two words: *Ayur* meaning "life" and *Veda* meaning "knowledge or science" i.e., "Science of Life".

It is the system of health care, both physical and mental, and therefore Ayurveda is based on the interaction of the body, mind and spirit. Health, in Ayurveda, has been defined as a well-balanced metabolism, plus a happy state of being. It is one of the most noted systems of medicine in the world.

Ayurveda is based on the hypothesis that everything in the universe is composed of five basic elements, viz., space, air, energy, liquid and solid. They exist in the human body in combined forms like *Vata* (space and air) *Pitta* (energy and liquid) and *Kapha* (liquid and solid). *Vata, Pitta* and *Kapha* together are called *Tridosha*, the three pillars of life. It is believed that these three doshas are in balance with each other in the body but, in every human being one of them is dominating, which in turn is called as the *Prakruti* of that person. Each disease is characterized by a specific status of these three doshas.

Disease has been considered four-fold:

* ❋ Body
* ❋ Mind
* ❋ External factors
* ❋ Natural intrinsic causes

In Ayurveda, treatment is done by a salubrious use of drugs, diets and practices. Pharmaceuticals occupy an important place in Ayurveda. Ayurvedic medicine makes use of vegetables, animals, minerals, metals and even physical forces. Utilization of plants is mentioned in *Rigveda* and *Atharvaveda*. *Charaka Samhita* (900 BC) is the first recorded treatise on Ayurveda. It consists of eight sections divided into 150 chapters, and describes 341 plants used in medicine. Charaka Samhita, a famous scholar of the 9th century BC, is considered as the father of Ayurvedic medicine. The other treatise on Ayurveda is Sushruta Samhita (600 BC) with special emphasis on surgery. It has six sections covering 186 chapters and describes 395 medicinal plants, 57 drugs of animal origin and 64 minerals and metals as drugs.

Ayurvedic formulations are *Churana* (powder), *Kalpa* (paste), *Kashaya* (extract), *agava* (tincture), etc. Mostly all of them are polyherbal formulations. Some of the important herbs from Ayurveda are *Rauwolfia serpentina, Asparagus racemosus, Cassia angustifolia, Sesamum indicum, Withania, Aconitum,* pepper, etc.

Ayurvedic methods of diagnosis are extremely simple. Ayurveda also stresses upon the use of a wholesome diet along with the use of drugs for the successful treatment of diseases. In Ayurveda, drugs are classified depending on their taste, attributes, potency, their taste after digestion and therapeutic effect.

Four types of therapies—elimination therapy, alleviation therapy, psychic therapy and surgery—are used for the treatment of diseases. In addition to single drugs, compound formulations are generally used by Ayurvedic physicians in the form of pills, powders, decoctions, infusions, linctus, alcoholic preparations, medicated ghee, fractional distillation, collyrium, etc.

Ayurvedic drugs are administered both externally in the form of ointment, dusting powder, collyrium, ear drops and eye drops, and internally as tablets, pills, powder, syrups, etc.

SIDDHA SYSTEM

It is one of the oldest systems of medicine in India. Its origin dates back to the pre-Vedic period and was practised even in the era of Mohenjo-daro (5000–3000 BC). The term "Siddha" means achievement and *siddhars* were considered as saintly figures.

The literature on Siddha system was written on palm leaves only in Tamil, therefore it is extensively practised in Tamil Nadu. It consists of 96 principles with at least 3500 formulae, which have been described by 18 siddhars in their work. The principles and doctrines of this system closely resemble that of Ayurveda. According to this system, the human body is a replica of the universe and all things in this universe are composed of five basic elements namely earth, fire, water, air and sky.

The food which is consumed by humans and the drugs that they use contain various proportions of these five basic elements, which have therapeutic action. As in Ayurveda, Siddha system also considers the human body as a conglomeration of the three humours, seven basic tissues and the water products of the body such as faeces, urine and sweat. The food is considered to be the basic building block of the human body which gets processed into humours, body tissues and waste products. The equilibrium of humours results in a healthy body, and its imbalance leads to disease.

Siddha system of medicine was achieved through the practice of yoga and meditation. Siddha medicine is actually a psychosomatic system of medicine and gives more importance to minerals and metals rather than using herbs and products of animals in pharmaceutical preparations. The plants garlic, neem and lemon are highly valued in this system. Hence, the medicines for the human body are prepared based on the theory of panchabutas like metals of gold, lead, copper, iron and zinc. For example, gold and lead are used for the maintenance of the body, whereas copper is meant for the preservation of heat in the body. Iron is the only metal attracted by the electric power of magnet, and zinc is used for generating electricity. All these metals are used after proper detoxification and these metals and minerals like salt and sulphur are used for the extension of life of humans on this earth. Siddha system is otherwise called "Agasthyar system" after the famous sage Agasthya.

UNANI SYSTEM

The Unani system of medicine originated from Greece.

The roots of this system of medicine go deep into the times of the well-known Greek philosopher–physician Hippocrates (460–377 BCE) who is credited for it.

The main framework of Unani medicine is based on the teachings of Hippocrates who is the father of medicine. He was then followed by a number of Greek scholars who enriched this system considerably. Aristotle Galen (131–210 CE), a well-known scholar, made valuable contributions to it.

This system of Greek origin was further carried to Persia (Iran), where it was enhanced by Arabian physicians like Rhazes (850–925 CE) and Avicenna (980–1037 CE).

In India, the Unani system of medicine was introduced by the Arabs.

Like Ayurveda, the Unani system of medicine is also based on ancient principle. The most important similarity is the principle of the four elements, which is identical to Ayurveda's Panchbhuta principle. Therefore Unani interacted intimately with Ayurvedic system. Unani practice is mainly confined to the Indian subcontinent.

According to the Unani system, the functions of the human body are regulated by four humours which include:

* Blood
* Phlegm
* Yellow bile
* Black bile

The imbalance of any of these humours produces a disease. Unani system emphasizes a correct diet. Also, it characterizes and diagnoses a disease by examining the pulse.

Unani medicine aims at combating disease and preservation and promotion of health through curative, preventive and promotive measures. For the treatment of diseases, medicines obtained from natural sources, e.g., plants, animals and minerals, are used in this system.

The Unani system lays great emphasis on the maintenance of proper ecological balance on one hand, and on keeping water, food and air free from pollution on the other.

The Unani system of medicine aims to treat the cause of disease and not its symptoms. The drugs are mostly polyherbal formulations and their collective effect is considered.

Unani medicines are not only cheap and easily available, but are also very effective and free from side effects. The Unani pharmaceutical preparations include powders, suspensions, syrups, electuaries, distilled medicinal waters and many other forms and are more than seventy in number.

HOMEOPATHIC SYSTEM

The word "Homeopathy" is derived from two Greek words, *Homois* meaning "similar" and *pathos* meaning "suffering".

In comparison to other traditional systems of medicine, Homeopathy is a newer one and was developed in the 18th century by Dr. Samuel Hahnemann (1755–1843), a German physician, pharmacist and chemist.

Homeopathy is the system of treatment based on the natural law of healing, "similia similibus curentur" which means "likes are cured by likes".

In the homeopathic system, the choice of drug depends on symptoms and the clinical condition of the patient.

According to the homeopathic medicinal system, until the potency governing the body of a human being is powerful and controls the functions of all organs, the person will not be affected by a disease. Any symptom of a disease cannot be complete without the governing power of the body.

For prescribing the medicine, it is essential that information about characteristics of elements, mental symptoms and other symptoms is collected and the medicine with more pronounced characters prescribed.

Homeopathic medicines used are in the form of mother tincture, small pills, powder and distilled water. The mother tincture is further diluted in terms of decimal. (Centesimal with 1000 C represented as 1 M and LM (1 in 50,000)). Various medicinal plants used in homeopathy are nux-vomica, *Thuja occidentalis, Colchicum autumnale, Aconitum napellus,* etc.

In homeopathy, diseases are not produced by the attack of microorganisms. Weak body potency is responsible for disease. If the symptoms of a disease are considered a manifestation of the body's own defence mechanism against the disease, the homeopathic treatment serves to stimulate such inherent defensive and curative processes. Homeopathic remedies have the distinct advantage that, they are without side effects.

AROMATHERAPY

It is one of the most ancient of healing arts, and traces its origin to 4500 BC when Egyptians used aromatic substances in medicines. Greeks also used plant essences for aromatic baths and scented massage. Ayurveda has also mentioned scented baths. Prof. Gantle Fosse, a French cosmetic chemist, coined the term "Aromatherapy" and described the healing properties of essential oils. Different essential oils from various parts of plants are massaged into the skin to treat a range of diseases as well as to have an effect on the mind and emotions. They are massaged into the skin or inhaled or taken as a bath. They have been shown to heal wounds, and promote formation of scar tissue, and to treat acne and skin problems, tension, rheumatism, poor circulation and also nervous disorders like headache, stress, insomnia, etc. Various essential oils used in aromatherapy are basil, pepper, calendula, caraway, eucalyptus, fennel, garlic, geranium, ginger, jasmine, juniper, lavender, rosemary, sandalwood, and so on.

REVIEW QUESTIONS

State whether the following sentences are true or false

1. Vata, pitta and kapha are considered as the three pillars of life.
2. The oldest homeopathic system was developed by Samuel Hahnemann.
3. A system of medicine that is native to a particular country is called Unani System.
4. The term kalpa means churnam.
5. Siddha system is mainly practised in Tamil Nadu.

Choose the correct answer

6. Traditional system of medicine practised in India is _____
   a. Homeopathy b. Ayurveda c. Allopathy
7. The term given to tincture is _____
   a. vata b. unani c. agava
8. The Unani system was introduced by _____
   a. Indians b. Arabs c. Chinese
9. The father of ayurvedic medicine is _____
   a. Charaka b. Avicenna c. Brown
10. The word aromatherapy was introduced by _____
    a. Fosse *et al.* b. Gantle *et al.* c. Gantle Fosse

Answers				
1. True	2. True	3. False	4. False	5. True
6. b	7. c	8. b	9. a	10. c

Answer the following

1. What is "Tridosha"?
2. What is "Prakruti"?
3. Mention the types of traditional systems of medicine practised in India.
4. What are the humours that regulate our body?
5. Comment on the following:
    i. Siddha medicine
    ii. Aromatherapy
    iii. Homeopathic system
6. Explain Unani system of medicine and its significance.
7. Discuss the basic principles of Unani and Ayurvedic systems of medicine.
8. Ayurveda is considered as a traditional system of medicine in India—Discuss.

NATURAL SOURCES OF DRUGS

3

A drug may be defined as any substance, which is used to cure, prevent or diagnose a disease. It also helps to alter the bodily function of humans or other animals.

In general there are two major types of drugs namely synthetic and natural drugs. Many drugs used in medicine today are developed by chemical synthesis. But a number of drugs are obtained from natural sources. The most important natural sources are

1. Plants
2. Animals
3. Marine organisms
4. Minerals and
5. Microbes

DRUGS OBTAINED FROM PLANTS

Plants have been used in the treatment of various diseases from time immemorial. The traditional Indian systems of medicine—Ayurveda, Unani and Siddha systems—are based on the use of plants and other natural substances. The majority of the crude drugs are derived from seed-bearing plants. Among the seed-bearing plants, the angiosperms have a good number of useful medicinal plants than the gymnosperms. The gymnosperms yield oils, resins and the alkaloid ephedrine.

In Angiosperms, both dicots and monocots provide useful drugs. Mostly, only certain parts of plants are used as drugs, but rarely the entire plant may be used. Thus a crude drug may be obtained from seeds, fruits, leaves, flowers, roots and barks of stem or wood. Many of the plant products are important therapeutic agents. These are represented by the alkaloids, cardiac glycosides, anthraquinones, flavonoids, mucilages and enzymes. Besides higher plants, some of the lower plants also provide a number of useful drugs.

Drugs can be obtained from many of the higher plants or lower plants.

Higher Plants

In the case of higher plants, drugs can be derived from

1. Entire plants, e.g., *Ephedra* and chirata
2. Roots, e.g., *Hemidesmus* and *Vetiveria*
3. Barks, e.g., *Cinchona* and *Cassia*
4. Flower buds, e.g., clove
5. Flowers, e.g., rose and lotus
6. Leaves, e.g., tulsi and pudina
7. Seeds, e.g., castor and pepper
8. Exudations, e.g., Gums and resins like *Acacia* and asafoetida

Drugs obtained from plants may belong to any one of the chemical classes like alkaloids, glycosides, oils, tannins, flavonoids, etc.

Lower Plants

Lower plants like fungi produce antibiotics, e.g., Penicillin from *penicillium notatum*. Ergot which is a fungal drug is a source of ergotamine. Algae produce agar and alginic acid and a green alga *Chlorella* produces an antibiotic chlorellin.

DRUGS OBTAINED FROM ANIMALS

Certain animal parts and animal products are used as drugs in therapeutics. Drugs obtained from animals can be classified as hormones, enzymes, and extracts of animal organs.

Hormones

They are secretions of endocrine (ductless) glands. Hormones are directly poured into blood. The following are some of the examples of hormones which are used as therapeutic agents:

i. *Thyroid hormones* The hormone thyroxin is prepared from the thyroid glands and can be obtained from sheep and pigs. It is used in the treatment of thyroid insufficiency.

ii. *Insulin* Insulin is secreted by the beta cells of the islets of Langerhans of pancreas. It is obtained from pancreas of cattle and pigs. Insulin is used in the treatment of diabetes.

iii. *Oxytocin* It is a hormone secreted by the posterior pituitary gland and can be obtained from cattle and pigs. It is used for induction of labour pain in pregnant women during childbirth.

iv. *Vasopressin* It is a hormone secreted by the posterior pituitary gland. It is also obtained from cattle and pigs. It is used in the treatment of intestinal paralysis, diabetes insipidus, excessive thirst, and passing large amount of urine.

Enzymes

They are biological catalysts produced by living organisms. Enzymes are protein in nature. The following are some of the important enzymes used in medicine.

i. *Pepsin* It is an enzyme of the gastric juice, obtained from the glandular layer of stomach of pigs used in dyspepsia caused due to deficient gastric secretion.

ii. *Rennin* It is a milk-curdling enzyme. It is obtained from the glandular layer of calf stomach, and is used to coagulate milk to make it more digestible.

iii. *Pancreatin* It is produced from pancreas, and can be commercially obtained from pig pancreas. It is used in the treatment of pancreatitis.

iv. *Hyaluronidase* It is a mucolytic enzyme. It is prepared from mammalian testes. It is used to increase the absorption of subcutaneous infusions.

Extracts of Animal Organs

Liver and stomach preparations and bile are the examples of this group. Liver and stomach are derived from healthy animals and converted into suitable preparations. It is used in the treatment of pernicious anaemia.

Bile is a natural secretion of the liver and it helps in the digestion of fats. These bile acids are obtained from ox bile. Bile acids are used in treating deficiency of biliary secretion.

The other useful animal drugs are as follows:

* Carmine, a colouring substance taken from cochineal insects.
* Cantharidin, an irritant constituent taken from cantharides insects.
* Wool fat and lanolin obtained from animals are used in certain formulations and in cosmetic industry.

DRUGS OBTAINED FROM MARINE ORGANISMS

In recent years a number of new drugs have been obtained from marine organisms. Sea weeds are used in drug industries as thickening, emulsifying and suspending agents.

Alginates taken from a brown alga *Laminaria* is used in adhesive formulations and as stabilizers, ingredients of ointment bases, suspending agents, etc.

Agar taken from algae such as *Gelidium* and *Gracilaria* is used in culture medium.

Cephalosporin-C, an antibiotic, is obtained from a marine fungus, *Cephalosporium acremonium*, which is used as an antibiotic in the treatment of various diseases.

Protamine is a protein obtained from the sperm and testes of salmon fish. It is used with insulin, so that the insulin is absorbed at a slow and steady rate.

Cod-liver oil is obtained from the fresh liver of cod, a fish. This oil contains growth-promoting substance and vitamins A and D and also fatty acids. It helps in the growth of children.

Spermaceti is a waxy solid taken from the head of the sperm of whale. It is used in the preparation of cold creams and also forms a base for many ointments.

DRUGS OBTAINED FROM MINERALS

Natural drugs are mostly derived from plants and animal sources, but a few of them are obtained from mineral sources. Nowadays these mineral sources are used as pharmaceutical aids very significantly and some of them are discussed in this chapter.

1. SILAJIT

Ancient Hindu literature recorded the following forms of the drug Silajit:

1. Blackish brown variety—iron Silajit
2. Blue variety—copper Silajit
3. White variety—silver Silajit
4. Red variety—gold Silajit

Source   It is an example of a herbo-mineral drug taken out of iron-rich rock fissures during hot weather.

Geographical distribution   It is produced naturally in mountainous regions of Vindhyas, Himalayas and other mountainous areas of India and Nepal. It is also found as a tar in the earth crust and is formed due to decomposition of vegetable substances.

Uses

1. The drug is used as a general tonic, stimulant and aphrodisiac.

2. Silajit of authentic quality forms an important ingredient of ayurvedic preparations, which is purely meant for strengthening the immune system of the human body.

2. FUELLER'S EARTH

The word Fueller's earth is derived from the ancient method of cleaning wool to remove oil and dust particles with the help of earth slurry.

Source   It is mined in an open quarry.

Its common name is floridin.

Geographical distribution   It is usually found in Surrey, Somerset, Hampshire, Dorset and Glaucestershire.

Description

Colour	White to yellowish grey.
Odour	Odourless
Taste	Tasteless powder
Solubility	It swells in water and becomes non-plastic in texture (like kaolin).

Chemical constituents   It contains 55% SiO_2, 16% Al_2O_3, 3.5% CaO, 2% MgO, 6% Fe_2O_3, 10% water with 50% montmorillonite and 18% silica.

Uses

1. It is used in the preparation of dusting powders due to its absorbent property.

2. It is used as a decolorizer for oils and other liquids and as clarifying and filtering agent.

3. It is also used for cleansing of woolen fabrics.

3. TALC

Source   It is a natural hydrated magnesium silicate and a fine variety of soapstone. It occurs as a massive mineral with a foliated structure and can be split into thin pieces.

The common names are purified talc, soapstone, talcum, French talc, French chalk, etc.

Geographical distribution   It is obtained from Italy, France, New York and also India.

Preparation   The powdered talc is usually purified for pharmaceutical use. The powder is purified by boiling with dilute hydrochloric acid and then decanting the liquid. The residue is washed with water several times and dried at 110°C. The mineral is flexible but not elastic and this character distinguishes it from mica.

Description

Colour	The mineral is a very fine powder with a white, greyish or greenish tint and possesses a pearly lustre. It is free of gritty particles, greasy to touch and adheres to skin.
Odour	Odourless
Taste	Tasteless
Size and shape	Minute particles of various sizes and shapes. The particles are irregular or sharply angular and often show jagged and laminated ends. The powder particles polarize brightly and often show bright colours between crossed nicols.
Solubility	The mineral is insoluble in water, dilute acids and alkali hydroxides.
Standard	Density 2.2 to 2.8.

Chemical constituents   Talc is a hydrated magnesium silicate. Chemically it is represented by $Mg_6(Si_2O_5)_4 \cdot (OH)_4$. The native mineral contains 1 to 2% iron oxide and traces of aluminium oxide. The greyish or greenish tint of the talcum powder is due to its iron oxide content.

Uses

1. It is used as a filter aid for filtration and clarification of cloudy liquids, as lubricant in preparation of pills and tablets and as a dusting powder.

2. It is used in the paint and varnish industry, as a filler for paper and as heat and electric insulator.

4. KAOLIN

Source   The native clay is derived from decomposition of the felspar (potassium aluminosilicate) or granite rock and contains silica (47%), alumina (40%) and water (13%). Kaolin is a purified native hydrated aluminium silicate free from gritty particles.

The common name is China clay.

Geographical distribution   It is found in south-eastern United States, England, France, and India.

Preparation   It is obtained by powdering the native kaolin, elutriating and collecting the fraction, which complies with the requirements of particle size.

Types   The following are the different types of kaolin available in the market.

1. *Heavy kaolin*   It is the purified hydrated aluminium silicate powder free from gritty particles by elutriation. Heavy kaolin with a particle size up to 60 μ is used in the preparation of kaolin poultice.

2. *Light kaolin*   It is similar to heavy kaolin, but is limited to particle size and contains suitable dispersing agent. Light kaolin with a particle size less than 10 μ is used in pharmacy.

3. *Natural light kaolin*   It is also a native hydrated aluminium silicate free from gritty particles, but devoid of dispersing agent.

Description

Colour	It is white, but often gets tinged to grey, yellow or brown due to impurities.
Odour	Almost odourless.
Size	10 μ to 60 μ
Standards	Fusion point: 1700° to 1800°C
Density	2.3
Solubility	It is insoluble in water and mineral acids but absorbs small amount of water. It is not affected by dilute hydrochloric acid, but is decomposed by conc. sulphuric acid after prolonged boiling.
Identification	Heating kaolin on charcoal block with cobalt nitrate, forms a blue mass due to the presence of alumina.

Uses

1. It is mainly used as an adsorbent. Oral administration of the drug helps in the treatment of enteritis, dysentery and food poisoning.

2. It is applied externally as a dusting powder and also as clarifying agent during filtration.

3. It is used as a filler in the paper, rubber, ceramics, cement and fertilizer industries.

4. It is also used in anticaking preparations, cosmetics, insecticides and paints and as a source of alumina.

5. ASBESTOS

Source   It is a fibrous variety of mineral silicates, which occur in two forms, i.e., Serpentine and Amphibole. The most commonly available fibrous form is serpetine, which is also known as chrysotile [$Mg_6(S_{14}O_{10})OH_8$], whereas the other form amphibole is subdivided into the following four types.

1. Anthophyllite
2. Amosite
3. Actinolite
4. Tremolite

Asbestos is mined in open quarries.

The common name is Amianthus.

Geographical distribution   It is mainly distributed in Canada, Italy, Australia and Rhodesia.

Description

Colour	White to slightly greenish mineral.
Odour	Odourless

It can be easily broken into long, fine and flexible fibres. It can be spun and woven to form a coarse cloth. It is unaffected by heat and alkalies. Chrysotile is attacked by acids, whereas amphiboles are acid-resistant. The fibres appear as long, delicate and transparent when mounted in cresol and are highly refractive, and polarize brightly on a dark field when examined with crossed nicols.

Chemical constituents   It is a double silicate of calcium-magnesium with a varying amount of iron which actually gives the due colour to asbestos.

Uses

1. Asbestos is used as a filtering medium for caustic alkalies and for making bacterial filters.
2. It is also used as a heat-resistant insulating material for ovens, steam pipes, fireproof gloves, fireproof clothing and for packing round boilers.

Caution   Constant exposure to asbestos is carcinogenic.

6. PREPARED CHALK

Source   Chalk is a native form of calcium carbonate, available in most of the areas of the world and is freed from most of the impurities by elutriations.

Common names are chalk, crela, Paris white, whiting, English white, etc.

Collection and preparation   Chalk is mined in open quarry, pulverized by elutriation. The water content is removed and the remaining insoluble chalk is allowed to settle forming flat cakes, called "whiting". It is then purified by dilute acid for pharmaceutical use.

Colour	Generally colourless, white earthy and soft to touch.
Odour	Odourless
Taste	Slightly acrid taste.
Shape and size	Various shapes and sizes of foraminiferous shells.
Solubility	Chalk is insoluble in water and it effervesces with acids.

Chemical constituents   Chalk contains 96% calcium carbonate, 0.5% magnesium carbonate, 0.5 to 1% silica, traces of iron, manganese and aluminium oxides.

Uses

1. It is used as a dietary supplement, an antacid, a dusting powder and an anti-diarrhoeal.

2. It is used as an abrasive in toothpowders and toothpastes and in face powders.

3. It is also used in the manufacture of antibiotics and other pharmaceuticals.

7. KIESELGUHR

Biological source   It is a natural diatomaceous earth consisting of siliceous skeletons of fossil diatoms of the family Bacillariaceae of the group algae.

The common names are diatomaceous earth, Celite, Super-col, industrial earth, etc.

Geographical distribution   A large quantity of diatomaceous earth is found in Denmark, West Germany, Algeria, Kenya, U.S.A. (Virginia and California), Scotland and Ireland.

Preparation   Kieselguhr is normally mined in an open quarry. The large blocks containing 30–40% moisture are arranged in such a way and dried in the air. The blocks containing

5–10% moisture are then pulverized to form fine powder and subsequently graded. The powder is purified further by treating with dilute hydrochloric acid, washed thoroughly with water and finally dried for pharmaceutical use.

Description

Colour	Brownish grey to white-coloured light powder.
Odour	Odourless
Taste	Tasteless
Size	Diatoms vary in size from 5–500 μ.
Shape	Elongated and circular or triangular in shape.

Kieselguhr is very smooth and it adheres to the skin after rubbing. It absorbs moisture, but does not swell when mounted in cresol.

Chemical constituents   Diatomite contains 75–90% silica, 1 to 5% aluminium oxide, 1.5% calcium oxide, 1.5% magnesium oxide and 5% iron oxide.

Uses

1. It is used as a filler aid and for clarification and decolorization of liquid.
2. It is used in the manufacture of toothpowder, face powder and nail polishes.
3. Filter candles are also made from diatomite.

8. BENTONITE

Sources   Bentonite is a natural mineral deposit of colloidal hydrated aluminium silicate (clay). It was formed by the weathering of volcanic ash.

The common name is Wilhinite.

Geographical distribution   It occurs in many parts of the world including Italy, Canada, South Africa and mid-western US. The finest variety of bentonite is obtained from a mine near Fort Benton in Wyoming, USA, and hence it got its name bentonite.

Collection and preparation   The mineral is obtained from the mines in the form of soapy lumps. The lumps are dried by steam heat and are then powdered and sifted for pharmaceutical use. Mounted in alcohol or glycerin, bentonite is seen to be composed of small fragments, about 50 to 150 μ in diameter with numerous minute particles about 1.0 to 2.0 μ in size interspersed amongst them. The Wyoming variety is generally preferred for pharmaceutical use.

Description

Colour	Bentonite of commerce is a pale greenish grey or pale buff (yellowish or pinkish tint), very fine powder, free of gritty particles.
Odour	Odourless
Taste	Earthy in taste
Solubility	Bentonite is insoluble in water and organic solvents. When added to water, it swells to about 12 times its volume.

Standards

Loss on drying	5 to 12%
pH of 2.0% solution	9 to 10.5
Arsenic	> 3.0 ppm
Swelling power	< 12 times

Identification test

1. Bentonite mounted in 1% solution of safranin in 70% alcohol permanently stains deep red in colour.

2. Bentonite also stains deep blue with 1% solution of methylene blue in 95% alcohol.

Chemical constituents Bentonite consists largely of montmorillonite. $(Al_2O_3 \cdot 4SiO_2H_2O)$. Analysis reveals about 60% SiO_2, 20% Al_2O_3, 6–10% water, about 2–3% MgO, 3% Fe_2O_3 with traces of one or two other elements.

Uses

1. Bentonite is used as suspending and emulsifying agent and base for ointment, plasters and creams.

2. In the cosmetic trade it is used to make bases for lipsticks, rouges and depilatories.

DRUGS OBTAINED FROM MICROBES

The microbes or microscopic organisms from which drugs are obtained, include viruses, bacteria and rickettsiae. These organisms produce immunizing substances such as vaccine and toxoid, which give active or passive immunity to organisms when administered.

Vaccines

Vaccines are suspensions of living, dead or attenuated microbes, which give immunity to the body when administered. Depending upon the source, the vaccines are classified into three groups.

1. Viral vaccines from virus.
2. Bacterial vaccines from bacteria.
3. Rickettsial vaccines from Rickettsiae.

Viral vaccines   Viral vaccines are used against smallpox, rabies, influenza, polio, measles and mumps. Viral vaccines contain live, attenuated or killed virus.

i. *Smallpox vaccine*   It contains living virus of vaccinia (cowpox), which has been grown in the skin of a vaccinated bovine calf. The vaccine is also available in dried form. It is used as immunizing agent and is prophylactic against smallpox infection.

ii. *Rabies vaccine*   It is a sterile preparation of killed, fixed virus of rabies obtained from duck embryos which have been infected with fixed rabies virus. It is available in dried form. It is used in the prevention of rabies in persons bitten by a rabid animal.

iii. *Influenza virus vaccine*   It is a sterile aqueous suspension of suitably inactivated influenza virus. It is prepared from extra-embryonic fluid of chick embryo infected with influenza virus.

iv. *Poliomyelitis vaccine*   It is of two types—poliovirus vaccine inactivated and poliovirus vaccine live oral. The poliovirus vaccine inactivated type is a sterile suspension of inactivated poliomyelitis virus strains, and the other type is a preparation of one or a combination of strains of live, attenuated polioviruses. These are used as active immunizing agents against poliovirus.

v. *Measles vaccine*   It contains live attenuated Rubeola (measles) and Rubella (German measles) viruses. The viruses are grown on cultures of either bird's embryo tissue or human diploid cell tissue.

vi. *Mumps vaccine*   It is prepared from a specific strain of mumps virus, which is grown in cell cultures of chicken embryo tissue. The vaccine is available in lyophilized form and is used for immunization against the mumps disease.

Bacterial vaccines   Bacterial vaccines consist of suspensions of killed or attenuated pathogenic bacteria and it is cultivated in a suitable broth medium.

Typhoid vaccine, cholera vaccine, plague vaccine and BCG vaccine are the important bacterial vaccines, which are used as active immunizing agents.

i. *Typhoid vaccine* It is a sterile suspension containing killed selected strain of typhoid bacillus and *Salmonella typhi*. It is used for producing immunization against typhoid fever.

ii. *Cholera vaccine* It is a sterile suspension of killed cholera vibrios called *Vibrio cholerae* in saline. It is an active immunizing agent for producing immunity against cholera.

iii. *Plague vaccine* It is used to produce immunity against the plague disease. It is a sterile suspension of killed strain of plague bacilli *Yersinia pestis*.

iv. *Pertussis vaccine* Whooping cough is caused by *Bordetella pertussis*. Pertussis vaccine is a sterile suspension of killed *Bordetella pertussis*. It is used as an immunizing agent against whooping cough disease.

v. *BCG vaccine* It is a dried, living culture of the Bacillus Calmette Guerin (BCG) strain of *Mycobacterium tuberculosis* var *bovis*. BCG vaccine is an active immunizing agent against tuberculosis.

Toxoids Toxoids are also microbial products used to produce active immunity against tetanus and diphtheria diseases. The waste products of bacteria are treated with formaldehyde and their toxic properties are reduced, but their antigenic property is retained. These products are called fluid toxoids. It is precipitated with alum or aluminium hydroxide to produce toxoid. It is treated for tetanus and diphtheria disease.

A combination of diphtheria and tetanus toxoid with pertussis (whooping) vaccine is known as triple antigen or DPT. It is used to produce immunity in young children against diphtheria, tetanus and whooping cough. The vaccines and toxoids are used as inoculations to stimulate the production of antibodies against pathogenic microorganisms.

Rickettsial vaccines Rickettsiae are a group of very small gram-negative microbes, intermediate in size between the average bacteria and the large virus. They cannot be grown in artificial media and therefore require chick embryo or monkey kidney tissue cultures for their growth. The rickettsial vaccine is a sterile suspension of the killed rickettsial organisms, which are obtained by culturing in yolk-sac membrane of the developing embryo of the domestic fowl. It is used for producing active immunity against typhus fever.

REVIEW QUESTIONS

State whether the following sentences are true or false

1. Asafoetida is an example of resin.
2. The useful part of *Hemidesmus* is the leaf.
3. Ergotamine is a fungal drug.
4. Insulin is used in the treatment of diabetes.
5. Oxytocin induces labour pain in pregnant women.
6. The milk-curdling enzyme is pepsin.
7. Carmine dye is obtained from cochineal insects.
8. Agar is obtained from red alga.
9. Protamine is a carbohydrate.
10. Rubella is used against German measles.

Choose the correct answer

11. DPT refers to _____
    a. triple antigen   b. diphtheria   c. BCG vaccine
12. Alginates are produced by _____
    a. diatom   b. laminaria   c. gelidium
13. Tuberculosis is treated with _____
    a. triple antigen   b. toxoids   c. BCG vaccine
14. The plague bacilli is _____
    a. *Yersinia pestis*   b. *Yetisia molli*   c. *Salmonella typhi*
15. Spermaceti is obtained from _____
    a. cod fish   b. whale   c. salmon fish
16. Cephalosporin is an example of _____
    a. antibiotic   b. vaccine   c. cold cream
17. *Bordetella pertussis* causes _____
    a. typhoid   b. whooping cough   c. cholera

18. The beta cells of the Islets of Langerhans of pancreas produce _____
    a. insulin   b. pepsin   c. rennin

19. *Acacia* species yield _____
    a. latex   b. oil   c. gum

20. The morphology of the useful part of *Cinchona* is the _____
    a. flower   b. bark   c. leaves

Answers				
1. True	2. False	3. True	4. True	5. True
6. False	7. True	8. True	9. False	10. True
11. a	12. b	13. c	14. a	15. b
16. a	17. b	18. a	19. c	20. b

Answer the following

1. Define the term vaccine.
2. What is triple antigen? Give an example.
3. What are crude drugs? Cite an example.
4. Give a note on the following:
    i. Marine drugs
    ii. Viral drugs
5. Write down the use of hormonal drugs.
6. Explain the various crude drugs obtained from microbes.
7. Write an essay on animal drugs and their importance in the life of human beings.
8. Describe drug-yielding plants and their significance.
9. "Vaccines are highly useful in giving immunity to man."—Justify.
10. Give an elaborate account on natural sources of crude drugs.

CLASSIFICATION OF CRUDE DRUGS

4

The term "crude drug" generally refers to the products obtained from plants, animals and minerals in a raw form. The word "crude" is also used in relation to the natural product that has not been improved in its value by any process or treatment. The main sources of crude drugs are higher plants, microbes, animals, marine organisms and minerals. In order to understand the nature of the crude drugs, they must be classified in a particular sequence and it is known as system of classification of drugs.

In pharmacognosy, the crude drugs may be classified according to their

* Taxonomy
* Morphology
* Chemical nature of active constituents
* Therapeutic applications

Each of these systems of classification has its own merits and demerits. Therefore, none of these systems give a total profile of natural drugs.

SYSTEMS OF CLASSIFICATION OF CRUDE DRUGS

1. *Morphological classification* is more helpful in the identification and detection of adulteration of drugs.
2. *Taxonomical classification* is more suited to study the evolutionary development of drugs and also the name of the organism from which the drug is derived.

3. *Therapeutical or pharmacological classification* is more useful to get the therapeutic utility of the drugs.

4. *Chemical classification* of drugs helps to identify the presence of chemical constituents that are present in the drugs and the activity of drug is due to its chemical constituents.

These systems of classification are discussed in detail in the following sections.

TAXONOMICAL CLASSIFICATION

It is otherwise called biological classification. Taxonomical classification is based on the principle of natural relationship or phylogeny among plants or animals. The drugs are classified according to the plants or animals from which they are obtained in phylum, class, order, family, genus, species, sub-species, etc. Crude drugs of plant origin are classified on the basis of one of the accepted systems of botanical classification. Although botanically speaking this system of classification looks appropriate and whole plants are rarely used as drugs, in majority of the cases a part of the plant is used as a drug, for example cinnamon bark. This system also does not show any relationship between the chemical constituents and biological activity of the drugs. Thus it is of no significance from identification point of view.

The taxonomical classification of a crude drug derived from a dicot plant is given below:

Phylum	Spermatophyta/Angiosperms
Class	Dicots
Order	Rubiales
Family	Rubiaceae
Genus	*Cinchona*
species	*officinalis*

MORPHOLOGICAL CLASSIFICATION

In this system of classification, the crude drugs are classified according to the part of the plant used as drug, like roots, stems, leaves, barks, flowers, etc. The drugs are divided into organized and unorganized drugs. The morphological classification of organized and unorganized drugs is shown in Tables 4.1 and 4.2.

The drugs obtained directly from the parts of the plants and containing cellular tissues are called organized drugs, e.g., rhizomes, leaves, stems, roots, flowers, fruits, seeds, barks, etc.

The drugs which are prepared from plant or plant parts by some physical processes such as incision, drying or extraction with a solvent, and which do not contain any cellular plant tissues are called unorganized drugs, e.g., Aloe juice, gums, mucilage, dried latex, etc.

Table 4.1   Morphological classification of organized drugs

Part of the plant	Name of the drug
Seeds	Castor, neem, cardamom, mustard, almond, nutmeg
Leaves	*Adhatoda* (vasaka), senna, *Eucalyptus*, tea, *Digitalis*, pudina, belladonna, tulsi
Barks	*Cinchona*, cinnamon, *Cassia*, kurchi
Wood	Sandalwood, red sandals
Roots	*Rauwolfia*, aconite, Ipecac, *Hemidesmus*
Rhizomes	Turmeric, ginger, *Acorus*, *Glycyrrhiza*
Flowers	Saffron, rose, pyrethrum
Flower buds	Clove
Fruits	Coriander, fennel, cardamom, cumin, amla, tamarind, *Capsicum*
Entire plants	Ergot, *Ephedra*, *Vinca*, yeast, *Datura*
Hairs and fibres	Cotton, hemp, jute, silk

Table 4.2   Morphological classification of unorganized drugs

Category	Name of the drug
Dried latex	Opium, papain, gutta percha
Dried juices	*Aloe*, kino and red gum
Dried extracts	Agar, gelatin, pectin
Gums	*Acacia*, tragacanth, guar gum, Indian gum
Resins	Benzoin, colophony, asafoetida, myrrh
Oleoresins	Balsam of Tolu, balsam of Peru, copaiba
Fixed oils	Arachis oil, castor oil, sesame oil, coconut oil, cod-liver oil, olive oil
Volatile oils	Turpentine, rosemary, dill, camphor, lemongrass, *Eucalyptus*
Fats	Lard
Waxes	Beeswax, wool fat
Animal products	Shark-liver oil, lactose, shellac, honey, cod-liver oil
Fossil organisms and minerals	Bentonite, kaolin, talc

The morphological system of classification is much useful for practical study, especially when the chemical nature of the drug is not clearly understood. This classification also has the drawback in that chemically and therapeutically unrelated drugs are grouped together.

PHARMACOLOGICAL (THERAPEUTICAL) CLASSIFICATION

Grouping of crude drugs according to their therapeutic use is called pharmacological or therapeutic classification of drugs (Table 4.3). Regardless of morphology, taxonomical status or chemical relationship, the drugs are grouped together, provided they exhibit similar pharmacological action. Thus drugs like digitalis, squill and strophanthus having cardiotonic action are grouped together. Similarly drugs like cascara bark, senna leaves and castor oil are grouped as purgative drugs, because of their common pharmacological action. Similarly, *Cinchona*, nux-vomica and *Swertia* are grouped as bitters.

The advantage of this system of classification is that even if we do not know the chemical constituents of a particular drug, it can be grouped according to its therapeutic use.

Table 4.3 Classification of drugs based on pharmacological action

Pharmacological action	Name of the drug
Anticancer	*Vinca, Podophyllum*
Anti-inflammatory	Colchicum corm and seed, turmeric
Antiamoebic	Ipecac root, kurchi bark
Anthelmintic	*Artemisia*, Male fern, Chenopodium oil
Antiasthmatic	*Ephedra*, vasaka, *Tylophora*
Antispasmodic	Belladonna, *Datura*
Astringent	Tannic acid, myrrh, *Myrobalan*
Analgesic	Opium, *Cannabis*
Bitter tonics	Nux-vomica seed, gentian root, chirata
Carminatives and flavours	Cinnamon bark, cardamom seed, nutmeg fruit, clove flower bud, asafoetida, tulsi, ginger, vanilla, coriander, caraway
Purgatives	Cascara bark, senna leaf, aloe juice, castor oil
Expectorant	Benzoin, glycyrrhiza, vasaka
Antiexpectorants	Stramonium leaves

(Contd.)

Table 4.3   (Continued)

Pharmacological action	Name of the drugs
Antitussives	Opium
Peptic ulcer treatment	Liquorice, raw banana
Central Nervous System (CNS) stimulants	Coffee, tea
CNS depressants	Belladonna, opium
Hallucinogenics	*Cannabis*, poppy latex
Tranquilizer	Rauwolfia root
Antirheumatics	*Aconite, Cardiospermum*
Antimalarials	*Cinchona*
Immunizing agents	Vaccines, toxoids
Sweeteners	Honey, liquorice root and rhizomes

The main drawback of this classification is that a drug having different action on the body is placed in various classes according to its therapeutic use. For example, *Cinchona* can be grouped in antimalarial drugs because of the presence of quinine, but it can also be put under the group of drugs affecting heart because of the antiarrhythmic action of quinidine.

CHEMICAL CLASSIFICATION

Crude drugs are divided into different groups according to the chemical nature (Table 4.4), since the pharmacological activity and therapeutic significance of crude drugs are due to the presence of certain chemical constituents in the drug. Plants and animals synthesize chemical compounds such as fats, carbohydrates, proteins, volatile oils, alkaloids, gums, resins, etc. and some of these are pharmacologically active constituents. A single active constituent may be isolated from the crude drug and used as a medicinal agent and thus it is known as the active principle. More than 75 pure compounds derived from higher plants are used in modern medicine. For example, the important traditional active plant principles are atropine, ephedrine, digitoxin, theobromine, quinine, caffeine, papaverine and colchicine. These active constituents are differentiated from the inert compounds like starch, cellulose, lignin, cutin, etc. The active constituent may be present in a very low concentration in the drug. The chemical classification of drugs is thus based upon the grouping of drugs with identical chemical constituents. The crude drugs containing alkaloids are grouped together, regardless of their morphology and taxonomical relationship.

Table 4.4   Chemical classification of drugs

Chemical constituents	Name of the drugs
Carbohydrates and derived products	
Gums	*Acacia*, tragacanth, guar gum
Mucilages	Plantago seed
Others	Starch, honey, agar, cotton
Glycosides	
Anthraquinone	*Aloe*, cascara, senna
Saponins	Liquorice, *Dioscorea*
Cyanophore	Wild cherry
Alkaloids	Nux-vomica, ergot, *Cinchona*, *Datura*
Tannins	*Myrobalan*, pale catechu, ashoka
Volatile oils	Peppermint, clove, *Eucalyptus*, garlic
Lipids	Castor oil, beeswax, cod-liver oil, kokum butter
Resins and resin combinations	Colophony, balsam of Tolu
Vitamins and hormones	Yeast, shark-liver oil, insulin
Proteins and enzymes	Casein, papain, trypsin

The chemical system of classification is more useful to the study of drugs from natural sources and gives logical reasoning of the biological activities, which are attributed to the chemical entities in the drugs.

When crude drugs are used as such as in herbal and traditional drugs, it may be a combination of various chemical constituents which exert a therapeutic effect.

REVIEW QUESTIONS

State whether the following sentences are true or false

1. Anticancer property is noted in *Podophyllum*.
2. The therapeutic action of any crude drug is due to its chemical component.
3. Turmeric is an example of anti-inflammatory drugs.
4. *Rauwolfia* and chirata are used as purgatives.
5. Liquorice root is a natural sweetener.
6. Insulin is an enzyme.
7. Plantago seed is the source of mucilage.
8. Myrobalan contains high percentage of alkaloids.

Choose the correct answer

9. The analgesic drug is _____
   a. benzoin   b. opium   c. catechu
10. Grouping of crude drugs based on their therapeutic use is called _____ classification.
    a. taxonomical   b. chemical   c. pharmacological
11. High percentage of lipid occurs in _____
    a. maize   b. kokum butter   c. *Cinchona*
12. The raw products obtained from any organism is called _____
    a. synthetic drug   b. natural drug   c. adulterated drug
13. Chirata is a _____
    a. bitter tonic   b. expectorant   c. flavour
14. Antimalarial drug quinine is derived from the bark of _____
    a. *Tectona*   b. *Cinchona*   c. *Eucalyptus*
15. Immunizing agents are _____
    a. enzymes   b. pesticides   c. vaccines
16. Peptic ulcer is cured by _____
    a. raw banana   b. vasaka   c. belladonna

Answers				
1. True	2. True	3. True	4. False	5. True
6. False	7. True	8. False	9. b	10. c
11. b	12. b	13. a	14. b	15. c
16. a				

Answer the following

1. What is active principle? State its significance.
2. Differentiate between organized drugs and unorganized drugs.
3. Give the sources of the following:
    i. Purgative
    ii. Expectorant
    iii. Sweetener
4. Write a note on carminative drugs.
5. Enumerate the chemical and therapeutic classification of drugs. Add a note on their merits and demerits.
6. With suitable examples bring out the morphological classification of crude drugs.
7. Write an essay on various systems of classification of crude drugs.

COLLECTION AND PROCESSING of CRUDE DRUGS

5

Crude drugs need to undergo suitable preparation for being marketed.

The reasons for such preparation of crude drugs are as follows:

* To stabilize them in transport and storage.
* To ensure the absence of foreign organic matter and substitutes.
* Market preparation of crude drugs also takes care of pharmaceutical values.

The following methods are used in the preparation and processing of crude drugs.

* Collection and harvesting of crude drugs
* Drying
* Garbling
* Packing prior to marketing to meet the standard pharmacopoeial requirements

COLLECTION AND HARVESTING

Harvesting of medicinal plants needs proper planning. Plant materials, animal products and other forms from which crude drugs are extracted are collected in such a way that the parts must contain maximum concentration of the active constituents. Wild plants have high concentration, and collection of wild plants has a number of drawbacks—it is unscientific, labour-oriented and undependable. Collection by untrained collectors leads to the danger of wrong identification and collection of poisonous crude drugs.

Methods of Plant Collection

The collection process is initiated usually when plant collectors are free and when there is a demand for crude drugs. Generally, the existing environmental conditions are not taken into consideration.

The following are some of the important points to be borne in mind during collection.

- ✤ Underground parts like roots, corms, bulbs, rhizomes, tubers, etc. should be collected at the end of flowering and fruiting, that is, during July–August and March–April.
- ✤ All the underground parts collected should be cleaned and dried in shade or in ovens at 30°C, e.g., *Gloriosa*, garlic, etc.
- ✤ In case these parts are collected during summer, their survival and regeneration are affected during the next growing season.
- ✤ Fruits and flowers should be collected individually only after maturity.
- ✤ Annual stems should be cut 5–10 cm above the ground, after the plant has begun to flower.
- ✤ Leaves are collected at a young stage, as the leaves usually contain a high content of chemical constituents during this stage, e.g., tulsi, pudina, etc.
- ✤ Specific parts of flowers like petals (lotus), whole flowers (*Vinca*) and flower buds (clove) are collected separately.
- ✤ Fruits and berries are harvested in early autumn, e.g., cardamom.
- ✤ Barks are collected in spring or early summer when the cambium is very active, as it is easy to detach them from the stem during this season.
- ✤ Sap from tree is collected in spring as it rises, e.g, silver birch.
- ✤ Milky latex is collected by squeezing the stems over a bowl using gloves, e.g., *Taraxacum*.

Methods of Harvesting

Harvesting is an important operation, as it reflects upon the economic use of the crude drugs. The following are the important do's during harvesting.

- ✤ Skilled workers should do this operation.
- ✤ Selective collection of drugs is very essential.
- ✤ Hand-picking is done for leaves, fruits, etc.
- ✤ Mechanical devices like diggers or lifters should be used for underground drugs.

* All aerial parts are harvested by binders.

* Seed strippers are used for flowers, seeds and small fruits.

* Cochineal insects are collected from branches of cacti by brushing.

* Mowers are used for peppermint.

* Sometimes reaping machines are also used for harvesting.

DRYING

Proper drying preserves the drugs for a long time and also maintains better pharmaceutical value. The purpose of drying is to remove sufficient amount of moisture, so as to get good quality of crude drugs. In addition, drying helps the drugs to resist microbial growth and also helps in proper grinding of the drugs. Before drying, special methods like fermentation, slicing and cutting the materials into small pieces are utilized to enhance better drying.

Methods of Drying

Depending upon the type of chemical compounds present in the drugs, the following two methods of drying may be practised—natural drying or sun drying and artificial drying.

1. Natural drying In order to retain the natural colour as in clove, senna and also the volatile principles of the drug (e.g., peppermint), the crude drugs are dried in the shade. Gums, resins and some fruits and seeds can be dried directly in sunlight.

2. Artificial drying It includes oven-drying (tray-drying), vacuum-drying and spray-drying.

Oven-drying In this method hot air is used to dry the drugs, e.g., cinchona bark, tea, etc.

Vacuum-drying Drugs that are sensitive to high temperature are dried by this process, e.g., tannic acid and digitalis leaves.

Spray-drying Drugs that are highly sensitive to atmospheric conditions and also to vacuum-drying are dried by this process. Spray-dryers are used for quick drying of drugs, e.g., papaya latex, pectin, etc.

De-humidifying In this method herbs are dried with the help of an instrument called dehumidifier. The dehumidifier is placed in a sealed small room, in which the herbs are hung in loose bunches or placed on mesh trays, to suck water out of the plants. In this process herbs will dry very quickly as no heat is used. It is an expensive method, but very effective.

GARBLING

The process of removal of dirt and foreign organic materials from the drug is called garbling or dressing. The unwanted stems, bark, rootlets, iron, etc., should be removed to get better quality.

PACKING

The following aspects are to be considered while packing the drugs.

* Morphological and chemical nature of drugs.
* Its ultimate use.
* Climatic conditions during transportation and storage.

The materials used for packing different drugs are as follows:

* Goatskin is used for packing *Aloe*.
* Colophony and balsam of Tolu are packed in kerosene tins.
* Asafoetida is packed in closed containers.
* Senna and *Vinca* leaves are pressed and wrapped.
* Roots and seeds can be packed in gunny bags.

STORAGE OF DRUGS

A number of drugs absorb moisture during their storage and become susceptible to microbial growth. The excessive moisture also facilitates enzymatic reactions resulting in decomposition of active constituents, e.g., ergot. Exposure of drugs directly to sunlight also causes destruction of active chemical compounds. Shape also plays a very important role in preserving the crude drugs. Colophony is preserved well when it is in entire form, but if stored in powdered form, it gets oxidized. The fixed oil in the powdered ergot becomes rancid on storage. Therefore, to maintain a good quality, ergot drug should be defatted with lipid solvent prior to storage.

Lard should be preserved against rancidity by adding siam benzoin. In order to protect the drugs from destruction of atmospheric oxygen, they must be stored in air-tight containers. For certain drugs like shark-liver oil, papain, etc., they are stored in containers where air is replaced by an inert gas like nitrogen. Protection and preservation of drugs against insects or mould attacks are also very important during storage. Therefore, drugs should be dried properly before storage, and can also be treated with fumigants like methyl bromide, carbon disulphide and hydrocyanic acid. As high temperature decomposes the drug during storage, drugs should be stored at a very low temperature. Costly phytopharmaceuticals are preserved at refrigerated temperature in closed containers. Wooden boxes and paper bags should not be used for storage of drugs.

PROCESSING OF CRUDE DRUGS

Medicinal plants must be treated and modified in such a way that its specific curative substance can be put into practice. The active principles of the medicinal plants either individually or in groups can have a specific action on our organs. Different methods are followed to extract these substances from medicinal plants.

Methods of Preparation

The medicinal plants are subjected to the following processes for the preparation of different herbal dosage forms.

* Decoction
* Maceration
* Infusion
* Juice extraction

Decoction   This method is used for plants whose active principles are not soluble in standing cold or boiling hot water or difficult to extract. Therefore the crude drug is boiled in a specified amount of water for a defined time, cooled and filtered to get the decotion.

A decotion is generally prepared using woody parts like roots, rhizomes, barks and berries, but sometimes leaves and flowers may also be included. For decoction, fresh materials should be sliced or cut into small pieces, while dried crude drugs should either be powdered or well-bruised.

Maceration   This process is carried out for drugs whose active principles take very long time to dissolve in cold water, without heating. As a result their active constituents will be released in cold water, e.g., mucilage.

Infusion   The dilute solutions of the readily soluble constitutents of crude drugs are called infusions. An infusion is the simplest way to prepare the most delicate aerial parts of plants especially leaves and flowers or fruits for use as a medicine or as a revitalizing or relaxing drink. It is usually prepared using either a single herb or a combinations of herbs and may be drunk cold or hot. There are two types of infusions, namely, cold infusions and hot infusions.

Cold infusions   When the active principles of the drugs are volatile in nature or sensitive to heat treatment, cold infusions are preferred. These are prepared by immersing the drug in cold water for some time. During this time all the active principles present will be released into the water without heating, e.g., tamarind.

Hot infusions These are made by pouring boiling water over the herb or herbal parts and keeping the infusion in a covered container for 5–10 minutes to dissolve and then filtered. Herbal teas are generally prepared in this way, e.g., Chebula, tea, etc.

Juice extraction Extraction of juice is similar to preparing decoction of medicinal plants.

REVIEW QUESTIONS

State whether the following sentences are true or false

1. Nitrogen gas is used in the storage of crude drugs.
2. Methyl bromide acts as an antibiotic.
3. The specific action of any drug is due to its active principle.
4. Drying of drugs leads to microbial growth.
5. Pharmaceutical values of drugs depend upon market preparation of drugs.

Choose the correct answer

6. Rancidity of lard is prevented by adding _____
   a. bentonite   b. siam benzoin   c. kaolin
7. The removal of foreign materials from the drug is called _____
   a. grinding   b. garbling   c. drying
8. The drug aloe is packed in _____
   a. goat skin   b. gunny bags   c. tin
9. The use of dehumidifier is to suck out _____
   a. chemicals   b. mucilage   c. water
10. Vacuum drying method is used for drugs sensitive to _____
   a. high temperature   b. high pH   c. low temperature

Answers				
1. True	2. False	3. True	4. False	5. True
6. b	7. b	8. a	9. c	10. a

Answer the following

1. What is decoction?
2. Distinguish between maceration and infusion.
3. Comment on storage of drugs?
4. Write a note on collection of plant drugs.
5. How will you dry the crude drugs for market sales?
6. "Pharmaceutical values and long-term storage depend on proper drying and preservation of crude drugs." Explain.
7. Write an account on the method of collection, processing and storage of crude drugs.

PHYTOCONSTITUENTS OF THERAPEUTIC VALUE

6

INTRODUCTION

Phytoconstituents are the chemical substances of organic nature, which are formed in plants by the activity of their individual cells. The process by which plants are able to convert the simple chemical substances into complex organic compounds with the help of enzymes is known as biosynthesis. The products of biosynthesis like glycosides might serve as storage form of carbohydrates for plants, and volatile oils help in the process of pollination by attracting insects to the flowers. Carbohydrates, lipids, glycosides, tannins, volatile oils, resins, gums, alkaloids and polyphenols are the secondary metabolites produced as a result of biosynthetic pathways, and constitute phytoconstituents of therapeutic significance.

The medicinal value of any plant drug however depends on the nature of the chemical constituent present in it and is referred to as the active principle. The isolated chemical constituents of plants have various applications in medicine.

CARBOHYDRATES

Carbohydrates are plant products, which contain carbon, hydrogen and oxygen. The ratio of hydrogen to oxygen is the same as that is in water, i.e., two to one. Carbohydrates are widely distributed in plants and provide means of storage and transport of energy and are also the building blocks of the cell wall. Carbohydrates can be classified into three main groups:

*   Mono- and oligosaccharides (true sugars)
*   Polysaccharides (non-sugars)
*   Derived carbohydrates (gums, mucilage and pectin)

Mono- and Oligosaccharides

The term sugar refers only to those carbohydrates that have sweet taste and are soluble in water. These include mono- and oligosaccharides. Monosaccharides are subdivided on the basis of number of carbon atoms into, trioses, tetroses, pentoses and hexoses. Aldehydic (—CHO) or ketonic($>$C$=$O) group is the main functional group. They are crystalline substances that are soluble in water and practically insoluble in organic solvents like ether, chloroform and absolute alcohol. They are optically inactive and respond positively to Molisch's test. The sugars of medicinal importance among pentose include arabinose, rhamnose, ribose and xylose. Among hexoses, glucose, fructose, galactose and mannose are the important ones.

Oligosaccharides consist of two or more than two monosaccharide units. Disaccharides consist of two sugar units, e.g., sucrose, lactose and maltose and are used in pharmaceutical industry. Sucrose, commonly called as cane sugar, is present in fruit juices, sugar cane, etc. Sucrose upon hydrolysis yields glucose and fructose. It is used as a nutrient and in making simple and other official syrups, which serve as flavoured vehicles in pharmaceutical preparations.

Lactose is the sugar obtained from milk. It is commonly called milk sugar. On hydrolysis, it yields glucose and galactose. Maltose (malt sugar) is produced by the hydrolysis of starch during the germination of barley and other grains. On hydrolysis it gives glucose.

Polysaccharides

Polysaccharides consist of a number of monosaccharide units joined together in long chains. They are not sweet in taste, and therefore are termed as non-sugars and are insoluble in water. Starch, pectins, agar and dextrins are examples of polysaccharides. Starch is widely distributed in plants and is the essential storage form of energy. Starch contains amylase and amylopectin. Starch is located in the chloroplasts in the form of granules. In pharmacy, starch is used as an adsorbent, emollient and diluent.

Cellulose is widely distributed in the plant cell wall. It is a fibrous material of and together with lignin, it is responsible for the structural rigidity of plants. It consists of long chains of glucose units. Cotton fibres consist of almost pure form of cellulose. Oxidized cellulose and methyl cellulose are examples of cellulose derivatives used in pharmacy.

Dextrins are the partial hydrolysis products of starch and are soluble in water. White and yellow dextrins are the two varieties of dextrin used in baby foods as a source of easily digestible carbohydrate.

Derived Products of Carbohydrates

Gums, mucilage and pectin are derived carbohydrates and are composed of acid or ester forms. Gums, mucilage and pectin are used in pharmacy. Gums are polyuronides consisting of sugar and uronic acid units of varying composition. On hydrolysis, gums yield sugars like arabinose, galactose, glucose, mannose, xylose together with sugar acids like glucuronic acid and galacturonic acid. Gums are translucent, amorphous substances which are formed in plants as a result of injury. They readily dissolve in water giving a sticky solution. They are used as tablet binders, emulsifiers, suspending agents and as thickeners in various pharmaceutical preparations. *Acacia*, sterculia gum and tragacanth are the important examples of gums.

Mucilages do not dissolve in water and form a clear colloidal solution like gums but merely swell. Chemically, mucilages may be similar to gums but differ in the nature of sugar and acid residues. They may contain sulphate groups or their salts. Mucilages are used as suspending agents. Agar, plantago seed and linseed are examples of mucilage-containing drugs.

Pectins consist of methoxylated polygalacturonic acid. They are present in the inner portion of the rind of citrus fruits and in apples. They swell in water and form stiff jellies. Due to their jelling property, they are used in medicinal preparations.

GLYCOSIDES

Glycosides are the organic substances which on hydrolysis give rise to one or more sugar moieties along with a non-sugar compound. The sugar component of glycoside is called glycone and the non-sugar component is aglycone or genin. The sugars involved in glycosides are of different types, but the most commonly occurring sugar is β-D-glucose. The other sugars detected are galactose, rhamnose, digitoxose, cymarose, etc. The linkage between glycone and aglycone is called glycosidic linkage. The glycosidic linkage can be alpha (α) or beta (β) depending on the arrangement of atoms in it. In plants only the β form of glycosides are formed, although the α-linkage is found in nature in some carbohydrates such as sucrose, glycogen and starch. Beta linkage can be hydrolysed by the enzyme emulsion whereas maltase hydrolyses alpha glycosides.

Chemically, glycosides are the acetals or sugar ethers formed by interaction of hydroxyl group, each of non-sugar and sugar moiety, with a loss of mater molecule. The hydroxyl group of aglycone may be alcoholic or phenolic and in some cases from amines also.

Glycosides are widely distributed in nature and many glycosides are used in traditional and modern medicine due to their therapeutic actions. These are colourless, crystalline or amorphous substances, which are generally soluble in water and dilute alcohol with the exception of resin glycosides, but insoluble in organic solvents like petroleum ether, chloroform, etc. The aglycone moiety is soluble in organic solvents like benzene or ether. Glycosides are easily hydrolysed by water, mineral acids and enzymes.

Glycosides are classified based on the nature of the sugar, and the type of linkage existing between the glycone and aglycone part. But generally glycosides can be classified on the basis of the chemical nature of aglycone and the pharmacological activity into anthraquinone glycosides, saponin glycosides, cardiac glycosides, cyanophore glycosides and isothiocyanate glycosides.

Anthraquinone Glycosides

The anthraquinone glycosides are the ones whose aglycone component is a polyhydroxyanthraquinone derivative. The drugs having these glycosides possess cathartic activity, e.g., *Cascara*, *Aloe*, *Rhubarb* and senna. The common polyhydroxy anthraquinone derivatives present in these drugs are chrysophanic acid (chrysophanol) and rhein in *Rhubarb*, aloe-emodin in *Aloe* and senna, and rhein in senna. The laxative action of certain drugs is attributed to derivatives of anthraquinones.

The anthraquinone derivatives are often orange-red in colour and are observed in the medullary rays of *Rhubarb* and *Cascara*. Their presence in plants can be detected by Borntrager's test.

Saponin Glycosides

Saponins are an important group of glycosides. These are of complex nature and are widely distributed in higher plants. The saponins are characterized by their property of producing soaplike foam when shaken with water. Their aglycone component is called sapogenin which generally has a steroidal structure. Most of the saponins are highly toxic when injected into the body.

The most important saponin-containing drugs are *Quillaia*, *Glycyrrhiza* and *Senega*. Most of the saponins are neutral and soluble in water. Like other glycosides, saponins are hydrolysed to form a sugar, usually dextrose, and an aglycone known as sapogenin, which may be a steroid or a triterpene. Saponins have a high molecular weight.

Diosgenin is the sapogenin derived from *Dioscorea* and serves as a source of starting material for the synthesis of corticosterone and related compounds. Saponins have the property of destroying red blood cells by their haemolytic action. Glycyrrhizic acid is the aglycone of glycyrrhizin, the glycosidic principle of *Glycyrrhiza glabra*.

Cardiac Glycosides

Cardiac glycosides are considered to be the most important constituents used therapeutically. Cardiac, cardiotonic and cardioactive glycosides are used medicinally to strengthen the weakened heart and thus allow it to function more efficiently. The aglycones or genins of cardiac glycosides consist of steroid. The different genins may vary in the type of substituents. The more prevalent steroidal aglycones are known as cardenolides, which are 23-carbon steroids with a five-membered lactone ring.

Digitoxigenin is the aglycone of digitoxin, the main cardiac glycoside of *Digitalis* species. Although the fundamental cardiac activity of these glycosides resides in the genin, the activity is greatly modified by the sugar moiety. The sugar helps in increasing the solubility of the glycosides and helps in the fixation of heart muscles.

The examples of other cardiac glycosides are digoxin, lanatoside-C, quabain, gitoxin and strophanthin. The genins of these glycosides are quite similar to each other.

Scillarenin-A, the aglycone of cardiac glycosides of squill, contains a six-membered lactone ring. It is placed in the category of bufadienolides. The drugs containing cardiac glycosides are *Digitalis, Strophanthus* and *Squill.*

Cyanophore Glycosides

Cyanophore glycosides are also called cyanogenetic glycosides. They yield hydrocyanic acid (HCN) as one of the products of hydrolysis. These glycosides are mostly found in the plants of the family Rosaceae. The most widely distributed cyanophoric glycoside is amygdalin which is found in large quantities in bitter almonds. Kernels of apricots, cherries, peaches and plums also contain amygdalin. On hydrolysis, amygdalin yields two molecules of glucose, one molecule of benzaldehyde and one molecule of hydrocyanic acid.

Prunasin is another cyanophoric glycoside present in wild cherry bark. The drug, in syrup form, is used in cough preparations, to which it gives a pleasant taste and a sedative and expectorant property.

Isothiocyanate Glycosides

Isothiocyanate glycosides (glucosinolate compounds) are those glycosides the ones, whose aglycone contains isothiocyanate group. This group is represented by sinigrin, the principal constituent of black mustard. Sinigrin on hydrolysis yields allylisothiocyanate, potassium acid sulphate and glucose. Sinalbin from white mustard and gluconapin from rape seeds are other examples for isothiocyanate glycosides.

LIPIDS

Lipids are water-insoluble, oily, organic compounds soluble in non-polar organic solvents. The term lipid is used for fixed oils, fats, waxes and other fatlike substances, which are found in nature. Fixed oils and fats are of common occurrence in plants and serve as reserve food for them. Many drugs contain fixed oils and fats as their principal constituents. The fixed oils and fats are separated from crude vegetable drugs or crude animal drugs and are used as drugs in the refined state.

Fixed Oils and Fats

Fixed oils and fats differ from each other only in their physical appearance. The lipids which are liquid at normal temperature are called fixed oils. The lipids which are semi-solid or solid at normal temperature are called fats. Thus, a lipid in cold climate (temperate) is a fat and the same in hot climate (tropics) is a fixed oil.

Chemically fixed oils and fats are esters of glycerol, with long-chain fatty acids. These esters are termed as glycerides. They may be simple glycerides or mixed glycerides. Glycerides can be hydrolysed readily with aqueous sodium hydroxide in which case the component fatty acids are obtained as sodium salts (soap) along with glycerol. This process is known as saponification.

The fatty acids constituting the glycerides may be saturated or unsaturated. Commonly the glycerides of unsaturated fatty acids are liquid whereas those of saturated fatty acids are solid.

Fixed oils and fats when dropped on paper form a permanent stain (grease spot) and get fixed on the paper. Hence the term fixed oil. They are non-volatile in nature and this property of fixed oil differentiates them from volatile oils. Fixed oils and fats are insoluble in water and are lighter than it. Except castor oil, which dissolves readily in alcohol, fixed oils and fats are sparingly soluble in cold alcohol. Fixed oils and fats are soluble in many organic solvents like petroleum ether, diethyl ether, chloroform, etc.

Fixed oils and fats may be separated from the crude drug by extraction with organic solvents or by expression. Generally, seeds contain large quantities of oils and fats than any other part of the plant. Oils and fats, which are to be used internally for medicinal purpose, must be obtained by expression. Extraction with an organic solvent may also extract certain toxic principles that are present in seeds. To check the purity of oils and fats, pharmacopoeias prescribe certain tests. These are determination of acid value, saponification value, and iodine value, etc.

Oils such as arachis oil, castor oil, linseed oil, etc. are some of the oils used in pharmacy. Shark-liver oil is an animal product used for medicinal purposes. Coca butter, a fat, and lanolin, a purified fatlike substance obtained from wool of sheep, are other lipids used in pharmacy.

Waxes

Waxes are the esters of fatty acids with high molecular weight monohydric alcohols such as cetyl alcohol, cholesterol, etc. Waxes usually contain esters of free acids, hydrocarbons, free alcohols and sterols. The hydrocarbons and sterols are unsaponifiable. The esters in waxes are generally more resistant to saponification than the glycerides of oils and fats.

In many ways, waxes resemble fats, but waxes are more difficult to saponify and can be saponified only by alcoholic alkalies, whereas fats may be saponified by either aqueous or alcoholic alkali. Waxes are also differentiated from fats and oils by their low saponification values and very low iodine value due to the presence of saturated acids.

Waxes are variably viscous and solid substances with characteristic waxy lustre. They are insoluble in water, but soluble in many organic solvents. Waxes are obtained from both plant and animal sources. Carnauba wax is obtained from plants and beeswax, wool fat, spermaceti, etc., are obtained from animals. In plants, waxes are usually present on the outer cell walls of epidermal tissue. They prevent the loss of water from plants.

Waxes are used in pharmacy for making ointments, cerates, cosmetic creams, and plasters.

VOLATILE OILS

The flavouring oil constituents which evaporate on exposure to air at ordinary temperature are called volatile oils. As volatile oils are responsible for the essence or odour of the plant, they are also known as essential oils.

Volatile oils are found in various plant parts such as flower petals (rose), fruits (fennel), leaves (mint), fruit rind (orange), bark (cinnamon), etc. They are secreted in specialized secretory cells like glandular hairs, modified parenchyma cells, in tubes called vitae or in lysigenous or schizogenous cavities. Volatile oils occur widely in nature and are commonly found in the members of families like Rutaceae, Umbelliferae, Lamiaceae, Myrtaceae, Pinaceae and Asteraceae.

Volatile oils when fresh are colourless liquids or crystalline or amorphous solids. On keeping for a long time, they become darker in colour, especially when exposed to air and direct sunlight. Volatile oils should, therefore, be stored in tightly closed amber-coloured bottles in a cool, dry place. Volatile oils are slightly soluble in water, but are readily soluble in ether, alcohol and other organic solvents.

The purity of volatile oils is checked by determining their densities, refractive indices, optical rotations and boiling points. Volatile oils differ from fixed oils in that volatile oils do not leave a permanent stain on paper and cannot be saponified by alkalis.

Volatile oils are generally obtained by the process of distillation. Water distillation, steam distillation and direct steam distillation are the usual techniques employed. For example lemon oil is obtained by expression.

Volatile oils are used as flavouring agents in perfumes, and in various pharmaceutical preparations, and some of them are used for their therapeutic action. In some cases, volatile oils are used as insect repellents.

Chemically, volatile oil consists of terpenoidal mixtures of a wide variety of compounds. These compounds may be hydrocarbons, alcohols, ketones, aldehydes, phenols, ethers, oxides, esters, acids or aromatic or aliphatic compounds.

Hydrocarbon volatile oils contain the compounds called terpenes. These compounds do not contain any functional group and may be branched. Turpentine oil is an example of a hydrocarbon volatile oil drug obtained from *Pinus* species.

Alcohol volatile oils contain alcoholic compounds which may be acyclic or cyclic in nature. The most common acyclic alcohol found in volatile oils is geraniol. Menthol is an important monocyclic alcohol found in peppermint. Other drugs containing alcohol volatile oils are coriander oil, cardamom oil, rose oil and pine oil.

The aldehydes occurring in volatile oils are benzaldehyde, cinnamaldehyde, citral, vanillin and anisaldehyde. Important drugs containing aldehyde volatile oils are cinnamon and lemon peel.

Ketone volatile oils contain the compounds called terpene ketones. Some examples are camphor, caraway, spearmint, etc.

Phenolic volatile oils are present in drugs like thyme, clove, creosote and pine tar. They have antibacterial, antifungal and antiseptic properties. Some of them also have slight local anaesthetic and analgesic actions. The important phenol-containing volatile oils are eugenol, thymol and carvacrol.

Some phenols occur as ethers in volatile oils. Anethole found in fennel and anise are important examples. Safrole is another phenolic ether found in sassafras.

Eucalyptus oil is an oxide volatile oil, which contains cineole as the chief constituent. It is used as a flavouring agent.

Ester volatile oils contain esters of wide variety. Terpeneol acetate, geraniol acetate and borneol acetate are of common occurrence. Methyl salicylate is the ester constituent of wintergreen oil. Lavender oil and mustard oil are the other examples of this group.

RESINS AND RESIN COMBINATIONS

Resins are solid or semi-solid amorphous products of complex chemical nature. Resins are usually obtained as exudates from plants (except shellac) and are considered as end products of metabolism. Resins and related resinous products are produced in plants during normal growth or secreted as a result of injury to the plants. They usually occur in schizogenous or schizolysigenous cavities or in special tubes called resin ducts. These ducts are anastomosing and thus a single incision may drain the resin from a particular area of the plant. The cells lining the ducts possess a layer of slimy matter bounded by a fine cuticle. In some cases, numerous resin ducts are present, e.g., *Copaiba*. In case of turpentine, only a few resin ducts are normally present, but when the cambium is injured, new secondary wood formed contains very large number of ducts and such resin is referred to as a wound, or as a traumatic or pathologically produced resin.

The resin may continue to flow for sometime from the injured part or in some cases it may be necessary to inflict wounds at frequent intervals. Resins are usually non-crystallizable and translucent hard masses. On heating, they soften and finally melt. All resins are insoluble in water but soluble in organic solvents such as alcohol, ether and chloroform. Chemically, resins are complex mixtures of resenes (hydrocarbons), resin acids, resin alcohols (resinols), resin phenols (resinotannols) and resin esters. Examples of resin-containing drugs are resin or colophony, *Podophyllum* and *Cannabis*.

Resins often occur in homogeneous combination with other plant products. These are collectively called resin combinations. The resin combinations of pharmaceutical importance are oleoresins, oleogum resins and balsams.

The resinous exudation may consist almost entirely of resins, e.g., benzoin; or it may be associated with volatile oil, e.g., turpentine, *Copaiba*, or resin associated with gum, e.g., gum resin. If a considerable amount of volatile oil is present, the substance is called an oleogum resin, e.g., myrrh.

The resins or oleo-resins, which contain benzoic or cinnamic acid, either free or combined, are commonly called balsams, e.g., benzoin, balsam of Tolu, balsam of Peru and storax.

Oleoresins are homogeneous mixtures of resins and volatile oils. They may be liquid, semi-solid or solid depending on the amount of volatile oil in the mixture. The oleoresin-containing drugs are *Capsicum*, ginger and male fern.

Oleogum resins are combinations of resins, gums and volatile oils. Myrrh and asafoetida are the oleogum resins used in therapeutics.

Balsams are resinous mixtures containing varying amounts of cinnamic acid, benzoic acid or esters of these acids. The balsams containing free acids are partially soluble in water, but those containing esters of these are insoluble. Benzoin, storax and balsam of Tolu are the drugs belonging to this category.

Besides the above major combinations, resins may also occur in combination with glycosides. These are called glucoresins. Such glucoresins are found in *Podophyllum*.

The estimation of resins and resin combinations in crude drugs or their extracts is carried out by extraction (usually with alcohol) of the substance followed by weighing the residue obtained.

ALKALOIDS

Alkaloids are a complex and heterogeneous group of compounds. They comprise the most important class of natural products. Alkaloids occur in plants and animals. They produce a wide variety of therapeutic effects. Alkaloids also constitute the largest group of phytoconstituents as over 6,500 alkaloids have been known so far.

The word "alkaloid" is derived from the term alkali-like and was used originally to describe a group of bases of botanical origin. The alkaloids occur most commonly in plants of families like Apocynaceae, Leguminosae, Papaveraceae, Ranunculaceae, Rubiaceae and Solanaceae. Among monocotyledons, Liliaceae and Amaryllidaceae are the two most promising families.

Alkaloids can be generally defined as basic, nitrogen-containing heterocyclic compounds of plant origin having marked physiological activity. Alkaloids contain one or more nitrogen atoms, usually as part of a cyclic system, but alkaloids like ephedrine and mescaline contain nitrogen which is not present in cyclic form. Colchicine is an exception to the above definition as this alkaloid is not basic in nature. Alkaloids are considered as derivatives of pyridine, quinoline or isoquinoline and contain carbon, hydrogen, oxygen, and nitrogen, but a few are without oxygen. Mostly alkaloids are solid colourless crystalline products, but a few alkaloids which generally do not possess oxygen, e.g., nicotine, conine and spartein are volatile colourless liquids. Some alkaloids are coloured, e.g., berberine (yellow colour) and sanguinarine (red colour).

Most of the alkaloids are slightly soluble in water, but soluble in organic solvents like chloroform, ether, alcohol and benzene. Alkaloids combine with acids to form salts and are used in this form. A water-soluble alkaloidal salt or other compound is more useful than one insoluble in water.

In plants, alkaloids are found in various parts as in seeds (Strychnos), in fruits (pepper), in leaves (*Belladonna*), in roots (*Rauwolfia*), in rhizomes and roots (*Ipecac*), in corms (*Colchicum*) and in barks (*Cinchona*).

Alkaloids are biosynthesized from amino acids. They occur in cell sap in the form of water-soluble salts of such organic acids as malic, citric, oxalic, succinic, tartaric or tannic acid.

In pharmaceutical preparations, alkaloids are always present in the form of salts either as solid or in solution.

Classification of Alkaloids

Many different systems of classification are possible, as alkaloids show great variety in their source, chemical structure and pharmacological action. The chemical classification of alkaloids based on their carbon–nitrogen skeleton has been found to be most convenient. Some of the nitrogen-containing ring systems are pyrrolidine, pyridine, piperidine, quinoline, isoquinoline, tropane, indole, imidazole and purine.

Pyridine–piperidine alkaloids These are present in *Areca*, *Lobelia* and tobacco. Lobeline is obtained from *Lobelia* and is used as a respiratory stimulant. Its basic skeleton is a piperidine nucleus.

Tropane alkaloids Tropane is a bicyclic structure, in which both pyrrolidine and piperidine systems fuse through a common nitrogen atom. The alcohol corresponding to tropane is tropine. Its ester with organic acids like tropic acid or benzoic acid constitute tropane alkaloids. These alkaloids occur in the members of Solanaceae and Erythroxylaceae. Solanaceous drugs such as *Belladonna hyoscyamus* and *Stramonium* contain hyoscyamine, atropine and hyoscine. These alkaloids possess anticholinergic activity. Cocaine, obtained from coca leaves, is another tropane alkaloid, which is used as a local anaesthetic. While solanaceous alkaloids are esters of tropic acid, cocaine is an ester of benzoic acid.

Quinoline alkaloids The basic nucleus of quinoline alkaloids is obtained from the bark of *Cinchona* species. The major alkaloids of this group are quinine, quinidine, cinchonine and cinchonidine. Of these, quinine and its stereoisomer quinidine are of therapeutic importance. Quinine is an antimalarial drug, while quinidine is a cardiac depressant and is used particularly to inhibit auricular fibrillation.

Isoquinoline alkaloids The alkaloids of therapeutically important drugs like opium and *Ipecac* belong to this group. Opium alkaloids can be put into 2 groups on the basis of their structures, one group includes morphine, codeine and thebaine, and the other group includes papaverine and narcotine. Both the groups have the same biosynthetic intermediate, i.e., benzylisoquinoline.

Morphine and codeine are used as analgesics, while papaverine is a smooth muscle relaxant. Narcotine is used as a cough suppressant. The major alkaloid of the drug ipecac is emetine, which also has an isoquinoline skeleton. Emetine is an anti-amoebic drug. It also possesses expectorant and emetic properties.

Indole alkaloids   Indole alkaloids form the largest group of alkaloids and they possess diverse pharmacological action. These alkaloids have indole nucleus as part of their structure. Indole alkaloids are derived from *Rauwolfia*, *Vinca*, *Physostigma*, ergot, nux-vomica, strychnine and brucine.

Reserpine is the main alkaloid of the drug *Rauwolfia*. It has two nitrogen atoms and a complex structure. Reserpine alkaloid is used as antihypertensive and tranquilizer.

Vinca (*Catharanthus*) yields alkaloids such as vinblastine and vincristine which possess complex structures and are used for the treatment of leukaemia and Hodgkin's disease.

Ergot, the fungal drug, contains several indole alkaloids of medicinal importance. Ergometrine is an oxytocic and ergotamine is used as a specific analgesic in migraine.

Nux-vomica contains chief alkaloids, strychnine and brucine. Strychnine is used as a valuable tool in pharmacological and physiological research. It is also used as a rodenticide.

The indole alkaloid physostigmine taken from *Physostigma* is used in ophthalmology to treat glaucoma.

Imidazole alkaloids   Alkaloids containing imidazole ring are represented by pilocarpine which is the principal alkaloid of *Pilocarpus*. It is used in the treatment of glaucoma.

Steroidal alkaloids   Alkaloids which have a steroid nucleus with nitrogen attached to it are called steroidal alkaloids. Kurchi contains these types of alkaloids, conessine being the major one. The drug is used in amoebic dysentery.

Alkaloidal amines   These are alkaloids which do not contain nitrogen as part of the heterocyclic ring; the nitrogen is present outside the cyclic structure. Such alkaloids are categorized as alkaloidal amines. Ephedrine, the major alkaloid of *Ephedra* (gymnosperm), is an example. It is used as a bronchodilator.

Colchicine is an alkaloid obtained from colchicum. It is used in the treatment of gout.

Purine bases   A group of nitrogenous compounds with purine nucleus consists of caffeine, theophylline and theobromine. These compounds are very weak bases and form unstable salts with acids. These are called purine bases.

Tea leaves and coffee seeds are the sources of these bases. They stimulate the central nervous system and possess diuretic activities.

Many alkaloids in small quantities (10 to 100 µg) in neutral or slightly acid solution, form precipitates, often coloured, with certain reagents. These reagents are called alkaloidal reagents and are used to detect the presence or absence of alkaloids in a powdered crude drug, in an extract of the plant material or in a chemical mixture.

Mayer's reagent, Hager's reagent and Wagner's reagent are used for testing alkaloids. Besides the above-mentioned reagents, ammonium phosphomolybdate solution and tannic acid solution also give precipitate with alkaloids.

TANNINS

Tannins are complex phenolic compounds present in plants. Tannins are characterized by their ability to combine with proteins of animal hides and convert them into leather. Large quantities of tannin are found in oak galls and these serve as the source of tannic acid.

Tannins are classified into two groups on the basis of their ability to be hydrolysed. They are hydrolysable tannins and non-hydrolysable tannins.

Hydrolysable Tannins

Hydrolysable tannins can be readily hydrolysed by acids, alkalis or enzymes to yield phenolic acid and sugar. These are otherwise known as gallic acid tannins, as this class consists of gallic acid and related polyhydric compounds esterified with glucose. They produce a dark blue to black colour with ferric chloride, but are not precipitated with bromine water. They are also called pyrogallol tannins.

Non-hydrolysable Tannins

These are otherwise called condensed tannins. These are resistant to cleavage. When heated with acid, they are decomposed into red insoluble compounds known as phlobaphenes. On dry distillation, they yield catechol and therefore are also called catechol tannins or phlobatannins. They react with ferric chloride to give a green colour and are precipitated with bromine water. Condensed tannins occur in catechu, gambir, krameria and chestnut leaves. Tannins are also present in *Cinchona*, clove, cinnamon, *Hamamelis*, etc.

Tannins are usually amorphous in nature. They are soluble in water and form a colloidal solution in water. Tannins have an astringent taste. They produce blue or greenish black colour with ferric salts. They are precipitated from aqueous solution by gelatin, albumin or other proteins. Tannins also combine alkaloids to form tannates, most of which are insoluble in water, and they yield a bulky precipitate of phenazone.

Tannins when applied to living tissue combine with proteins to have astringent action. Many tannin-containing drugs are used therapeutically due to their astringent action. They are used in the treatment of burns, as they precipitate the proteins of exposed tissues to form a protective coating. They are also used in the tanning industry for converting animal hides to leather.

PHENOLS

Phenolic compounds constitute one of the largest groups of secondary metabolites, widespread in nature. Such compounds are composed of one or more aromatic benzene rings with one or more hydroxyl groups (C—OH).

This enormous class includes numerous plant compounds that are chemically distinct from terpenes. These compounds have diverse functions in plants such as protection against herbivores, pathogens, ultraviolet radiation and attraction to pollinators and seed dispersal. In some plants they (phenolics in the form of phytoalexins) are secreted by the root system to check the growth of nearby competitor plants.

Phenols are important constituents of some medicinal plants, and in the food industry, they are utilized as colouring agents, flavourings, aromatizers and antioxidants.

Some phenolics are water-soluble while others are soluble in organic solvents, and still others are insoluble polymers.

Classification of Phenolic Compounds

Depending upon the complexity and length of the side chain, phenolic compounds are categorized into the following five groups:

1. Simple phenols
2. Phenol carboxylic acids
3. Phenol propanes
4. Lignin
5. Flavan derivatives (flavonoids)

1. Simple phenols   Simple phenols contain an aromatic ring/skeleton bearing one or more hydroxyl groups. The ring may contain other additional substituents especially methyl groups and its glycoside. A common example of this group is vanillin taken from the pods of *Vanilla fragrans*, and is widely used in confectionary and in perfumery.

Capsaicin obtained from *Capsicum annuum* is used as a condiment and the drug is given internally in atonic dyspepsia and flatulence. It is used externally as a counter-irritant in the form of ointment, plaster, medicated wool, etc., for the relief of rheumatism, lumbago, etc.

Hydroquinone, arbutin, caffeic acids, ferulic acids, salicylic acid, etc., are also other examples of phenolic acids.

2. Phenol carboxylic acids This group of phenols includes simple phenols containing a carboxylic group as a substituent. These phenols have one carbon in their side chain. Common examples are gallic acid, protocatechuic acid, etc.

Protocatechuic acid is derived by β-oxidation of caffeic acid, a phenol propane. Onion bulbs also contain protocatechuic acid and catechol and are water-soluble.

3. Phenyl propanes Phenyl propanoids are naturally occurring phenolic compounds with an aromatic ring to which a three-carbon side chain is attached. They are derived biosynthetically from the aromatic protein amino acid, phenylalanine. Common examples of this group are coumarins, hydroxy cinnamic acids, cinnamic alcohols, isocoumarins and lignin.

In plants, four hydroxy cinnamic acids are common—ferulic, sinapic, caffeic and p-coumaric acids.

Coumarin occurs in over 27 plant families, e.g., *Melilotus*, Umbelliferae, etc. The double-ring phenolic compound is called coumarin, which imparts the distinctive sweet smell to newly mown hay. It is very common in many grasses and fodder crops. Coumarin is used as an anticoagulant that represses the synthesis of prothrombin, a plasma protein produced in the liver in the presence of vitamin K. Coumarin is converted into anticoagulant dicoumarin and is used in human medicines as blood thinners.

The phenylpropenes are another group of phenyl propanoids, which impart volatile flavours and odours to plants. These are usually isolated together with the volatile terpenes in the "essential oil" of the plants, e.g., eugenol from clove, anethole from anise and fennel.

4. Lignin Lignin is one of the most abundant organic substances present in plants. Lignin is widely distributed in the cell walls of fibres, tracheids, vessels of xylem and sclerenchymatous cells. It is a highly branched polymer of phenyl propanoid groups. Lignin gives tensile strength and helps in cementing of cell walls.

Lignin comprises three kinds of aromatic alcohols—coniferyl, coumaryl and sinapyl alcohol. Angiospermic lignin consists of coniferyl alcohol and sinapyl alcohol in almost equal proportion and traces of coumaryl alcohol, while conifer lignin contains coniferyl, coumaryl and sinapyl alcohol.

5. Flavan derivatives Flavan is derived from the word "flavus", which literally means 'yellow'. This group of phenols have flavan skeleton. It comprises two aromatic rings, 'A' and 'B'. Two rings are bridged by an oxygen-containing heterocycle (3-carbon bridge). The basic carbon skeleton of a flavonoid has 15 carbon atoms.

Most of the phenolic compounds belong to the flavonoids. Flavonols are colourless or yellow flavonoids found in leaves and many flowers. Quercetin is the yellow flavonol pigment of oak pollen. In some trees such as *Acer rubrum* (red maple) and *Quercus coccinea* (scarlet oak), colourless flavonols are converted into red anthocyanin as the chlorophyll breaks down. The flavonoids are all structurally derived from the parent substance, flavone. Flavonoids with glucose side chains are called glucoflavonoids or glucosides.

Flavonoids are mainly water-soluble compounds and they can be extracted with 70% ethanol. Bioflavonoids secreted by plants have allelopathic effects. Common examples are as follows: Catechin taken from *Centaurea biebersteinii* (Knapweed) has both antibiotic and antioxidant properties. It also occurs in *Camellia sinensis* (green tea). Catechin induces oxidation and cellular death of the cells, and is hence used in cancer treatment.

Classes of flavonoids

i. *Anthocyanins* These are coloured flavonoids that include scarlet, red mauve, purple and blue flower pigments and are also found in fruits, vegetables, cereal grains, leaves and other tissues. Man is naturally exposed to anthocyanins through the ingestion of fruits and vegetables.

ii. *Proanthocyanidines* These are mainly colourless and occur in heart woods and in leaves of woody plants.

iii. *Flavonols* These are mainly colourless co-pigments in both cyanic and acyanic flowers and also in wide leaves and is used as repellent.

iv. *Flavones* These are cream-coloured pigments of flowers.

v. *Glycoflavones* These are the flavonols in gymnosperms.

vi. *Chalcones and aurones* These are yellow flower pigments, which are occasionally present in other tissues.

vii. *Flavanones* These are colourless and present in leaf and fruits (*citrus*).

viii. *Isoflavones* These are also colourless and often found in root, and common only in the family Leguminosae. They serve as phytoalexins and have oestrogen effects, toxic for fungi.

Flavonoids differ from other phenolic substances in the degree of oxidation of their central pyran ring and very fundamentally also in their biological properties.

The quinines are another group of phenolic compounds. They are mostly coupled to other molecules, often to glucosyl residues, but to sulphate or acetyl residues too.

Many low molecular weight compounds, for example thymol, are used in medicine as antiseptics due to their toxicity.

ENZYMES

Enzymes are organic catalysts produced by animals and plants and are responsible for regulating all biochemical reactions inside the living cell. Hence they are called biocatalysts.

Enzymes have high molecular weight from 13,000 to 84,000. Enzymes play a vital role in the functioning of cells and activities of an organism. Therefore, enzymes are necessary to increase the efficiency of biochemical reactions to get the products quickly.

The term enzyme was introduced by Kuhne in 1878. Enzymes are made up of proteins. However, many enzymes contain a non-protein part attached to the protein part. The complete enzyme with the protein part and non-protein part is called a holoenzyme. The protein part of the enzyme is called apoenzyme. The prosthetic group or non-protein part may be a metal or an organic group. If the prosthetic or non-protein part is a metal, it is called activator or cofactor. Normally the metal component is firmly attached to the protein part of the enzyme.

Some enzymes along with their protein part have certain organic prosthetic groups instead of inorganic metal. These organic prosthetic groups are called co-enzymes. Enzymes are soluble in water and dilute alcohol, but they are precipitated in concentrated alcohol.

Classification of Enzymes

The classification of enzymes is based on the nature of the reaction catalysed by the enzymes. According to the International Union of Biochemistry (IUB), enzymes are classified into six major groups. They are as follows:

1. **Hydrolases** These enzymes carry out hydrolytic reactions.
2. **Transferases** These enzymes help in the transfer of the chemical group from one molecule to another.
3. **Lyases** These enzymes catalyse the addition of groups to double bonds or vice versa.
4. **Oxido-reductases** These enzymes catalyse the oxidation–reduction reactions.
5. **Isomerases** These are responsible for intramolecular rearrangements.
6. **Synthetases** These enzymes are otherwise called ligases and catalyse the condensation of two molecules coupled with the cleavage of pyrophosphate bond of ATP or similar triphosphate.

Properties of Enzymes

The properties of enzymes are as follows:

❋ An enzyme is a biocatalyst speeding up a reaction and changing the substrate into other products, without undergoing any change.

❋ Enzymes catalyse only specific reactions and in many cases only one reaction is catalysed by a given enzyme.

❋ Enzymes effectively catalyse dehydrations, oxidation–reduction reactions, hydrolytic reactions, aldol condensations, polymerizations and free-radical reactions.

❋ Enzymes exhibit all the colloidal properties.

❋ Enzymes are thermolabile and are inactivated at 0°C, and beyond 65°C, they get denatured.

❋ Enzymes show maximum activity between 35°C to 40°C.

Some Important Enzymes

1. Papain It is obtained from the latex of unripe fruits of papaya tree, botanically called *Carica papaya*. It belongs to the family Caricaceae.

Preparation The latex of unripe fruits of papaya is collected in trays, made up of aluminium. Potassium metabisulphate (5 g/kg of latex) is added to the collected latex and by passing through sieves, the waste materials are removed from the latex. The cleared latex is dried in a vacuum shelf drier at 55–60°C. It is also dried by spray-drying method. The dried latex is called papain. Papain may be a light brown or white amorphous powder. It is with typical odour and taste. It is a mixture of proteolytic enzymes, which shows maximum proteolytic activity between pH 5 and 6. Papain is soluble in water.

Uses

❋ It is used medicinally as an anti-inflammatory agent and in relieving the symptoms of episiotomy.

❋ It is used for degumming of silk fabrics in textile industry and in leather industry for dehairing of skins and hides.

❋ It is also used in clarification of beverages and as a meat tenderizer and is also employed in cheese manufacturing as a substitute for rennin.

2. Asparaginase It is extracted from the bacterium, *Escherichia coli*. An asparaginase enzyme without anti-leukaemic activity is also extracted from this bacterium during the purification process. L-asparaginase enzyme is also obtained from plants and animals, fungi and yeast in pure form. It catalyses hydrolysis of L-asparagine to L-aspartate and ammonia. It is a crystalline powder, white in colour and soluble in water. Each milligram of L-asparaginase contains 250 units.

Uses

❋ The enzyme interferes with the growth of malignant cells, which are not capable of synthesizing L-asparagine for their metabolism and hence it is used in chemotherapy of acute lymphocytic leukaemia.

✤ It is useful for the induction of remission in children with relapse of acute lymphocytic lymphoma.

✤ The enzyme also exhibits immuno-suppressive activity.

3. Rennin Rennin is a partially purified, milk-curdling, proteolytic enzyme. It is extracted from the glandular layer of the fourth or true-digesting stomach of the calf, *Bos taurus*, which belongs to the family Bovidae. It occurs as scales or powder and has peculiar odour and characteristic saline taste. The powder is yellowish white or greyish white in colour and hygroscopic in nature.

Preparation The glandular layers of the digesting stomach of calf are minced and macerated in sodium chloride solution (5%), followed by filtration. The filtrate is acidified with hydrochloric acid and then rennin is precipitated by saturating the filtrate with sodium chloride.

Rennin is separated, dried and powdered. It is also prepared microbiologically by controlled fermentation of *Endothia parasitica*, *Bacillus cereus* and *Mucor pusillus*.

Uses

✤ It is used to prepare junkets and cheese.

✤ It is also used to coagulate milk, thereby making the milk easily digestible for weak patients.

4. Trypsin It is a proteolytic enzyme derived from the pancreas of mammals like ox, *Bos taurus*. It belongs to the family Bovidae.

Preparation It is obtained by the alcoholic or aqueous acid extraction of its precursor trypsinogen and further conversion to crystalline trypsin. It is a crystalline or amorphous powder, which is yellowish white in colour, without any odour. It is insoluble in alcohol, chloroform and ether, but soluble in water. At pH 7.6 to 8, it shows maximum enzymatic activity. It is stable in dry air, but the solution should be freshly prepared due to its degradation. Calcium ions increase its stability.

Uses

✤ It does not act on living tissue due to the presence of inhibitors.

✤ It is used to enhance the proteolysis of different protein substrates, like blood clot, necrotic tissue, purulent exudates, etc.

Each mg of trypsin should contain not less than 2500 USP trypsin units.

5. Urokinase It is extracted from the kidney and urine or from kidney tissue cultures of humans. It is a lyophilized white powder, which is soluble in water. This enzyme is an activator of the endogenous fibrinolytic system that converts plasminogen to plasmin and degrades fibrinogen, fibrin clots and other plasma proteins.

Uses

✤ It is used to dissolve fibrin or blood clots in the anterior chamber of the eye and in acute massive pulmonary emboli.

6. Pancreatin   It is extracted from the pancreas of animals like hog, *Sus scrofa* var. *domesticus* (Family: Suidae), and ox, *Bos taurus*, (Family: Bovidae). Pancreatin contains different enzymes namely amylase, lipase and protease. It is an amorphous powder with white or buff colour. It has no taste and odour. It is soluble in water, but insoluble in alcohol and ether. It shows maximum activity in faintly alkaline medium. Each gram of pancreatin contains not less than 12,000 units of amylase activity, 1000 units of protease activity and 15,000 units of lipase activity.

PROTEINS

Proteins are very complicated nitrogenous organic substances of plant and animal origin. Proteins contain carbon, hydrogen, nitrogen, oxygen and sulphur. Few proteins also contain phosphorus and halogens. They are present in different forms in all the living cells of plants and animals. They also form the essential foodstuff, like carbohydrates and fats. Proteins also provide very important therapeutically active compounds such as hormones, enzymes, sera, antitoxins, etc.

In plants, protein is present in the form of solution in the cell sap and in a semi-solid state in the protoplasm. They are generally stored in the form of aleurone grains in plant cells and are solid proteins which constitute nitrogenous reserve food material, which is generally stored in the seeds and it can be easily extracted from the seeds. But in animals, proteins are present as structural material in the form of collagen (connective tissue), keratin (hair, wool, nail, feathers and horns), elastin (epithelial connective tissue), casein (milk) and plasma proteins.

Casein, gelatin, heparin and haemoglobin are pharmaceutically important proteins of animal origin. Proteins are the compounds of high molecular weight. They form a colloidal solution in water and are amphoteric in nature. They get easily denatured due to heat, changes in pH or by ultraviolet radiation. The ultimate products of complete hydrolysis of proteins are amino acids.

Depending upon the products of hydrolysis, proteins are classified as:

✤ Simple proteins—containing only amino acids, e.g., protamine sulphate.
✤ Conjugated proteins—containing amino acids and non-amino acids.
✤ Derived proteins.

There are several classes of proteins and are as follows:

Albumins These are soluble in water, insoluble in alcohol, coagulated by heat, and not precipitated by partial saturation with ammonium sulphate.

Globulins These are insoluble in water and alcohol, soluble in saline solutions, coagulated by heat, and precipitated by partial saturation with ammonium sulphate.

Some Important Therapeutic Proteins

1. Protamine sulphate It is a purified mixture of different simple proteins obtained from the sperms or mature testes of the fish, salmon, which belongs to the family Clupeidae or Salmonidae.

It is a hygroscopic crystalline powder with white or greyish yellow colour and an astringent taste. It is sparingly soluble in water and alcohol, and insoluble in ether and chloroform. For the preparation of protamine sulphate, the testes of salmon are isolated, frozen, then minced, washed with water, centrifuged and then dehydrated using solvents and dried under vacuum. It is then subjected to extraction by sulphuric acid and filtered, and the protamine sulphate-rich fraction is precipitated by adding cold alcohol. By dissolving in water, the fraction separates out as an oily layer on cooling. This layer is then separated and dissolved in hot water and refractionated using cold alcohol. The fraction is further solvent-dehydrated and finally dried under vacuum.

Protamine sulphate is used to counteract the effects due to heparin overdosage, like haemorrhage.

2. Heparin sodium It is a sterile preparation of active glycosaminoglycans, obtained from the lungs and intestinal mucosa of mammals.

Heparin acts as a natural anticoagulant in mammalian body, thereby maintaining the fluidity of blood. It is obtained as a white, hygroscopic powder without any odour or colour. It is soluble in water, saline, alcohol and acetone.

It is mainly used as an anticoagulant in vascular surgery and also in blood transfusions. It is used in venous and arterial thrombosis. It is also recommended for treating primary and post-operative thrombophlebitis, non-fatal pulmonary embolism, and immune thrombocytopenia.

3. Protein hydrolysate It is a sterile solution of short-chain peptides and amino acids and is nutritionally equal to that of plasma, fibrin, lactalbumin, casein or other proteins, from which it is derived.

It occurs as a yellowish to reddish amber transparent liquid and consists of not less than 50% total nitrogen in the form of α-amino nitrogen.

It helps to maintain the positive nitrogen balance under severe conditions of protein deficiency due to starvation, surgical operations of the gastrointestinal tract or severe illness. It is taken along with glucose or other carbohydrates to act as sources of energy. Protein hydrolysate injection is given by infusion into superficial vein or by catheter into the large central veins.

4. Collagen   It is obtained from animal source, which contains polypeptide substance. It forms the main constituent of skin, connective tissue and bones.

Collagen is used to stop bleeding in surgery. It is absorbed within 84 days. Normally, it is used on those wounds where surgical ligatures cannot be used.

5. Levodopa   It is naturally obtained from the seeds of *Vicia faba* L. It belongs to the family Fabaceae. It is available as crystalline powder without any odour or taste and is white in colour. It is sparingly soluble in water, but soluble in alkali, carbonates and mineral acids, and insoluble in alcohol, ether and chloroform.

Levodopa drug is used in the treatment of Parkinson's disease. It shows maximum effects in rigidity and hypokinesia. The drug is also used to suppress conditions like dysphagia, seborrhea, speech difficulties, etc.

6. Lectins   Lectins are glycoproteins or proteins that can agglutinate the cells and precipitate complex carbohydrates. They are isolated from seeds, barks, roots, eggs, fish and body fluids of organisms. Lectins are used for the following purposes:

* For isolation, purification and structural studies of molecules having carbohydrates.
* For blood grouping and erythrocyte polyagglutination studies.
* For mitogenic stimulation of lymphocytes.
* For histochemical studies of normal and pathological conditions.

Lectins containing natural sources are as follows: *Abrus precatorius* (abrin), *Canavalia ensiformis* (Concanavalin-A), *Dolichos biflorus*, *Phaseolus vulgaris*, the marine alga, *Codium fragile*, etc.

REVIEW QUESTIONS

State whether the following sentences are true or false

1. Cellulose consists of a long chain of glucose units.
2. Lactose is present in cane sugar.
3. The derived products of carbohydrates are latex and mucilages.
4. The sugars of carbohydrates have sweet taste.
5. Apple rinds and linseed fruits contain pectins.
6. The non-sugar component of glycoside is called aglycone.
7. The medicinal value of pectin is due to its jelling property.
8. Laxative action of rhubarb is due to the presence of anthraquinones.
9. Red blood cells in blood are destroyed by saponins.
10. Amygdalin is the glycoside that occurs in bitter almonds.
11. Cyanophore glycoside contains sedative property.
12. Squill contains scillarenin-A.
13. *Pinus* species are the source of turpentine oil.
14. Gum is obtained from *Cinchona*.
15. Myrrh is an example of oleogum resin.
16. Quinidine is a cardiac depressant.
17. Opium belongs to isoquinoline alkaloid.
18. Oak galls are the major source of tannins.
19. Hair contains keratin substance.
20. The protein part of the enzyme is called co-enzyme.

Choose the correct answer

21. Starch contains amylopectin and _____
    a. lactose   b. amylose   c. sucrose
22. *Aloe* and senna possess _____ property.
    a. cathartic   b. sedative   c. antimalarial

23. Sterculia secretes _____

    a. starch   b. saponin   c. gum

24. The members of Rosaceae contain _____

    a. cyanophoric glycoside b. steroidal glycoside   c. cardiac glycoside

25. The active principle of black mustard is _____

    a. digitalin   b. genin   c. sinigrin

26. Saponin-containing drugs are _____

    a. *Senega* and *Quillaia*   b. *Squill* and *Prunus*   c. apple and rose

27. Lactose is a _____

    a. monosaccharide   b. disaccharide   c. oligosaccharide

28. The petals of rose plant yield _____

    a. volatile oil   b. tannins   c. turpentine oil

29. Glucoresin is derived from _____

    a. *Podophyllum*   b. *Pinus*   c. Kurchi

30. Carbon–nitrogen skeleton is used for chemical classification of _____

    a. alkaloids   b. lipids   c. sugars

31. Tensile strength and cementing of plant cell walls are due to _____

    a. pectin   b. lignin   c. suberin

32. Reserpine is extracted from _____

    a. *Ephedra*   b. *Rauwolfia*   c. *Vinca*

33. The alkaloid vinblastine is used in the treatment of _____

    a. Parkinson's disease b. cancer   c. Hodgkin's disease

34. Ephedra belongs to the group _____

    a. Gymnosperms   b. Pteridophytes   c. Angiosperms

35. _____ is an antioxidant.

    a. *Atropa belladonna* b. *Cinchona officinalis*   c. *Camellia sinensis*

36. The dried latex of unripe fruit of papaya is called _____

    a. rennin   b. papain   c. zymin

37. _____ is a chemotherapeutic enzyme.

    a. urokinase   b. invertase   c. asparaginase

38. The natural anticoagulant in mammals is _____
    a. heparin   b. keratin   c. aspartin
39. Levodopa is obtained from the seeds of _____
    a. *Vicia faba*   b. *Coccus cacti*   c. *Acorus calamus*
40. The source of lectin is _____
    a. *Digitalis purpurea*   b. *Acacia catechu*   c. *Dolichos biflorus*

Answers				
1. True	2. False	3. False	4. True	5. False
6. True	7. True	8. True	9. True	10. False
11. True	12. True	13. True	14. False	15. True
16. True	17. True	18. True	19. True	20. False
21. b	22. a	23. c	24. a	25. c
26. a	27. b	28. a	29. a	30. a
31. b	32. b	33. c	34. a	35. c
36. b	37. c	38. a	39. a	40. c

Answer the following

1. What are monosaccharides? Cite an example.
2. Why are polysaccharides called non-sugars?
3. What are lipids? Give an example.
4. What are volatile oils? Give an example.
5. What are waxes? Cite an example.
6. Write notes on:
    i. Derived products of carbohydrates
    ii. Waxes
    iii. Lectins
7. Give a note on lipids.
8. Distinguish between fixed oils and fats.
9. Describe the therapeutic use of resins.
10. Explain tannins and their importance.

11. Explain lipids and their importance.

12. List out the properties of enzymes.

13. Explain the classification of phenols.

14. Give an account of glycosides and lipids?

15. Write a critical note on:
    i.   Saponin glycoside
    ii. Cardiac glycoside

16. Write an account on carbohydrates and their derivatives.

17. Write an essay on glycosides and their importance.

18. Explain the classification of glycosides.

19. Explain volatile oils and their therapeutic significance.

20. "Alkaloids are the largest group of phytoconstituents of pharmacological importance". Justify.

21. How will you classify alkaloids?

22. "Enzymes are the biocatalysts of all living organism". Justify.

23. Give an account of therapeutic proteins with suitable examples.

24. Write down the pharmaceutical use of gums and resins.

7

HISTOCHEMICAL TESTS FOR PHYTOCONSTITUENTS

To analyse the various phytoconstituents present in natural drugs, the following histochemical tests are carried out.

TESTS FOR CARBOHYDRATES

Carbohydrates are identified by the following microchemical tests.

Molisch's test A solution of carbohydrate is prepared in water containing α-naphthol. To this mixture, concentrated sulphuric acid is added along the sides of the test tube. A purple to violet coloured ring is formed in the junction below the aqueous layer. This indicates the presence of soluble carbohydrates, whereas with insoluble carbohydrates like cellulose, the reaction mixture is shaken to produce the same colour. Molisch's test is positive with soluble as well as insoluble carbohydrates.

Osazone formation test It is an important test to be carried out for identification of sugars. An aqueous solution of sugar is heated with phenylhydrazine hydrochloride and sodium acetate. This results in yellow crystals of osazone, indicating the presence of sugar.

Test for pentoses A small portion of extract is heated with an equal amount of hydrochloric acid containing a little phloroglucinol. A red colour is formed indicating the presence of sugars.

Benedict's test A small portion of test solution is added to 5 ml of Benedict's reagent and mixed well. The reaction mixture is then boiled for 2 minutes and allowed to cool. If a red, yellow or green precipitate is formed, it indicates the presence of sugar.

Fehling's solution test The test solution is heated with equal volume of Fehling's solution A and B. This produces a brick-red precipitate of cuprous oxide in case of reducing sugars such as all monosaccharides and many disaccharides like lactose, maltose, cellobiose and gentiobiose. Non-reducing sugars like sucrose and trehalose are boiled with acid to convert it into reducing sugars and then the test is performed.

Furfural test A sample of carbohydrate is heated with a drop of syrupy phosphoric acid to convert it into furfural. A disc of filter paper is moistened with a drop of 10% solution of aniline in 10% acetic acid. The disc is then placed over the mouth of the test tube and the bottom of the test tube is heated for 30–60 seconds. A pink or red stain formed on the paper indicates the presence of sugar.

Test for soluble starch

i.   Starch forms a deep blue colour with iodine.

ii. Starch gives a canary yellow colour on heating with 5 per cent KOH.

TESTS FOR GLYCOSIDES

Presence of glycosides in natural drugs can be detected by the following tests.

Test for Anthraquinone Glycosides

Borntrager's test A sample of powdered drug is boiled with dilute sulphuric acid and filtered. To this, ether or chloroform is added and again filtered. The lower organic layer is separated and to this, aqueous ammonia or sodium hydroxide is added slowly to give pink to red or violet colour after shaking the mixture to indicate the presence of anthraquinones. Borntrager's test gives negative result in the case of anthranols.

Tests for Cardiac Glycosides

The following tests are carried out for the identification of cardiac glycosides.

i. Keller–Kiliani's test One gram of powdered sample is boiled with 10 ml of alcohol (70%) for 2 to 3 minutes and filtered. To the filtrate, 5 ml of water and 0.5 ml of lead acetate solution are added and shaken well and then the filtrate is separated. The filtrate is then heated with equal amount of chloroform to give the extract. The extract is dissolved in glacial acetic acid and allowed to cool. A few drops of ferric chloride and concentrated sulphuric acid are added to it to form a reddish brown layer at the junction of the two layers and the upper layer turns bluish green indicating the presence of cardiac glycosides (digitoxose).

ii. Legal's test A few drops of pyridine and sodium nitroprusside are added to the extract of the drug and made alkaline. A pink or red colour is obtained.

iii. Baljet test A few drops of sodium picrate solution is added to a transverse section of the drug. The change of yellow colour to orange indicates the presence of glycosides.

iv. Liberman Burchard's test To a solution of the drug in glacial acetic acid, one drop of concentrated sulphuric acid is added. The colour change from rose to red or violet and blue to green is due to the presence of steroid molecule. Colours are slightly different from compound to compound.

Test for Cyanogenic Glycoside

Sodium picrate test A few drops of dilute sulphuric acid are added to the sample placed in a conical flask containing water. A strip of filter paper soaked in sodium picrate solution is suspended at the neck of a cork. The flask is gently heated to give brick-red colour on the strip to indicate the presence of cyanogenic glycoside.

Test for Saponin Glycoside

A drop of sodium bicarbonate solution is added to the aqueous extract and shaken well and then allowed to stand for few minutes to get a honeycomblike mass.

TESTS FOR LIPIDS

The tests for identification of fixed oils, fats and waxes are given below:

Detection of Sesame Oil

Two ml of sesame oil is shaken with 1 ml of 1 per cent solution of sucrose in concentrated hydrochloric acid and kept for 5 minutes. A pink or red colour formed in the acid layer indicates the presence of sesamol in sesame oil.

Detection of Castor Oil

To 1 ml of the sample is added half its volume of acidified petroleum ether (60–80°C) and then shaken well. To this a drop of ammonium molybdate solution is added. A white turbidity indicates the presence of castor oil.

Detection of Cottonseed Oil by Halphen's Test

Two ml of cottonseed oil is added to 2.5 ml of alcohol and 2.5 ml solution of sulphur in carbon disulphide (1%). The mixture is heated to form a pink or red colour indicating the presence of cottonseed oil.

Detection of Cholesterol

0.5 g of hydrous wool fat is dissolved in chloroform and to it is added 1 ml of acetic anhydride and two drops of sulphuric acid. Formation of a deep green colour indicates the presence of cholesterol.

TESTS FOR VOLATILE OILS

The presence of volatile oil in natural drugs can be detected by the following methods:

1. One ml of the sample is treated with phenylhydrazine hydrochloride. A red-coloured phenylhydrazone of cinnamic aldehyde indicates the presence of volatile oil in the drug.

2. One ml of volatile oil is dissolved in 5 ml of alcohol and then a drop of $FeCl_3$ is added to it. It gives a pale green colour due to the presence of eugenol.

3. To a thin section of the drug, a drop of alcoholic solution of Sudan III is added to form globules of red colour, indicating the presence of volatile oil in the sample.

4. To a thin section of the drug, a drop of tincture alkana is added to give a red colour to indicate the presence of volatile oil.

5. A thick section of the drug is treated with concentrated acid to give a dark green colour to indicate the presence of volatile oil.

TESTS FOR TANNINS

The following tests are carried out to identify tannins in the natural drugs.

Goldbeater's skin test   A membrane is prepared from the intestine of an ox called goldbeater's skin, which is dipped in 2% hydrochloric acid and rinsed with distilled water. It is then placed in a test solution of tannin for 5 minutes. The membrane is washed with distilled water and kept in ferrous sulphate (1%) solution to produce a brown or black colour on the skin to show the presence of tannins.

Heavy metal test   When a sample of tannin solution is added to heavy metals, it results in precipitation.

Gelatin test   To a solution of tannin (0.5–1%), aqueous solutions of 1% gelatin and 10% sodium chloride are added. A white buff-coloured precipitate is formed indicating the presence of tannin.

FeCl₃ test

i. A solution of tannin is treated with FeCl₃ to produce a blue-black colour indicating the presence of gallitannins and ellagitannins.

ii. A brownish green precipitate is formed with FeCl₃ test if condensed tannins are present.

Phenazone test 5 ml of an aqueous extract of crude drug is heated with 0.5 g of sodium acid phosphate and allowed to cool and then filtered. To the filtrate, phenazone (2%) solution is added to produce a bulky coloured precipitate.

Catechin test (Matchstick test) A matchstick dipped in the plant extract is dried and moistened with conc. hydrochloric acid. Warming on a flame, the matchstick wood turns pink or red colour due to the formation of phloroglucinol.

Chlorogenic acid test A small amount of the extract is treated with aqueous ammonia to give a green colour on exposure to air. Formation of green colour indicates the presence of chlorogenic acid in the extract.

Vanillin–hydrochloric acid test A small amount of drug is treated with vanillin–hydrochloric acid (vanillin 1 g, alcohol 10 ml and conc. hydrochloric acid 10 ml), to give a pink or red colour to indicate the presence of phloroglucinol.

TESTS FOR ALKALOIDS

The following chemical tests are used to detect the presence of alkaloid in the samples.

Dragendorff's reagent It is prepared by dissolving Bismuth nitrate (8 g) in nitric acid (20 ml) and separately dissolving potassium iodide (27.2 g) in water (50 ml) and mixing the two solutions to make up the volume to 100 ml. When this solution is treated with a drug, it gives a reddish brown or orange precipitate with alkaloids.

Wagner's reagent Wagner's reagent is prepared by dissolving 1.3 g of iodine and 2 g of potassium iodide in water to make up 100 ml. A sample of drug when treated with the prepared iodine solution it gives a brown or reddish brown precipitate with alkaloids.

Mayer's reagent (Potassium mercuric iodide solution) When the sample is treated with Mayer's reagent, it gives a creamy white or pale yellow precipitate, except with alkaloids of purine group and a few others.

Hager's reagent (Saturated solution of picric acid in cold water) A sample treated with Hager's reagent forms a yellow crystalline precipitate indicating the presence of alkaloid.

Tannic acid test The test solution is treated with a freshly prepared aqueous solution of tannic acid (5%, w/v). It gives a precipitate with most of the alkaloids.

TESTS FOR FLAVONOIDS

Mg–HCl Reduction Test (Shinoda Test)

A mixture containing a piece of magnesium ribbon and conc. hydrochloric acid are added to the plant extract. It gives a red colour which indicates the presence of flavonoids, flavanones, flavonols and xanthone. Sometimes pink, purple or red-magenta colour may also be formed in different derivatives of flavonoids.

FeCl₂ test   To the test solution, FeCl₂ solution is added to get a green to black colour. The colour change denotes the presence of flavonoids.

Lead acetate solution test   To the sample, 10% lead acetate is added to form yellow precipitate of flavonoids.

Sodium hydroxide solution test   To the test solution, 10% sodium hydroxide is added to produce yellow colour.

Alkali test   Flavonoids dissolve in alkalis giving yellow colour solution, which becomes colourless when acid is added to it.

Mineral acid test   To the test sample, conc. sulphuric acid is added. Formation of yellow-orange colour denotes the presence of flavonoids.

TESTS FOR RESINS

The following tests are carried out for resins.

1. To a solution of powdered resin (0.1 g) in acetic acid (10 ml), one drop of conc. sulphuric acid is added in a dry test tube. A purple colour changes to violet indicating the presence of resin.

2. Alcoholic solution of colophony turns blue litmus red due to the presence of diterpenic acid.

3. Alcoholic solution of balsam of Tolu (1 g) gives green colour with ferric chloride due to toluresinotannols.

4. To a petroleum ether solution of powdered colophony, twice its volume of dilute solution of copper acetate is shaken well. The colour of the petroleum ether layer changes to emerald green due to the formation of copper salt of abietic acid.

5. To an alcoholic solution of colophony, water is added. It becomes milky white due to the precipitation of chemical compound resins.

6. To a petroleum ether solution of benzoin (0.2 g), 2–3 drops of sulphuric acid are added in a china dish. Sumatra benzoin produces reddish brown colour, while Siam benzoin gives purple-red colour on rotating the dish.

7. A sample of asafoetida drug is boiled with hydrochloric acid (5 ml) for some time. It is then filtered, and ammonia is added to the filtrate. A blue fluorescence is formed, indicating the presence of resin.

8. To the fractured surface of the asafoetida drug, a drop of sulphuric acid is added. A red colour is obtained, which later changes to violet on washing with water.

TESTS FOR PROTEINS

The following are the colour reactions which indicate the presence of proteins in the samples.

Gelatin test

1. Gelatin is precipitated by trinitrophenol and solution of tannic acid, but not with alum, lead acetate or acids, indicating the absence of chondrin.

2. Gelatin gives a white precipitate with mercuric nitrate and on warming turns to brick-red colour.

3. Gelatin produces ammonia when heated with soda lime.

Biuret reaction   To the sample of alkaline solution of protein (2 ml), a dilute solution of copper sulphate is added. A red or violet colour is formed with peptides containing at least two peptide linkages. A dipeptide does not respond to this test.

Millon's reaction   Millon's reagent (mercuric nitrate in nitric acid containing a trace of nitrous acid) usually gives a white precipitate with a protein solution, which in turn becomes red on heating. This reaction is a characteristic feature of phenols, (e.g., phenolic amino acid tyrosine).

Xanthoproteic reaction   A sample of protein forms a yellow colour when warmed with conc. nitric acid. This colour changes to orange when the solution is made alkaline, and the colour is due to nitration of aromatic ring present in phenylalanine, tyrosine and tryptophan.

Ninhydrin test   To an aqueous solution of protein, an alcoholic solution of ninhydrin is added and then heated. Red to violet colour is formed due to the presence of protein.

Nitroprusside test   Proteins containing sulphur group when mixed with nitroprusside solution result in red colour.

Lead sulphide test   To alkaline solution of sulphur-containing proteins, lead acetate is added to produce a black precipitate.

REVIEW QUESTIONS

State whether the following sentences are true or false

1. Glycosides are identified by furfural test.

2. Molisch's test gives positive result to carbohydrates.

3. Osazone formation with sample indicates the presence of sugars.

4. Legal's test is a very common test to identify glycosides.

5. Cardiac glycosides are identified with Keller–Kiliani's test.

6. Saponin glycoside forms a honeycomb with sodium bicarbonate.

7. Formation of precipitate with heavy metals indicates the presence of tannins.

8. Matchstick test is otherwise called catechin test.

9. A saturated solution of picric acid in cold water is called Mayer's reagent.

10. Presence of yellow precipitate with lead acetate denotes flavonoids.

Choose the correct answer

11. Sample that gives canary yellow colour with KOH is _____

    a. starch   b. lipid   c. enzyme

12. Cyanogenic glycosides are tested with _____

    a. sulphuric acid   b. sodium picrate   c. nitric acid

13. Substance that gives pink or red colour with Halphen's test indicates the presence of _____

    a. castor oil   b. sesame oil   c. cottonseed oil

14. A section of a drug treated with concentrated acid gives dark green colour to indicate the presence of _____

    a. lipid   b. volatile oil   c. tannin

15. Goldbeater's skin is prepared from the intestine of _____

    a. cow   b. ox   c. zebra

16. Tannins are identified with _____

    a. phenazone test   b. Legal's test   c. iodine test

17. A reagent prepared with iodine (1.3 g) and potassium iodide (2 g) in water is called _____

    a. Hager's reagent   b. Wagner's reagent   c. Fehling's reagent

18. Shinoda test is used for testing _____

    a. lipids   b. waxes   c. flavonoids

19. Balsam of Tolu gives green colour with ferric chloride due to the presence of _____

    a. diterpenes   b. trinitrophenol   c. toluresinotannols

20. Formation of red or violet colour with ninhydrin test indicates the presence of _____

    a. protein   b. enzymes   c. sugars

Answer

1. False	2. True	3. True	4. False	5. True
6. True	7. True	8. True	9. False	10. True
11. a	12. b	13. c	14. b	15. b
16. a	17. b	18. c	19. c	20. a

Answer the following

1. Give an account of various histochemical tests to identify glycosides and lipids.

2. How will you test the presence of tannins, carbohydrates and alkaloids in the given drugs?

3. List out the histochemical tests carried out for resins and flavonoids.

8

DRUGS CONTAINING CARBOHYDRATES

The drugs containing carbohydrates and derived products, and having therapeutic values are discussed in this chapter.

1. GUM ARABIC

Biological source The gummy exudates from the stem of *Acacia arabica* L. and *Acacia senegal* Willd of the family Mimosae (Leguminosae) are used as drug.

The common names of the source tree are babul tree; gum acacia; karuvelam maram; pesin maram.

Geographical distribution The plant is an evergreen thorny tree. It is found wild in Africa, Morocco, Sri Lanka, Sudan and India. In India, it is distributed in Punjab, Rajasthan, Tamil Nadu and in the Western Ghats. A major part of the world supply comes from Sudan.

Collection The gummy exudate is produced as a result of injury to the plant. After the rainy season, the bark of the tree is cut transversely to expose the cambium. Due to injury, tears of gum get collected on the cut portion of the trunk. Within 20–30 days, the tears of gum collected on the surface may be picked up, and the bark pieces and foreign matter are removed. The collected gum is dried in the sun to remove the moisture.

During the process of drying, numerous cracks appear on the outer surface of the lumps of tears and the gum is bleached. The tears graded on the basis of appearance are packed, and exported.

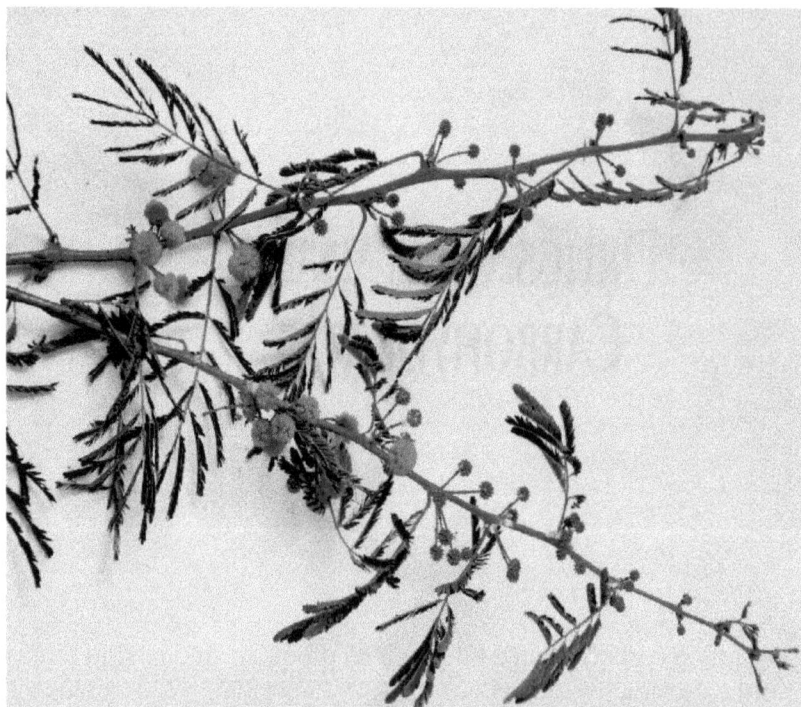

Figure 8.1 *Acacia arabica*

Characteristics of the gum

Colour	Tears are pale yellow to light brown or red. Gum from Sudan is more yellow to pinkish in colour. Powdered gum is light brown in colour.
Size and shape	Round or irregular with glossy surface
Odour	Odourless
Taste	Bland and mucilaginous
Surface appearance	Transparent and glossy with numerous fine cracks and brittle in nature.
Solubility	Freely soluble in water and the solution is viscous and acidic. It is insoluble in alcohol and organic solvents.

Chemical constituents The major component is arabin, a mixture of calcium, magnesium and potassium salts of arabic acid. Enzymes such as oxidase, pectinase and peroxidase are present.

Adulterants Tragacanth gum, starch, dextrin, gum ghatti and sterculia gum.

Uses

1. It is used as a very good binding agent in the preparation of pastilles, lozenges and compressed tablets.

2. It is an excellent emulsifying agent for fixed oils, volatile oils and also for liquid paraffin.

3. It is used to cover burns, sore nipples, etc.

4. In combination with gelatin, it is used to form conservates for microencapsulation of drugs.

5. It is also used for a variety of drug formulations.

Tests for identification

1. To the aqueous solution of gum, add a few drops of dilute solution of lead acetate. If no precipitate is formed, it shows the absence of tragacanth gum and agar.

2. Mix 1 ml of solution of gum with 4 ml of H_2O, boil, cool, and add 2 drops of N/10 iodine. Absence of colour indicates absence of starch and dextrin.

3. To 5 ml aqueous solution of gum, add 0.1 g of borax. A translucent mass is formed.

4. Gum arabic gives positive results for Fehling's test of reducing sugars and Molisch's test of carbohydrate.

2. TAMARIND

Biological source Ripe fruits of *Tamarindus indica* Linn. of the family Caesalpineae (Leguminosae).

The common names of the source plant include West Indian tamarind; Puliamaram; Imhil.

Geographical distribution The plant is a very big perennial tree, indigenous to tropical Africa. It is commonly grown and cultivated in India and the West Indies.

Collection and preparation The fully ripe fruits are collected from the plants by beating or by knocking with sticks. The collected fruits are dried and the woody brittle epicarp of the fruits is removed by breaking. The fleshy mesocarp which is reddish brown in colour is used as drug. The seeds are removed manually. To maintain the quality, salt and preservatives are added to the pulp. The preserved pulp is marketed.

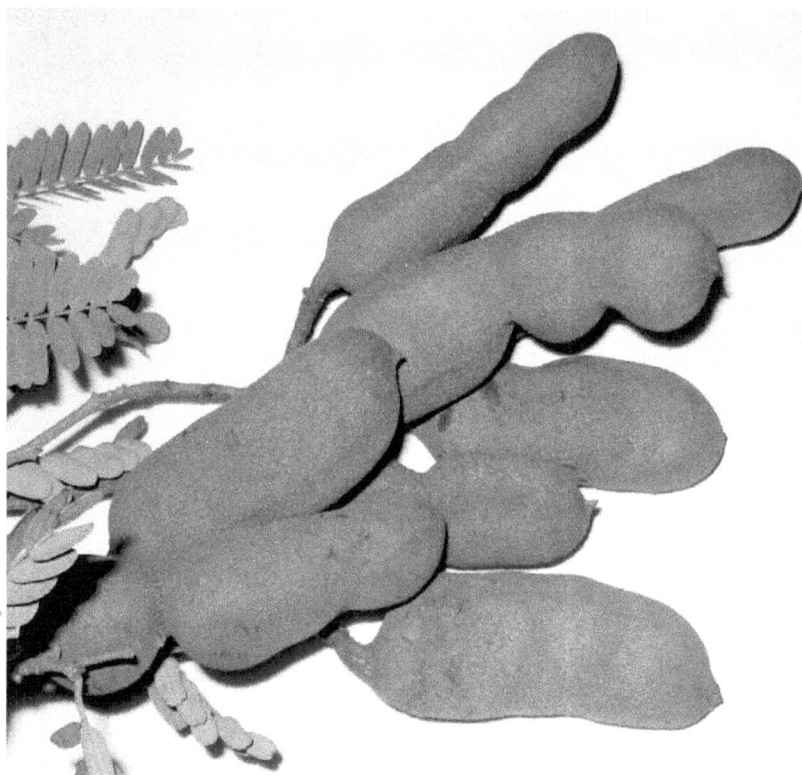

Figure 8.2 *Tamarindus indica*

Characteristics of the pulp

Colour	Reddish brown to pale brown
Odour	Pleasant aroma
Taste	Sweet and sour (acidic)
Size	Legume 5–20 cm length and 2.5–3 cm width.

Chemical constituents The fruit pulp contains 30–40% invert sugars, 10% tartaric acid, about 8–9% sodium potassium tartrate. In addition, pectin and small quantities of malic and citric acids are present. The seeds of tamarind contain 70–80% carbohydrates.

Uses

1. It is used as a mild laxative and pleasant acid refrigerant.
2. It is used for culinary purposes.

3. SODIUM ALGINATE

Biological source   Sodium alginate is derived from brown algal seaweeds, e.g., species such as *Laminaria hyperborea, L. digitata, Macrocystis pyrifera, L. saccharina, Ascophyllum nodosum*, etc. which belong to the class Phaeophyceae (brown algae).

The common names of the product include algin and sodium polymannuronate.

Geographical distribution   Seaweeds are found in US, Japan, Iceland, Canada, UK, Australia, China, South Africa, and coastal lines of Saurashtra in India.

Collection and preparation   The fully grown seaweeds are harvested and cleaned thoroughly with water. The seaweeds are dried and macerated. A pasty mass of macerated seaweeds is made using dilute sodium carbonate solution. It is further diluted with soft water to remove the insoluble matter. The diluted clear liquid is treated with calcium chloride to precipitate the salt as calcium alginate in the form of a gel.

The liquor is separated from the precipitate by roller or expeller press. It is further purified by repeated washing to get wood-pulplike product. It is then treated with hydrochloric acid to remove calcium and the pulp of alginic acid is again roller-pressed and then neutralized with sodium carbonate to give sodium salt. The alginic acid content on dry solid basis varies from 20–35% in different varieties of brown algae.

Characteristics of algin

Colour	White or cream-coloured powder
Texture	Coarse powder
Odour	Odourless
Taste	Tasteless
Solubility	Readily soluble in water to form a viscous and colloidal solution, and insoluble in alcohol, chloroform, ether and aqueous acids with pH below 3.

Chemical constituents   Alginic acid contains residues of L-glucuronic acid and D mannuronic acid. It is precipitated below pH 3.0.

Uses

1. It is used in the preparation of creams, pastes and for thickening and stabilizing emulsions.
2. It is a very good suspending agent in soft drinks.

3. It is used as binding and disintegrating agent in tablets and lozenges.
4. It is used in the food industry for the preparation of jellies, ice creams, etc., in the textile industry, and in the manufacture of cosmetics and paints.
5. In the pharmaceutical industry, it is used as a sizing agent and also in dental impression preparations.
6. Alginate fibres are used as absorbable haemostatic dressings.

Identification

1. One per cent solution of water forms a heavy gelatinous precipitate with dilute sulphuric acid.
2. The aqueous solution of sodium alginate produces copious precipitate with calcium chloride solution.

4. BAEL

Biological source The unripe fruits of *Aegle marmelos* Corr. belonging to the family Rutaceae are used as drug.

The common names of the source tree are Indian bael; bael fruit; bel; vilvam; etc.

Geographical distribution The tree is native to India and distributed throughout India. It is found growing in Myanmar and Sri Lanka.

Figure 8.3 *Aegle marmelos*

Collection and preparation The fruits ripen in the month of December. Therefore, the unripe and half-ripe fruits are collected and dried in the shade. They are washed thoroughly, and the hard epicarp is peeled off to cut the pulp of the fruits into transverse slices, for marketing.

Characteristics of unripe/ripe fruits

Colour	Unripe berry is green in colour and the ripe fruit is yellowish brown in colour. The pulp is pale red in colour.
Shape	Subglobose of 10–20 cm diameter with smooth but hard and woody epicarp
Odour	Slightly aromatic
Taste	Mucilaginous and astringent

Chemical constituents The fruit consists of 0.5% marmelosin. In addition the pulp also contains 11–17% carbohydrate, 20% tannins, 1% protein, volatile oil, vitamin A and vitamin C. It also contains alkaloids such as isopentylhalfordinol and *o*-methylhalfordinol.

Uses

1. It is used in habitual constipation, chronic diarrhoea, dysentery, and dyspepsia.

2. It is used as a digestive, appetizer and tonic.

5. STARCH

Biological source Starch is obtained from the grains of cereals such as *Oryza sativa* L., *Zea mays* L. and *Triticum aestivum* L. of the family Poaceae (Gramineae) and the tubers of potato (*Solanum tuberosum* L.) of the family Solanaceae.

The common name of starch is amylum.

Geographical distribution Cereals are commonly cultivated in both tropical and sub-tropical countries.

Preparation of starch The procedures used for the preparation of starch depend upon the type of source material and also the type of starch to be produced.

i. **Preparation of maize starch** The mature grains of maize are collected and washed thoroughly. The grains are soaked in water for 3–4 days at 50°C in the presence of sulphur dioxide to prevent fermentation. The softened grains are passed through a mill to break the grains and to separate the endosperms from the outer coatings of the grains. Water is added to the broken grains to make the embryos float on the surface so that they can be separated. The separated embryos are used to prepare corn oil or germ oil by expression method.

The water that is used to soften the grains dissolves most of the minerals, soluble proteins and carbohydrates present in the grains. Therefore, this water contains all these content and is called corn steep liquor, which is used as a culture medium for the production of microbes. The starchy material is further filtered through sieves to remove the remaining gluten and other debris. The starch settles at the bottom. With repeated washing and centrifugation, starch gets separated, which is dried in flash dryers and packed for marketing.

ii. **Preparation of rice starch**   The starch is obtained from the broken rice pieces that are formed during polishing of rice grains. The pieces of rice are soaked in 0.4% aqueous solution of sodium hydroxide to soften the tissue and to dissolve the insoluble protein gluten. The soaked material is crushed and the solution is further centrifuged. It is then washed with water and dried at 60°C and powdered. After packing, it is sent for marketing.

iii. **Preparation of potato starch**   The fresh potatoes are washed to remove the mud and other dirt. The cleaned potatoes are crushed into slurry and filtered to remove vegetable matter. After filtration, the milky liquid that contains the starch is purified by centrifugation and washed further. It is then dried and pulverized, packed and marketed.

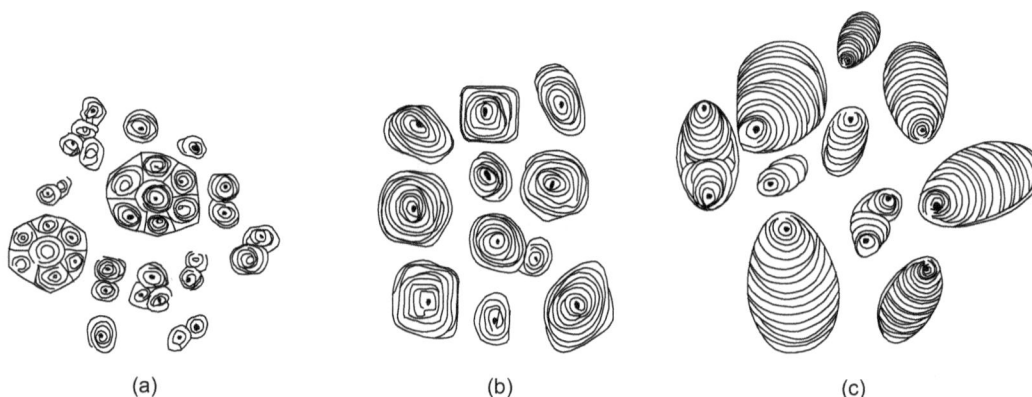

(a)　　　　　　　　　(b)　　　　　　　　　(c)

Figure 8.4   Starch grains (a) Endosperm of rice grain (b) Endosperm of maize (c) Tuber of potato

Characteristics of starch

> *Appearance*　　　Starch occurs as irregular, angular powder or a white mass.

Colour	Rice and maize starch grains are white, whereas potato starch is slightly yellow.
Odour	Odourless
Taste	Mucilaginous
Structure	Starch granules contain a central or eccentric hilum with concentric rings or striations, and differ in size and aggregation.

Size and shape Starch grains vary depending upon the source.

Rice starch Granules may be simple or compound and the simple granules are polyhedral in shape and 2–12 microns in diameter, whereas, compound granules are ovoid with 12–30 μ × 7–12 μ in size.

Maize starch Granules polyhedral or round, 10–30 μm in diameter and the hilum is distinct at the centre.

Potato starch Granules mostly simple, sub-spherical or somewhat flattened and irregularly ovoid in shape. Size varies from 30–110 μm in diameter. Hilum is eccentric, placed near the narrow end with well-marked concentric striations. Few granules are compound, and 2 or 3 are fused together.

Solubility Starch is insoluble in alcohol and cold water. On boiling, starch forms a viscous solution, and on cooling, it forms a translucent jelly. Rice starch is slightly alkaline, maize starch is neutral, and potato starch is slightly acidic.

Standard of starch required for pharmaceutical use

a) *Loss on drying* Not more than 15% for rice and maize, and not more than 20% for potato starch.

b) *Ash value* Not more than 0.6% in case of rice starch, and not more than 0.3% for maize and potato starches.

Chemical constituents It consists of a mixture of two different polysaccharides, amylose (β-amylose 25%) and amylopectin (α-amylose 75%) in 1:2 proportion. Amylose is more water-soluble whereas amylopectin is water-insoluble, but swells in water and is responsible for gelatinization of starch. Amylose gives blue colour with iodine, while amylopectin gives bluish black colour.

Identification

1. 1 g of starch is boiled with 15 ml of water and cooled. A translucent viscous jelly is formed.

2. The starch jelly turns deep blue colour with iodine solution. The blue colour disappears on warming and reappears on cooling.

3. It gives positive results to Fehling's solution, and for Molisch's test for carbohydrate.

Uses

1. It is the starting material for commercial production of glucose, dextrose, and dextrin.

2. Mucilage of starch B.P.C., zinc starch dusting powder B.P.C., and zinc oxide paste I.P. are used in pharmaceuticals.

3. Glycerin of starch is used as an emollient and as a base for suppositories.

4. It is used as a disintegrant in the preparation of tablets and pills, and also as a filler and binder.

5. It is used internally as a mild astringent, nutritive, demulcent, and protective agent.

6. It is used as an important ingredient of all talcum powders due to its absorbent property.

7. Starch is also used as an antidote for iodine poisoning.

8. It is a diagnostic aid in the identification of crude drugs.

9. Starch is industrially used for sizing of paper and cloth.

6. AGAR

Biological source Agar is obtained from red algal species of *Gelidium cartilagineum*, *G. amansii*, *G. conferoides*, *Pterocladia lucida*, *P. capillacea* of the family Gelidaceae and species of *Gracilaria* of the family Gracilariaceae.

Some common names of agar include vegetable gelatin; agar–agar; Japanese Isinglass; etc.

Geographical distribution Commercially used agar is produced in Japan, U.S.A., New Zealand, Australia, Spain, India and Korea.

Collection and preparation These seaweeds live at the bottom of the sea and grow flourishingly with the support given in the sea. Therefore, bamboos are spread in the sea and the algae grow well around this support. Algal collection is made usually in the months of May and October. At the time of harvest the bamboo poles are removed from the ocean and the algae are scraped out from the poles. The collected seaweeds are dried and beaten, and shaken well to remove the sand, shells and other unwanted materials. The dried materials are washed thoroughly with water to remove the salt. To bleach and dry, the materials are exposed to sunlight.

The bleached material is boiled with dilute acidified water for 5–6 hours. The boiled hot extract is strained through the cloth. On cooling, the extract forms a jelly and is further passed through wire netting under pressure to form strips of agar. The excess of moisture from the agar strip is removed by freezing, thawing and drying at about 35°C.

Characteristics of Agar

Colour	Yellowish to white
Odour	Odourless
Taste	Mucilaginous
Shape	Strips, sheets, flakes or coarse powder
Size	Sheets are 45–60 cm long and 10–15 cm wide, whereas strips are 4 mm in width.
Appearance	Transparent, or translucent, lustrous and slender, brittle on drying.
Solubility	It is insoluble in cold water, but forms a gelatinous mass after cooling the hot solution. It is slowly soluble in boiling water to form a viscid solution and insoluble in organic solvents.

Chemical constituents   It consists of two different polysaccharides—agarose and agaropectin. Gel property of agar is due to the presence of agarose, and the viscosity of the agar solution is due to the presence of agaropectin. Agarose is composed of D-galactose and 3, 6 anhydro L-galactose units, whereas agaropectin is a sulphonated polysaccharide with galactose and uronic acid units, and partly esterified with sulphuric acid.

Chemical test

1. To 0.2 g of agar solution in water, add tannic acid solution. No precipitate is formed, showing the absence of gelatin.

2. To agar powder, add a solution of ruthenium red to show pink-coloured particles to indicate the presence of mucilage.

3. Agar gives positive results to Fehling's solutions, and Molisch's reagent.

4. Boil 1.5 g of agar in 100 ml of water and cool to room temperature. It forms a stiff jelly.

5. Warm the agar with a solution of potassium hydroxide. It produces canary yellow colour.

Uses

1. It is used as a bulk laxative and as an emulsifying agent.

2. It is used in the preparation of jellies, confectionery items and also nutrient culture media in the laboratories.

3. It is used as surgical lubricants and also in the preparation of vaginal capsules and suppositories.

4. It is used as a thickening agent in food industries, especially in dairy and confectionary products, in meat canning, sizing for silks and papers, in the dyeing and printing of fabrics and textiles.

REVIEW QUESTIONS

State whether the following sentences are true or false

1. The major component of gum arabic is arabin.
2. The fruit of Bael is medicinally used.
3. The source of agar is the species of *Gelidium*.
4. Starch is used as an antidote for iodine poisoning.
5. The epicarp of tamarind fruit is used as drug.

Choose the correct answer

6. Gel property of agar is due to the presence of _____
   a. amylopectin   b. agarose   c. amylose
7. The chemical constituent marmelosin is present in _____
   a. tamarind   b. bael   c. custard apple
8. The botanical name of potato is _____
   a. *Solanum tuberosum*   b. *Solanum lycopersicum*   c. *Solanum melongena*
9. Tamarind belongs to the family _____
   a. Rutaceae   b. Mimosae   c. Caesalpineae
10. Gum arabic is adulterated with _____
    a. asafoetida   b. tragacanth   c. balsam

Answers				
1. True	2. True	3. True	4. True	5. False
6. b	7. b	8. a	9. c	10. b

Answer the following

1. Describe the biological source, distribution, collection, characters of any three carbohydrate-containing drugs.

2. Give a detailed account of the botanical name, collection, preparation, chemical constituents and use of sodium alginate and agar.

3. How will you differentiate between the following on the basis of characters:
   i. Maize starch and potato starch
   ii. Gum arabic and tamarind

4. Write the chemical tests for identification of agar, gum arabic and starch.

5. List out the pharmaceutical use of the following drugs:
   i. Agar
   ii. Sodium alginate
   iii. Gum arabic

DRUGS CONTAINING GLYCOSIDES

9

Glycosides are compounds which on hydrolysis give rise to one or more sugars (glycones) and a compound of non-sugar (aglycone or genin). Glycoside-containing drugs are discussed in this chapter.

1. SENNA

Biological source   The leaflets from the plants of *Cassia angustifolia* Vahl. of the family Caesalpineae (Leguminosae) are used as drug.

The common names of the plant include Indian senna; Tinnevelly senna; Nilavarai; Swarnamukhi; Sena-ka-patta.

Geographical distribution   Tinnevelly senna is widely cultivated in Tirunelveli, Madurai, Ramanathapuram and Tiruchirappalli districts of Tamil Nadu, Cudappa district of Andhra Pradesh, Mysore, Kutch in Gujarat, Jammu and some parts of Pakistan.

Collection and preparation   The plant is a small shrub bearing paripinnate compound leaves. The leaves are harvested before flowering by hand. The collected leaves are dried in the shade for 8–10 days till they become yellowish green in colour. The pods and other unwanted stalks are separated by sieves. The leaves are then graded according to size and compressed into bales and marketed. Indian senna leaves are in great demand in the U.S.A. market.

Figure 9.1 Twig of Indian Senna

Extraction of sennoides   Different methods are used for the extraction of sennoides from the leaves. But the common method used is as follows: The powder drug is either treated with 80% acetone or 90% methanol for 6 hours and then for 3 hours with cold water. This results in an extract containing about 17–20% sennoides.

Morphology of leaflets of Indian senna

Colour	Yellowish green
Odour	Slight
Taste	Slightly bitter and mucilaginous
Size	7–8 mm in width and 2.4–6 cm in length.
Shape	Leaflets lanceolate, entire, apex slightly acute, margin entire, base asymmetrical with epidermal trichomes on both sides of the leaflet, which are unicellular, conical, thick-walled and with a warty cuticle.

Chemical constituents   Senna leaves consist of anthraquinone glycosides such as sennosides A, B, C and D, which account for its purgative property. In addition, senna leaves contain rhein-anthrone 8-glucoside, rhein 8-diglucoside, aloe-emodin 8-glucoside, and aloe-emodin-anthrone diglucoside. Besides, the leaves contain two naphthalene glycosides namely tinnevellin glycoside and 6-hydroxy musizin glycoside. Senna also contains flavonoids like kaempferol, its glucosides (kaempferin), isorhamnetin; a sterol and its glucoside, mucilage, calcium oxalate, resin, phytosterol, salicylic acid, myricyl alcohol and chrysophanic acid.

Uses

1. It is used in habitual constipation.

2. It is also used as a purgative due to its laxative property and as cathartic.

Chemical test

1. A little drug extract is treated with 5N NaOH and sodium hyposulphate. On heating, red colour appears.

2. *Borntrager's test for anthraquinones*   Senna leaves are boiled with dilute sulphuric acid, filtered and the filtrate is extracted with ether or chloroform and shaken well. The organic layer is separated to which ammonia is added slowly. The ammoniacal layer shows pink to red colour due to the presence of anthraquinone glycosides.

2. ALOE

Biological source   *Aloe* is the dried juice of the leaves obtained from *Aloe vera* L. (*Aloe barbadensis* Mill.) of the family Liliaceae.

The common names of the plant are chotru kathazhai; ghirtakumari aloes; curacoa aloe; Karibolan; musabhar.

Geographical distribution   There are about 180 species of *Aloe* distributed widely. *Aloe barbadensis* is indigenous to Northern Africa. It is widely found in West Indies, Kenya, Socotra, Zangibar, Europe and in many parts of India. It is commonly cultivated in most of the Indian gardens. It is a typical perennial succulent xerophytic herb with thick, fleshy leaves having marginal spines. Leaves are arranged in rosette formation. The leaves are strongly cuticularized and waxed.

Collection and preparation   For collection of juice, the leaves are cut in March–April in V-shaped manner or a single incision is sufficient to draw out all the juice from the entire leaf. The juice of the *Aloe* is collected in a tin vessel. The juice is evaporated in copper vessel on open fire, poured into cans or tins, allowed to solidify and exported.

For marketing, it is packed in gourds, and sent to the market under the trade name Barbados or Curacao aloe. *Aloe vera* is also known as hepatic aloe or liver aloe.

Figure 9.2 *Aloe vera*

Characteristics of aloe juice

Colour	Ranges from yellowish brown to chocolate brown in colour.
Odour	Iodoform-like strong odour
Taste	Nauseating and bitter
Fracture	Waxy, with uneven fractured surface
Appearance	Opaque mass.

Uses

1. It is mainly used as a purgative and is given for constipation.

2. It is used as an ingredient of compound tincture of benzoin.

3. The ointment of aloe gel is used in the treatment of skin irritation and burns caused by sun.

4. It is also used to cure many skin diseases, ulcerative skin conditions, wounds, burns and snake bites.

5. It is used as a hair tonic and to treat enlarged spleen, and keratosis, and also for painful inflammations.

Chemical test 1 g of aloe leaf powder is boiled with 10 ml of water and filtered with the help of kieselguhr. The filtrate is used for bromine test and Schoenteten's reaction.

1. **Bromine test**   To a small quantity of the above filtrate, freshly prepared bromine solution is added. A pale yellow precipitate of tetrabromalin is formed.

2. **Schoenteten's reaction (Borax test)**   The above filtrate is treated with borax and shaken well till the borax dissolves. A few drops of this solution are added to a test tube nearly filled with water. A green fluorescence is formed.

3. **Borntrager's test**   1 g of powdered drug of aloe is boiled with 5 ml of 10% sulphuric acid for 2 minutes and filtered while hot. Cool the filtrate and shake gently with equal volume of benzene. Allow the benzene layer to separate from the lower layer. Using a pipette, the benzene layer is transferred to a clean test tube. To this add about half its volume of 10% ammonia solution, shake well and allow the layer to separate. The lower ammoniacal layer becomes pink in colour due to the presence of anthraquinones.

4. **Nitrous acid test**   To 5 ml of aqueous solution of aloe, add a few crystals of sodium nitrite and 2–3 drops of dilute acetic acid. Pink or purple colour is formed.

3. DIGITALIS

Biological source   The dried leaves that are used as drugs are obtained from *Digitalis purpurea* Linn. of the family Scrophulariaceae.

Some common names of the plant are fox glove leaves and folia digitalis.

Geographical distribution   Mostly cultivated in US, England, Southern and central parts of Europe, Germany, France and India. In India, it is grown in Kashmir, Kishtwar at 2000–2300 m altitude in Darjeeling and Nilgiris.

Collection   *Digitalis* is a biennial herb of about 1–2 m height. It grows wildly and is also cultivated for its leaves. The leaves are collected in the early afternoon from September to November by hand. Leaves collected in the first year of growth of the plant contain maximum amount of active constituents. The leaves are washed and dried immediately after harvest at a temperature below 60°C to keep up the potency of the drug. If drying is done rapidly, the characteristic green colour of the leaves is retained as such. Dried leaves containing not more than 5% moisture are packed in air-tight containers. Usually a desiccating substance like silica gel or calcium oxide is kept in the container to absorb moisture.

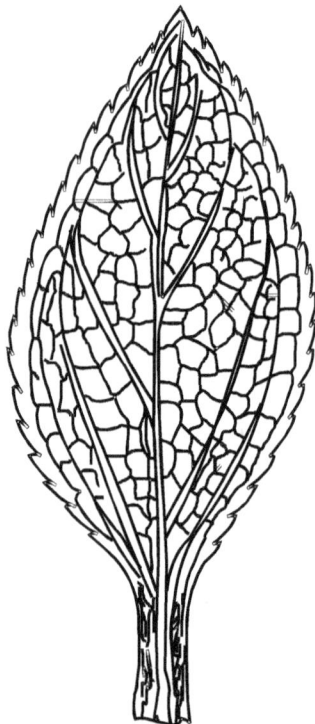

Figure 9.3   Leaf of *Digitalis purpurea*

Characteristics of dried leaves

Colour	Upper surface of the leaf is dark green and lower surface greyish green in colour with more prominent veins.
Odour	Slight
Taste	Bitter
Size	10–40 cm long and 4–8 cm width.
Shape	Ovate-lanceolate to broadly ovate, with crenate or serrate or dentate margin, apex obtuse or rounded, base decurrent with winged petiole.
Texture	Pubescent on both sides but lower surface is more hairy.

Chemical constituents   The drug contains both primary and secondary cardiac glycosides. Purpurea glycosides A and B and glucogitaloxin are the primary non-hydrolysed glycosides, all possessing at C-3 of the aglycone a linear chain of 3 digitoxose sugar moieties terminated by glucose. The primary glycosides are less stable and less absorbed than the secondary glycosides.

The secondary glycosides include digitoxin, gitoxin and gitaloxin, all possessing at C-3 of the aglycone a linear chain of 3 deoxy sugars. The purpurea glycoside A and purpurea glycoside B constitute the principal active constituents of the fresh leaves. The main cardioactive principles of the drug are digitoxin and gitoxin.

Uses

1. It is used as a cardiac stimulant and tonic.

2. The drug stimulates cardiac muscles to increase the systole of heart ventricle and normalizes the heart frequency.

3. In this way the drug is useful in congestive heart failure, atrial flutter and atrial fibrillation.

Storage *Digitalis* should be stored in a well-closed, well-filled container in a cool place away from light. Care must be taken to ensure that leaves do not contain more than 5% moisture, because it leads to destruction of glycoside and finally loss of cardiac activity on storage.

Chemical test Boil about 1 g of finely powdered digitalis drug with 10 ml of 70% alcohol for 2–3 minutes. The extract is filtered and used for the following test:

1. **Legal's test** The extract is dissolved in pyridine. Sodium nitroprusside solution is added to it to make it alkaline. Pink or red colour is produced.

2. **Baljet test** To a solution of digitalis leaves, sodium picrate solution is added. Yellow to orange colour is formed due to the presence of cardiac glycosides.

3. **Keller–Kiliani's test** To an extract of the drug in glacial acetic acid, a few drops of ferric chloride and conc. sulphuric acid are added. A reddish brown colour is formed at the junction of 2 layers, and the upper layer turns bluish green.

4. INDIAN SQUILL

Biological source The dried slices of the bulbs used as drugs are derived from *Drimia indica* (syn. *Urginea indica* Kunth.) of the family Liliaceae.

The common names of the plant include scilla; sea onion; jangli piyaz; nari-vengayam; kantena.

Geographical distribution It is distributed in the coastal areas of Konkan and Saurashtra and in the dry hills of the lower Himalayas at an altitude of 1500 m, and in Western Himalayas, Bihar and Chotanagpur.

Figure 9.4   Bulb of Indian squill

Collection   The plant is a glabrous perennial herb with underground bulb. The bulbs are collected after flowering in autumn. The uprooted bulbs are cleaned and cut into quarters, which are further sliced longitudinally. It loses about 80% of its fresh weight when sun-dried. The bulbs are then packed in bags or barrels for marketing.

Characteristics of slices of bulbs

Colour	The slices of bulbs are slightly yellowish to white.
Odour	Slight and characteristic
Taste	Bitter, mucilaginous and acrid
Size	30–60 mm in length and 3–8 mm broad
Shape	The slices are united in groups of 4–8, which are curved.
Entire bulb	15 cm in diameter
Slices	Translucent, and become tough and flexible, when they absorb moisture.
Surface of the bulb	Longitudinally ribbed
Fracture	Brittle

Chemical constituents   It contains 3% cardiac glycosides. In addition, it contains mucilage (40%) and calcium oxalate. Scillaren A and scillaren B are the major cardiac glycosides of the drug.

Uses

1. It is used as a cardiotonic, stimulant and also as a powerful expectorant, in chronic bronchitis.

2. It is used as a diuretic in small doses, but in large doses, it is an emetic and cathartic.

3. It is also used for asthma and possesses anti-cancer activity against carcinoma.

Chemical test Leaf mesophyll tissue stains red with alkaline colarin solution, and reddish purple with 0.1 M iodine solution.

5. DIOSCOREA

Biological source The dried tubers are taken from both cultivated and wild species of *Dioscorea deltoidea* Wall, *D. composita* and other species of *Dioscorea* of the family Dioscoreaceae.

The common names of the plant are yam; rheumatism root; colic root; poomikilangu; pitalu; kodikilangu; etc.

Geographical distribution The plant grows in the wild in north-western Himalayas from Kashmir and Punjab to Nepal and China up to an altitude of 1000–3000 m. In India it is cultivated in Jammu and Kashmir, Tamil Nadu, West Bengal, Maharashtra, Karnataka and in parts of Himachal Pradesh. It is also grown in U.S.A. and Mexico.

Collection The plant is a climber with tuberous roots. Fresh tubers are dug out from two-year-old crops. Before harvesting, deep ploughing is done in the dormant season and during this season the diosgenin content is found to be high. The dug-out rhizomes are washed to remove the mud and dried. The rhizomes lose 50% of their weight after drying.

Figure 9.5 *Dioscorea deltoidea* twig

Characteristics of tubers

Colour	Pale brown
Odour	Odourless
Taste	Bitter
Size	Varies with the source

Chemical constituents   Dioscorea rhizome contains 75% starch, but is non-edible due to the bitter taste. The chief active constituent is diosgenin, a steroidal sapogenin (4–6%). Tubers also contain an enzyme sapogenase. In addition it also contains rich glycosides and phenolic compounds.

Uses

1. Diosgenin is used as a precursor for synthesis of several sex hormones, corticosteroids and oral contraceptives.

2. It is used in the treatment of rheumatic arthritis.

6. LIQUORICE

Biological source   The dried, peeled or unpeeled roots and stolons derived from *Glycyrrhiza glabra* Linn. and other species of *Glycyrrhiza* of the family Leguminosae are used as drugs.

Some of the common names of the plant are liquorice root; mulethi; sweet wood; athimadhuram.

Geographical distribution   Spain, Sicily, England, Russia, Iran, Siberia, India, Italy, Iraq, Arabia, Greece and Asia Minor are the main countries which produce liquorice on a large scale. In India liquorice is cultivated in Punjab, tracts of Himalayas, Sind, Peshawar valley, and Andaman and Nicobar islands.

Figure 9.6   *Glycyrrhiza glabra* (Stolon)

Collection and processing of liquorice   *Glycyrrhiza* is a perennial herb, and grows to a height of 1 m. The drug is collected from 3–4-year-old plants, after planting when they show sufficient growth. Collection is made during autumn season. Roots and rhizomes are dug out from the soil, and the buds, and rootlets are removed and washed with water. The roots and rhizomes are cut into small pieces and dried first in the sun and then in shade. For the preparation of extracts and tinctures, only the unpeeled part is preferred, but for direct use, the peeled root or rhizome is used.

Processing of liquorice   For commercial purpose, besides peeled and unpeeled liquorice, the drug can be obtained in the following three forms.

i. Solid liquorice extract prepared by boiling roots with water.

ii. Aqueous extract dried and moulded into sticks and

iii. Granulated or spray-dried liquorice prepared from the aqueous extract.

The roots and rhizomes of liquorice are cut into small pieces by hand or using a shredding machine and extracted with boiling water. The extract is then cooled and the supernatant is decanted. The extract is evaporated at a controlled temperature. The concentrate is transferred to finishers for further evaporation and the process is continued till the extract contains 15–20% moisture and becomes a paste. The paste is then drawn out while still hot into wooden cases lined with paper. The product is moulded into blocks or sticks.

Characteristics of rhizome

Colour	Peeled liquorice is pale yellow in colour, whereas unpeeled rhizome is yellowish brown or dark brown outside and yellow colour inside.
Odour	Faint and characteristic
Taste	Sweet
Size	5–30 cm length and 1–2 cm diameter
Shape	Cylindrical, branched or unbranched, straight, peeled or unpeeled with longitudinal wrinkles.
Fracture	Fibrous in the bark and splintery in wood

Chemical constituents   It contains a triterpenoid saponinlike glycoside, glycyrrhizin (6–14%), which is 50 times sweeter than sugar. The amount of principle sweetening agent glycyrrhizin in *Glycyrrhiza* varies from 4–13% depending upon the geographical source of the drug and its time of harvest. In addition liquorice also contains sucrose (5%), starch (30%) and flavonoids (4%) like liquirtin, isoliquirtin, liquiritigenin, isoliquiritigenin, rhamnoliquirtin and various 2-methyl isoflavones which give yellow colour to liquorice.

Uses   Therapeutic uses of liquorice can be considered under the following headings:

   i.   **For respiratory tract**   Liquorice is used in a number of herbal syrups for diseases of respiratory tract. These include cough, chest congestion, and asthma, where it is used as an expectorant.

   ii. **For circulatory system**   Liquorice increases blood sodium and decreases blood potassium. The drug is used in Addison's disease because of glucocorticoid action, and causes oedema. Glycyrrhizin reduces blood cholesterol. The antipyretic effect of glycyrrhizic acid has been found equivalent to sodium salicylate.

   iii. **As anti-inflammatory**   Glycyrrhizic acid and two derivatives have anti-inflammatory and anti-arthritic activity. *Glycyrrhiza* is also used in tuberculosis. It is used in cosmetics and the glycyrrhizic acid is also used as material for the cosmetic gel, for skin and provides refreshing feeling, excellent transparency and consistency, together with anti-inflammatory value.

   iv. **For liver and urinary tract**   In Chinese medicine glycyrrhiza is used for jaundice and the drug increases the urinary volume and excretion of bilirubin.

   v.   **For gastrointestinal tract**   Liquorice is used as laxative due to increased absorption due to the presence of saponin. The drug is also useful in the treatment of stomach ulcers and duodenal ulcers.

   vi. **As sweetening agent**   Liquorice is used as a sweetening agent to mask the taste of such bitter drugs as aloes, quinine and others. Commercially liquorice is added in chewing gums, chocolate candy, cigarettes, smoking mixtures, and chewing tobacco. It is also added to beer to increase foam.

Chemical test   On addition of 80% sulphuric acid, the thick section of drug or powder shows orange-yellow colour.

7. BRAHMI [NIRBRAHMI]

Biological source   The fresh leaves and stem of the plant of *Bacopa moniera* Linn. (syn. *Herpestis monniera* H.B.K.) of the family Scrophulariaceae are used as drugs.

The common names of the plant include jalbrahmi (because of its habitat in wet, marshy and damp places); safed chamni; nir-brahmi; barna.

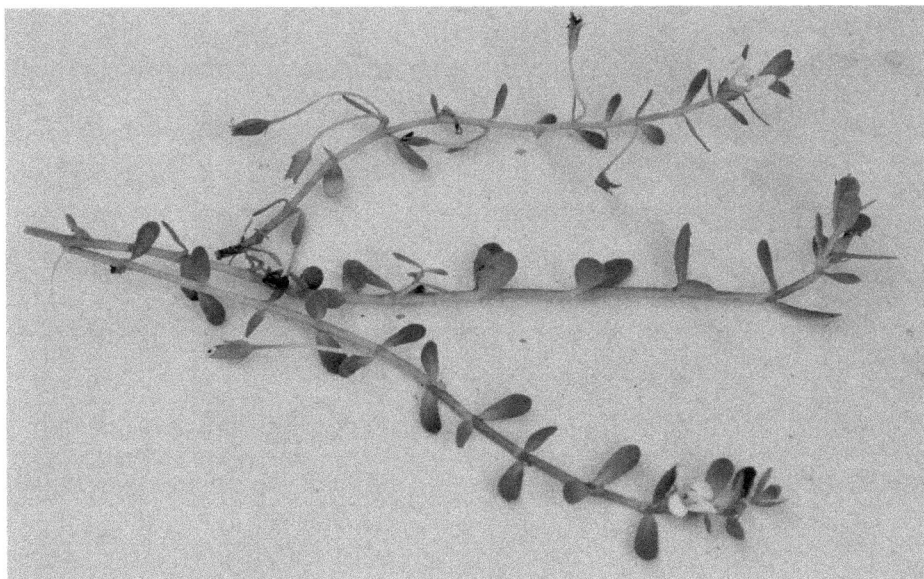

Figure 9.7 *Bacopa monniera*

Geographical distribution   The plant is a creeping succulent herb, rooting at the nodes, mostly found growing in wet habitats. It is distributed throughout India in wet, damp and marshy places.

Collection   The entire plant is collected and washed thoroughly to remove the mud particles. The drug may be used fresh or may be dried in the shade, packed and sent for marketing.

Characteristics of leaves

Colour	Green
Odour	Odourless
Taste	Bitter
Size	0.6–2.5 cm length and 0.5 cm width
Shape	Leaves are obovate–oblong, entire, with broad apex, sessile with entire margin.

Chemical constituents   *Bacopa* contains alkaloids like brahmine, herpestine and a mixture of three other alkaloids such as D-mannitol and saponin bacosides A and B. In addition it also contains betulic acid, stigmasterol and hersaponin.

Uses

1. It is used as a nervine tonic in the treatment of asthma, epilepsy, hoarseness and insanity.

2. It is aperient and also diuretic.

3. The alcoholic extract of the entire plant has anti-cancer activity.

8. BRAHMI

Biological source The entire herb of *Centella asiatica* urban. (syn. *Hydrocotyl asiatica* L.) of the family Umbelliferae (Apiaceae) is used as drug.

The common names of the plant include mandukparni, Indian pennywort, vallarai, etc.

Geographical distribution The herb grows throughout wet, damp and marshy places of India, Sri Lanka, Pakistan, Indonesia and Madagascar up to an elevation of 1500 m. It is a herbaceous creeper, rooting at the nodes with bigger leaves and long petioles.

Figure 9.8 *Centella asiatica*

Collection The entire plant or the leaves are collected before flowering and washed thoroughly. The drug is used either fresh or in dried form. The cleaned herbs or the leaves are dried in the shade, powdered, and packed for sale.

Characteristics of leaves

Colour	Green
Odour	Slight
Taste	Bitter

Size   2.5 cm length and 2 cm width

Shape   Simple, entire, crenate, orbicular, reniform and glabrous on both sides.

Chemical constituents   The drug contains triterpenoid saponin glycosides in the form of α-amyrin derivatives called asiaticoside and madecassoside. They yield asiatic acid and madecassic acid on hydrolysis. It also contains brahmic acid, arabinose, glucose and rhamnose. Besides saponin, the drug also contains flavonoids like kaempferol and quercetin.

Uses

1. The drug is used as a nervine tonic.

2. It shows sedative, spasmolytic, anti-anxiety, anti-stress, and anti-bacterial action.

3. The drug is also employed in skin diseases, leprosy, eczema and syphilis.

4. In India, the drug is used as a brain tonic.

9. SWEET ALMOND

Biological source   The ripe seeds of *Prunus amygdalus* Baill var. *dulcis* Schneider of the family Rosaceae are used as drug.

The common names include almond, badam, badama, etc.

Geographical distribution   The plant is a tree, which grows to a height of about 5 m and is native to Persia and Asia minor. It is cultivated in Italy, Spain, Sicily, Morocco, Portugal and South France. In India, it is cultivated in Kashmir.

Figure 9.9   *Prunus amygdalus* var. *dulcis* (Seed)

Collection and preparation   The fruits are collected and dried in the shade for few days. The drupes (fruits) are broken using machines. The seeds are gathered and graded according to the size and packed. The packed seeds of badam are sent to market for sales. The drug is also available in the form of powder called almond powder or badam powder.

Characteristics of seeds

Colour	Brown
Odour	Odourless
Taste	Sweet and slightly bitter.
Size	20 mm in length, 12.5 mm width and 10 mm in thickness.
Shape	The seeds are oblong-ovoid in shape and flattened and rounded at one end and pointed at the other end, with markings on testa. Kernel consists of 2 large oily planoconvex cotyledons.

Chemical constituents   Sweet almonds contain about 40–50% bland oil, 20% proteins, mucilage and an enzyme, emulsin, and a colourless crystalline bitter cyanogenic glycoside known as amygdalin (2–4%). It also contains 0.5% volatile oil. Amygdalin on hydrolysis decomposes into benzaldehyde (80%) and hydrocyanic acid (2–6%). Sweet almonds contain also a little sucrose, gum and asparagin. Almond oil chiefly contains olein with a small proportion of linolein.

Uses

1.   Almond oil is used as a demulcent skin lotion.

2.   It is also used in the preparation of amygdalin, almond water, in perfumery and in the form of liquors.

3.   Sweet almonds are highly nutritive, hence used as a non-starchy food for diabetic patients.

10. MUSTARD

Biological source   The drug is constituted by the dried ripe seeds, which are obtained from the species of *Brassica nigra* (L) Koch. or *B. juncea* HK.f and T. and their varieties of the family Cruciferae (Brassicaceae).

The common names include black mustard; brown mustard; true mustard; sarson; kadugu; rai; etc.

Geographical distribution   The plant is largely cultivated in Europe, south-western Asia, USA and India. In India, mustard is grown in Uttar Pradesh, Bihar, Bengal and Tamil Nadu as rabi crop.

Collection   The plant is an annual erect herb with yellow flowers and siliqua fruit. When fruits mature, the aerial parts of the plant are cut and dried partially. The partially dried herb is thrashed or sometimes beaten to collect the seeds. The gathered seeds are then dried in the sun. The seeds are separated from other foreign materials by sieving, and marketed.

(a)

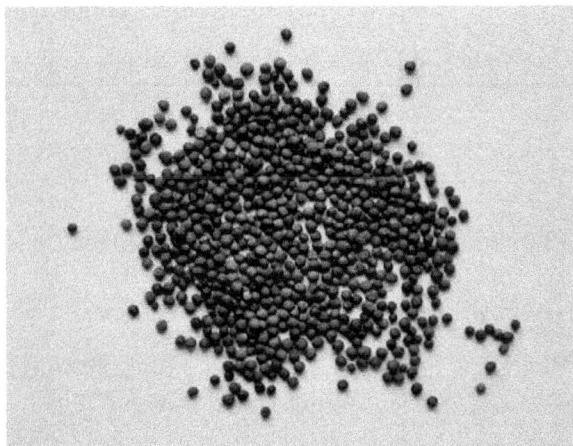

(b)

Figure 9.10   *Brassica nigra* (a) Twig   (b) Seeds

Characteristics of seeds

Colour	Black or dark brown or reddish brown.
Odour	Seeds odourless, whereas crushed seeds give pungent smell.
Taste	Bitter
Size	1 mm diameter
Shape	Spherical to globular

Chemical constituents   Black mustard seeds contain about 30–35% fixed oil, 20% proteins and 0.7–1.3% volatile oil. The seeds also contain 4% isothiocyanate glycoside, which is a major active constituent called sinigrin (potassium myronate) and its amount varies from 0.7–13%. In addition the seeds also contain mucilage and a little quantity of acid sinapine sulphate. Sinigrin is hydrolysed in the presence of water to form allyl isothiocyanate, called volatile mustard oil.

Uses

1. It is used as a condiment.

2. It is used externally as counter-irritant and rubefacient in the form of poultice or plasters.

3. It acts as emetic in large doses.

4. Other preparations of mustard include:

    ❉ Mustard spirit which contains 2% volatile mustard oil in alcohol.

    ❉ Mustard liniment which contains volatile mustard oil, camphor, castor oil and alcohol.

    ❉ Black mustard oil which is the oil expressed from black mustard, and used as mild rubefacient.

Chemical test   When treated with alkali, mustard powder gives bright yellow colour.

11. PSORALEA FRUIT

Biological source   The dried ripe fruits derived from *Psoralea corylifolia* L.of the family Leguminosae are used as drug.

The common names include bavchi; malaya tea; bawchang seed; babchi; Karpo–karishi; etc.

Geographical distribution   It grows throughout India as a weed in wastelands. It is found in the states of Maharashtra, Madhya Pradesh and Tamil Nadu. It is also seen in Sri Lanka.

Collection   The plant is an erect annual herb growing to a height of 60 cm to 1 m. Fruits are minute. The mature pods are picked from the plants and dried. The dried pods are graded, packed, and marketed. The pods are also powdered, packed and sent for sale.

Figure 9.11 *Psoralea corylifolia* twig

Characteristics of fruits and seeds

Colour	Dark brown to black
Odour	Seed odourless, whereas the powder is pungent and characteristic after crushing.
Taste	Bitter, unpleasant and acrid.
Size	Fruits are 4 mm long and 2–3 mm broad
Shape	Pods ovoid to oblong, compressed, glabrous, closely pitted seeds with campylotropous ovule, non-endospermic, smooth, oily and free of starch, and the seeds are kidney-shaped.

Chemical constituents *Psoralea* fruits contain essential oil (0.25%), resin (8.9%), brown fixed oil (10%), a pigment, raffinose, and coumarin compounds, psoralen, isopsoralidin, psoralidin and corylifolin.

Uses

1. The drug is used in the treatment of leucoderma, leprosy, psoriasis and inflammatory diseases of the skin.

2. It is given orally as well as locally applied externally in the form of an ointment or paste.

3. The seeds of *Psoralea* are used as stomachic, diuretic, diaphoretic, anthelmintic and laxative.

Chemical test

1. A small quantity of drug is dissolved in alcohol to which 3 times propylene glycol, 5 times acetic acid and 40 times water is added. A blue fluorescence is observed under ultraviolet light.

2. The drug shows yellow fluorescence in UV light when dissolved in alcohol and little sodium hydroxide solution is added to it.

12. PICRORRHIZA

Biological source   The dried roots and rhizomes of *Picrorrhiza kurroa* Royle of the family Scrophulariaceae are used as drug.

The common names of the plant are Indian gentian; kadu; katuka; kutki; kutkikuru; kitchli; etc.

Geographical distribution   It is a perennial herb with woody rootstocks, grows wild at an altitude of 3000–5000 m in alpine Himalayas ranging from Kashmir to Sikkim, Punjab, Uttar Pradesh, Himachal Pradesh and also in China. The rhizome is covered with dry leaf bases.

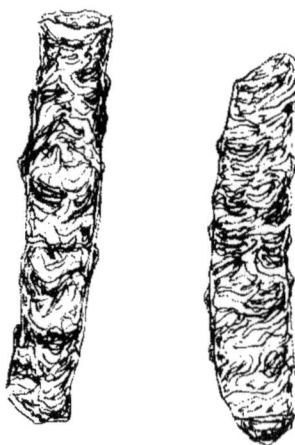

Figure 9.12   *Picrorrhiza* rhizome

Collection   The drug is collected from the naturally growing plants during the period July–August. The large underground woody rootstocks are dug out and washed thoroughly. The cleaned rootstock are cut into small pieces. The rootlets and other unwanted materials are removed and dried.

Characteristics of rhizomes

Colour	Rhizomes deep greyish brown in colour, externally white, internally blackish with whitish wood
Odour	Slightly unpleasant
Taste	Very bitter
Size	3–5 cm in length and 0.5–1 cm in diameter.
Shape	Cylindrical and straight or slightly curved pieces with longitudinal wrinkles and annulations at the tip
Fracture	Tough

Chemical constituents   The drug contains irridoid glycosides picroside I, picroside II and kutkoside, which are responsible for the activity of the drug.

Uses

1. *Picrorrhiza* is used as a valuable bitter tonic, anti-periodic, febrifuge and stomachic.
2. It is a laxative in large doses.
3. Alcoholic extract of the root has antibacterial effect.
4. It removes kidney stone and is used as an emetic, abortifacient, antidote for dog-bite and externally used for skin diseases.
5. The drug is useful in the treatment of jaundice and hepatitis and is also used as a substitute for gentian.

13. CHIRATA

Biological source   The entire herb of *Swertia chirata* Buch Ham. of the family Gentianaceae is used as drug.

The common names are chirayata; east Indian balmony; kirata-tkta; nelabedu; nilavembu.

Geographical distribution   The plant is indigenous to temperate Himalayas and is found growing wild in Kashmir, Simla, Mussoorie Hills, Madhya Pradesh, Khasi Hills and Meghalaya at an altitude of 1300–3500 m. In addition to India, the herb is distributed in Nepal and Bhutan.

Figure 9.13 Chirata twig

Collection The plant is an erect annual herb that grows to a height of 1 m. The plant is small with quadrangular stem. The drug is collected from wild-grown plants during the flowering stage, and dried. The plant flowers from July to October and is collected for medicinal purposes, when the capsules are fully formed. The harvested plants are washed and dried in the shade to retain their active principle. The dried plants are made into sheaves weighing 1 kg and marketed.

Characteristics of leaves, flowers and fruits

Colour	Leaves, flowers and fruits with yellowish tinge, stem is round and purple at the base but angular and yellowish to brown above.
Odour	Odourless
Taste	Extremely bitter
Size	Stems are about 6 mm thick and 1 m in length
Shape	Cylindrical, glabrous and quadrangular at the apex; leaves are sessile, thin, opposite decussate, unequal, broad at the base, ovate-lanceolate, margin entire, apex acuminate and glabrous.

Chemical constituents Chirata contains a bitter yellow acid known as ophelic acid and two bitter glycosides, chiratin and amarogentin. Amarogentin is a phenol carbonic acid ester of sweroside, a substance related to gentiopicrin. It also contains yellow crystalline substance used in dyeing. The two alkaloids reported in chirata are gentianine and gentiocrucine.

Uses

1. Chirata is used extensively in Ayurvedic medicine as a bitter tonic and as an ingredient of an ayurvedic preparation "sudarshan powder" used in chronic fever.

2. It has been used since ancient times as a febrifuge.

3. It is used as a remedy for scanty urine and also used in epilepsy and administered either as an infusion or a tincture.

4. It is used industrially for dyeing cotton cloth.

5. It is free of constipative effect and also prescribed in dyspepsia.

14. KALMEGH

Biological source   The dried leaves and tender aerial shoots of the plant *Andrographis paniculata* Nees. of the family Acanthaceae is used as drug.

The common names are king of bitters; chiretta; green chirata; andrographis; kiryat; nilavembu; kirata; siriyanankai; etc.

Geographical distribution   Kalmegh is an erect annual herb reaching a height of 1–3 ft. The herb is found throughout the plains of India, specifically in Maharashtra, Karnataka, Uttar Pradesh, Tamil Nadu, Andhra Pradesh, and Madhya Pradesh. It is also cultivated in Assam and West Bengal.

Figure 9.14   *Andrographis paniculata*

Collection   Flowering starts in the period September to December, and collection of the drug is made in November and December after the flowering season. The aerial parts of the plants are harvested, washed and cleaned to separate the foreign materials. The aerial parts are dried and packed for sale.

Characteristics of leaves, stem and flowers

Colour	Leaves are dark green, stem quadrangular and flowers are red in colour.
Odour	Odourless
Taste	Strongly bitter taste
Size	Leaves 7 × 2.5 cm, flowers 1–5 cm in length.
Shape	Leaves lanceolate, petiolate, with entire margin and apex acuminate.

Chemical constituents   The drug contains kalmeghin (0.85–2.5%), a bitter crystalline diterpene lactone such as andrographolide flavonoids and phenols.

Uses

1. It is used as a bitter tonic, an antidote and anthelmintic.

2. Green leaves are used as stomachic, and in the treatment of dysentery, dyspepsia, fever and in weakness.

3. It possesses antityphoid and antibiotic activities.

REVIEW QUESTIONS

State whether the following sentences are true or false

1. The laxative nature of senna is responsible for its purgative property.

2. Digitalis belongs to the family Bignoniaceae.

3. Bromine test is used to test the purity of *Aloe*.

4. The dried stems of *Dioscorea* are used as medicine.

5. Glycyrrhizin is used as a sweetening agent.

Choose the correct answer

6. The drug used as nervine tonic is _____

   a. pudina   b. senna   c. brahmi

7. The drug used in Ayurvedic "Sudarshan powder" is _____

   a. chirata   b. myrobalan   c. mustard

8. The main cardioactive principles of *Digitalis purpurea* are _____

   a. gitoxin and digitalinin   b. digitoxin and gitoxin   c. digitoxin and digitin

9. The biological source of Indian squill is _____

   a. *Dioscorea deltoidea*   b. *Drimia indica*   c. *Centella asiatica*

10. The bitter cyanogenic glycoside present in almond is _____

    a. sinigrin   b. myosin   c. amygdalin

Answers				
1. True	2. False	3. True	4. False	5. True
6. c	7. a	8. b	9. b	10. c

Answer the following

1. What are cardiac glycosides? Write the biological source, geographical distribution, method of collection, chemical constituents, characters, use and storage of *Digitalis*.

2. Define bitter glycoside. Explain any two anthraquinone glycoside-containing drugs known to you.

3. Write notes on:

   i. Chirata

   ii. Indian Senna

   iii. Mustard

   iv. Kalmegh

4. Give the therapeutic use of *Aloe* and brahmi.

5. Give the botanical name, collection, processing, characteristics, chemical constituents and uses of liquorice.

6. Give the pharmaceutical importance of the following drugs:

   i. Picrorrhiza

   ii. Chirata

   iii. Indian squill

    iv. Dioscorea

    v. Mandukparni

    vi. Sweet almond

    vii. *Psoralea*

    viii. Black mustard

7. What are the chemical tests carried out for identifying the following drugs: *Aloe, Digitalis, Psoralea.*

DRUGS CONTAINING LIPIDS

10

Lipids are the substances derived from plants or animals and consist of fixed oils, fats and waxes. The primary function of oils and fats is to store energy and in addition, they are also used in medicine and other industries.

FIXED OILS

Fatty oils are the reserves of food materials of plants (arachis oil, sesame oil, castor oil, rice bran oil, cocoa butter) and animals (cod-liver oil, shark-liver oil). The lipids that are liquid at 15.5°C to 16.5°C are called fixed oils (fatty oils) while those which are solid or semi-solid at that temperature are called fats. These are esters of glycerol with higher long-chain and have common characteristics.

1. RICE BRAN OIL

Biological source   Rice bran is the cuticle layer that occurs between the grain and the husk of the paddy and encloses the embryo and endosperm of the seeds of *Oryza sativa* Linn. and other paddy varieties of the family Gramineae (Poaceae).

The common names of the product include rice oil; oryzae oil.

Geographical distribution   The plant is a semi-aquatic herb, which is cultivated in tropical and sub-tropical countries.

Collection and method of preparation   Rice bran is obtained as a by-product during polishing of rice in a rice mill. It is removed after dehusking of paddy.

The quality of rice bran oil depends upon the time which elapses between milling of the rice and removal of oil from the separated bran. Therefore, the solvent extraction plant for rice bran oil should be constructed near the rice mill, so as to process the bran oil quickly. Rice bran contains an active enzyme called lipase, that increases the free fatty acid content on storage. The oil derived from fresh bran is of good quality with good flavour and contains low free fatty acid.

Rice bran is found in extremely small pieces, and is also hard-textured. It is impermeable to solvents. Therefore, it is dried, cooked and flaked before solvent extraction. It is then pressed and extracted with solvent in special continuous immersion extractors.

Description Rice bran oil is an example of non-drying oil, which is difficult to bleach. It is not affected by temporary heating to 160°C.

Colour	Dark golden yellow colour
Odour	Odourless
Taste	Acrid with ghee taste
Appearance	Transparent
Solubility	Oil is insoluble in water and soluble in common fat solvents.

Standards

Acid value	4 to 5
Iodine value	99 to 108
Saponification value	181 to 189
Hydroxyl value	5 to 14
Refractive index	1.470 to 1.473
Specific gravity	0.916 to 0.921
Thiocyanogen value	69 to 76

Chemical constituents Rice bran oil contains 20–25% of saturated and 80–85% of unsaturated fatty acids as glycerides. The main fatty acids present in this oil are oleic (40–50%), linoleic (30–40%) and palmitic acids (12–18%). It also contains antioxidants like tocopherols. Due to the presence of antioxidant, its keeping quality is good.

Uses

1. Rice bran oil is used in the manufacture of cosmetics and as an emollient.

2. It is an edible oil and used in the preparation of vegetable ghee.

2. GROUNDNUT OIL

Biological source Groundnut oil is expressed from the seed kernels of *Arachis hypogaea* Willd. of the family Fabaceae (Leguminosae)

The common names are peanut oil; monkey nut oil; earth nut oil; kadalai ennai.

Geographical distribution Groundnut plant is a low-growing annual herb, 0.3–0.6 m tall. A native of Brazil, it became widely distributed in south and Central America, West Africa, Philippines, China, Japan, Peru, Malaysia, Australia, Nigeria, Sudan, Thailand, Argentina and India.

The principal producing states of India are Andhra Pradesh, Gujarat, Tamil Nadu, Karnataka, Uttar Pradesh, Madhya Pradesh, Punjab and Rajasthan.

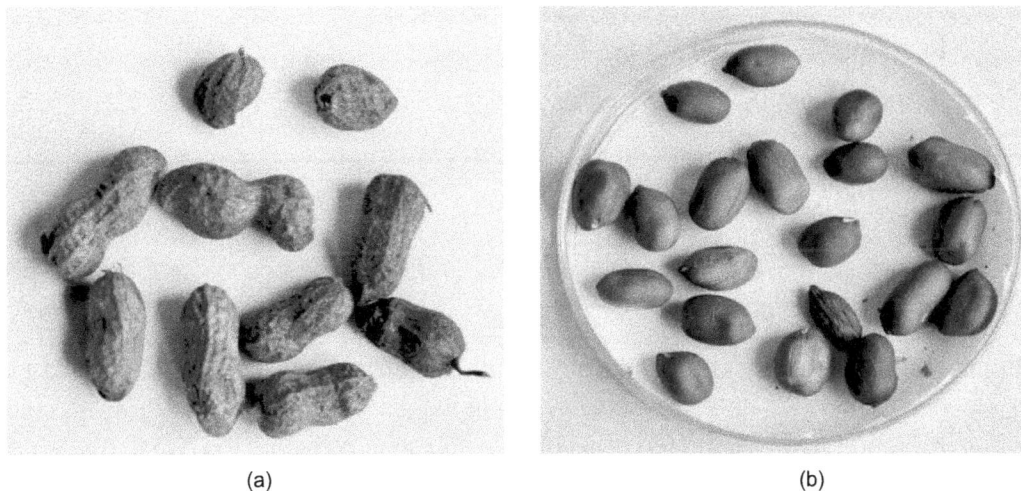

(a)   (b)

Figure 10.1 *Arachis hypogaea* (a) Pods (b) Seeds

Collection and extraction of oil The crop is harvested by hand and the fruits are separated from the uprooted plants by picking or flailing. Harvesting and shelling are also carried out mechanically. The fruits are washed to remove the mud particles and other unwanted earthern materials, and dried in the sun. The shells are decorticated either by machine or hand. The cleaned kernels are cold-pressed to get the oil, which is golden yellow in colour, with a faint agreeable odour, while the hot-pressed oil has a reddish yellow tinge. Hot expression is more common and involves the pre-heating of kernels with steam to allow the lipids to come out from the cells, whereas in cold expression, the kernels are pressed without resorting to steaming. Pre-cooking, results in higher yield, but the quality is inferior. To remove the impurities, the oil is subjected to the refining process.

Characteristics of oil   It is a non-drying fixed oil belonging to the oleo-linoleic acid group.

Colour	Cold-pressed oil is golden yellow in colour, whereas hot-pressed oil is a pale yellow-coloured liquid.
Odour	Faint and characteristic.
Taste	Bland taste and nutlike
Solubility	It is miscible with solvent ether, petroleum ether, chloroform, carbon disulphide, benzene and very slightly soluble in alcohol.

Standards

Acid value	Not more than 0.5
Iodine value	85 to 105
Saponification value	185 to 195
Refractive value	1.467 to 1.470
Unsaponifiable matter	Not more than 1%

The oil is lighter than water having 0.911–0.920 wt. per ml.

On exposure to air, the oil thickens very gradually and becomes rancid. It is sterilized by dry heat process.

Chemical constituents   The peanut oil contains principal glycerides of fatty acids such as oleic acid 56%, linolenic acid 25%, palmitic acid 6–12%, stearic 3.1%, arachidic acid 3.4% and higher saturated acid. Arachin and conarachin are the two important proteins found in the seed. The yellow colour of the oil is due to the presence of carotenoid pigments chiefly β-carotene and lutein. The kernels contain about 45% of fixed oil, 20% of proteins and are rich in thiamine content.

Uses

1.   Principally the oil is used as an edible oil and as a substitute for olive oil.

2.   It is used as a solvent for intramuscular injection in pharmaceutical aid, for manufacture of margarine, soap, paints, liniments, plasters and ointments.

3.   Due to its non-drying property, it is used as a valuable lubricant.

4.   Peanut oil cake is a valuable livestock food.

3. CASTOR OIL

Biological source   Castor oil is a semi-drying fixed oil extracted from the seeds of *Ricinus communis* Linn. of the family Euphorbiaceae.

The common names of the product are ricinus oil; oleum ricini; oil of palma christ; aamanakku oil; vilakkennai.

Geographical distribution   The plant is an annual woody shrub, reaching up to 2–3 m height. It is indigenous to India, which is the second largest producer of castor seed in the world. It is extensively cultivated in temperate, tropical and sub-tropical countries of the world. Castor-growing countries are India, Brazil, Russia, China, Thailand, U.S.A., Romania and West Indies. In India, it is largely cultivated in Gujarat, Karnataka, Andhra Pradesh and Tamil Nadu.

(a)   (b)

Figure 10.2 *Ricinus communis* (a) Twig (b) Seeds

Collection and preparation of oil   The mature seeds are collected, graded, cleaned, washed and sun-dried. The dried seeds are decorticated through a decorticator. The rollers with sharp cutting edges break the testa without injuring the kernels. The kernels are cold-expressed by suitable hydraulic press with a pressure of 2 tonnes per square inch, to get about 30% oil at room temperature. The extracted oil is known as **cold-drawn oil,** which is used medicinally.

The oil is then refined by steaming, filtration and bleaching. Therefore, the cold-drawn oil is steamed at 80°C to destroy the enzyme lipase and a toxic protein albumin called ricin. It is then bleached and de-acidified with sodium carbonate to remove free fatty acid. Finally, it is treated with animal charcoal to remove the impurities present in it by adsorption. This refined oil is known in commerce as 'cold drawn oil'. It is filled into containers and sold.

The remaining oil from the seeds is extracted by hot expression or solvent extraction methods. This oil is not suitable for medicinal purposes and is of industrial significance.

Characteristics of oil

Colour	Cold-drawn oil is a pale yellow or almost colourless, clear viscid liquid.
Odour	Slight or characteristic nauseating odour.
Taste	Acrid taste with a nauseating after-taste.
Appearance	Transparent and viscous liquid.
Keeping quality	Excellent keeping qualities and does not turn rancid unless subjected to excessive heat. When exposed to air, it gradually thickens, darkens in colour and develops a strong odour and taste.
Solubility	The oil is soluble in alcohol and this is an exceptional property which is lacking in all other fixed oils. It is miscible with all organic solvents.

Standards

Weight per ml	0.945 to 0.965 g
Acid value	Not more than 2
Iodine value	81 to 90
Saponification value	176 to 187
Optical rotation	$+3.5°$ to $+6.0°$
Solidifying point	$-10°$ to $-18°C$
Refractive index	1.467 to 1.479
Viscosity	6.8 poises
Specific gravity	0.958 to 0.969
Acetyl value	About 150

Chemical constituents Castor oil chiefly composed of mixture of fixed oil 45–55%, proteins 20%, (consisting of globulin, albumin, nucleoalbumin, glycoprotein, ricin (a toxalbumin)), ricinine alkaloid and some enzymes. The fixed oil mainly contains triglycerides of ricinoleic acid about 87% and also isoricinoleic 2%, stearic 1%, dihydroxy stearic acid in traces, linoleic 3%, oleic 7% and palmitic acid 2%. Hydrolysis of triricinolein of in triglycerides by lipases in the duodenum releases free ricinoleic acid and its stereoisomers, which are actually responsible for purgative action because of its irritating action on the membranes of small intestine.

Uses

1. Castor oil is a purgative and emollient and is used for lubrication commercially.
2. It is used for external application for its emollient effect.
3. The hydrogenated castor oil is used as a useful ointment base, and also in the preparation of flexible collodion U.S.P, to prepare undecylenic acid, which is a fungistatic preparation, to manufacture lipsticks, perfumed hair oil, and hair fixers.
4. In case of food poisoning, the use of castor oil in hospitals is more popular.
5. For intestinal examination, the oil is administered as a preliminary step.
6. It is used in textile industry, coating, surfactants, cosmetics, paints, varnishes, grease, polishes, printing ink, etc.
7. It is used in abortifacient paste, and ricinoleic acid is used in contraceptive creams and jellies.
8. It is the chief raw material for the production of sebacic acid.
9. Castor oil is also used as a substitute for spermaceti, beeswax and carnauba wax.

Identification Test

1. An equal volume of ethanol added to the oil results in a clear liquid. On cooling at 0°C and on storage for 3 hours, the liquid remains clear.
2. It can be mixed with half its volume of light petroleum ether (40–60°C).

4. SESAME OIL

Biological source Sesame oil is a semi-drying fixed oil obtained from the seeds of different cultivated varieties of *Sesamum indicum* Linn. of the family Pedaliaceae.

The common names of this oil are gingelly oil; benne oil; till oil; sesamum seed oil; nallaennai.

Geographical distribution The plant is an annual herb, 1 m in height. Sesame is indigenous to India and widely cultivated in China, Japan, West Indies, Africa and in Southern United States.

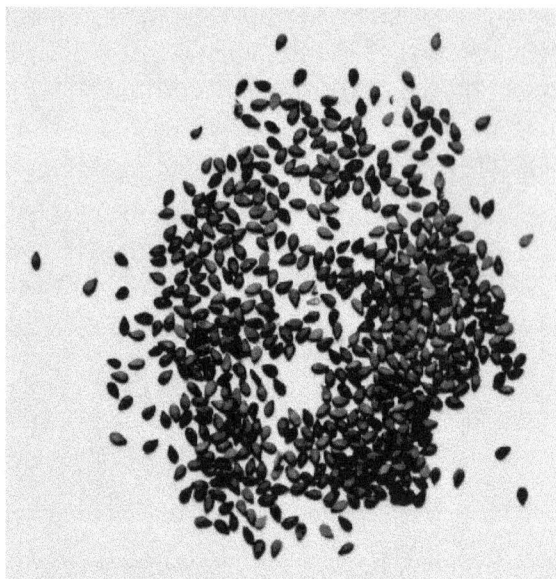

Figure 10.3 *Sesamum indicum* (Seeds)

Collection and method of preparation of oil   The seeds are very small in size. There are two varieties of seeds, white or yellowish white and red or brownish black. The colour of seeds partly determines the quality and the quantity of the fixed oil present. White or yellowish white seeds give the best quality oil, whereas, dark-coloured seeds yield a larger quantity of fixed oil. Commercially these varieties are referred to as "white" and "black" sesame seeds.

The white varieties of seeds are more desirable and used for pharmaceutical purpose. The mature seeds are gathered, cleaned, if necessary washed and sun-dried. The seeds are hot-pressed at room temperature to obtain the oil. Subsequently, the temperature and pressure are raised, when a yellow to dark amber-coloured oil is obtained. The crude oil is refined by the usual refining procedures by using caustic alkali, bleaching and deodorizing agents and packed for marketing.

Characteristics of sesame oil

Colour	Pale yellow liquid
Odour	Slight, characteristic odour
Taste	Bland taste
Solubility	The oil is slightly soluble in alcohol but miscible with all other organic solvents.

The oil does not solidify at 0°C, but on cooling to –4°C, it solidifies to a butterlike mass.

Standards

Acid value	Not more than 2
Free fatty acids	Not more than 1%
Iodine value	103 to 116
Saponification value	188 to 195
Refractive index	1.472 to 1.476
Specific gravity	0.916 to 0.919
Unsaponifiable matter	Not more than 1.5%

Chemical constituents   The seeds contain about 50% of semi-drying fixed oil. Sesame oil consists of a mixture of glycerides of higher fatty acids mainly oleic acid (43%), linoleic acid (43%), palmitic (9%), stearic (4%) and myristic acids. It contains a phenolic constituent known as sesamol, which is responsible for the excellent stability of the oil. In addition, the oil also contains lignan derivatives such as sesamin and sesamolin and also vitamins A and E.

Uses

1. The oil is nutritive, and is a laxative and demulcent having emollient properties.

2. The refined oil is non-allergenic and therefore, used as a vehicle in many pharmaceutical intramuscular oil injections of steroids, antibiotics and hormonal drugs.

3. The oil is also used in the preparation of liniments, plasters, ointments, soaps, cosmetics, iodized oil and oleomargarine.

4. The oil is more valuable in pharmacy as a solvent or as a suspending agent.

5. Sesamolin present in the unsaponifiable fraction of the oil is an effective synergist for Pyrethrum insecticide.

6. Sesame oil yields top quality black ink after burning.

7. It is also used in dysentery and urinary complaints.

Identification test

> **Baudonin's test** Two ml of sesame oil shaken with 1 ml of 1% solution of sucrose in concentrated hydrochloric acid produces pink to red colour due to the presence of the phenolic component, sesamol.

5. SUNFLOWER OIL

Biological source Sunflower oil is an example of a semi-drying fixed oil, extracted from the seeds of cultivated varieties of *Helianthus annuus* Linn. of the family Compositae (Asteraceae).

The common names are sunflower oil; suryakamal oil; suriya kanthiennai; arkakantha.

Geographical distribution The plant is an annual woody herb, reaching a height of 1.5 m. It is indigenous to North America. It is cultivated in countries like India, Egypt, Canada, Tanzania, South Africa, Argentina, Australia and Russia.

Figure 10.4 *Helianthus annuus*

Collection and preparation of oil The mature seeds are collected and dried in the sun or shade. Prior to extraction, extraneous materials such as twigs, leaves, stones, floral parts and tramp materials are removed by screening and passage over magnets. After cleaning, the seed coats (hull) are removed with the help of specially designed decorticating machines and thereafter the kernels are reduced to thin flakes. The oil is extracted by mechanical expression method.

Mechanical expression method involves the application of pressure to the oil-bearing tissues to squeeze out the fat. This is accomplished either by hydraulic pressing or screw pressing. In hydraulic press, the flaky material is usually placed in sacs or wrapped in strong cloth or filter cloth which holds back the residual mass as the oil strains through.

Hot expression is more common and involves the pre-heating of kernels with steam to allow the lipids to escape from the cells. In cold expression, the kernels are merely pressed without resorting to steaming. As a result of pre-cooking, the oil is rendered more labile, giving a higher yield but the quality is inferior. As the crude oils often contain impurities, they are subjected to a refining process to rid them of all impurities.

Characteristics of oil

Colour	Pale yellow
Odour	Odourless liquid
Taste	Bland agreeable taste
Solubility	It is soluble in chloroform, benzene and carbon tetrachloride, and insoluble in water and alcohol.

Standards

Acid value	0.6
Iodine value	125 to 136
Saponification value	188 to 194
Refractive index	1.466 to 1.468
Specific gravity	0.915 to 0.919
Unsaponifiable matter	Below 1.5%
Hydroxyl value	14 to 16
Thiocyanogen value	78

Chemical constituents   Sunflower seeds contain about 54–60% oil. The oil contains a mixture of glycerides, which are rich in unsaturated acids. It contains a large amount of linoleic acid (65%), oleic acid (21.3%), behenic acid (0.8%), arachidic acid (94%), stearic acid (1%) and palmitic acid (6.5%) and also vitamin E.

Uses

1. It is used as cooking oil and as a dietary supplement.

2. It is also used in the preparation of oleomargarine and on a large scale in the production of soaps.

6. KARANJA OIL

Biological source   It is a semi-drying fixed oil obtained from the seeds of the tree *Pongamia glabra* Vent. of the family Fabaceae (Leguminosae).

The common names are karanja oil; pungamara oil; nalatamala; kuranj; pungammaram.

Geographical distribution   The plant is a small, evergreen tree. It is found growing in India along streams and rivers, mainly in the states of Tamil Nadu, Andhra Pradesh, Kerala, Madhya Pradesh and Maharashtra.

Figure 10.5   *Pongamia glabra*

Collection and preparation of oil The seeds are removed from the legumes and cleaned to remove impurities. They are dried in sunlight and washed thoroughly to remove the mud particles. The seeds are used for extracting oil by expression method.

Characteristics of oil

Colour	Fresh oil is yellow-orange in colour, but darkens and becomes reddish brown on storage.
Odour	Peculiar pungent and disagreeable odour.
Taste	Bitter taste.

Standards

Acid value	6 to 7
Iodine value	89 to 90
Saponification value	181 to 182
Refractive index	1.4736
Specific gravity	0.9273
Acetyl value	21
Unsaponifiable matter	4.2%

Chemical constituents It is a semi-drying fixed oil, containing a mixture of fatty acids, such as palmitic acid (3.5–9%), stearic acid (2.5–9%), arachidic acid (2.2–5%), lignoceric acid (1–3.5%), oleic acid (44–75%), linoleic acid (10–18.5%) and behenic acid (4.0–5.5%).

The unsaponifiable matter of the oil contains β-sitosterol, while non-fatty components are composed of karanjin, karanja, chromene, pongapin and pongamol.

Uses

1. Pongamia oil is mostly used in the treatment of herpes, scabies, leucoderma and other skin diseases.

2. It is used in rheumatism, in the manufacture of toilet soap, laundry soap and also as a lubricant for dressing of leathers in tanning.

3. The oil cake is used as a remedy against white ants and also as manure in the fields.

7. LINSEED OIL

Biological source   Linseed oil is a drying fixed oil, extracted from the seeds of *Linum usitatissimum* Linn. of the family Linaceae.

The common names of the oil are flaxseed oil; alasi; arish vettu.

Geographical distribution   It is an annual, cultivated herb, 1 m high, and indigenous to India. At present, it is extensively cultivated as a source of oil in India, Afghanistan, Turkey, South America, US, Canada, England, Russia, etc., while as a source of fibre in Spain, Algeria, Egypt, Argentina, Scotland, Holland and Belgium.

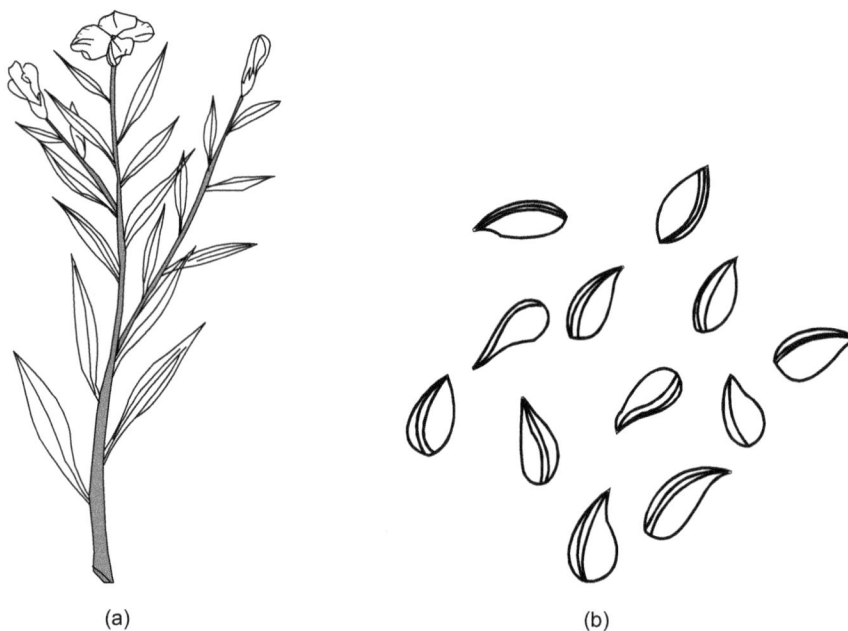

(a)   (b)

Figure 10.6   Linseed (a) Twig (b) Seeds

Collection and method of preparation of oil   Fully mature seeds from high oil-yielding variety are collected for oil extraction. The seeds are sieved to free the earthy matter and other impurities. The dried seeds are crushed in oil seed rollers, moistened and heated to 85–90°C in steam-jacketed troughs to soften the seed tissues. The softened, seed tissues are pressed through hot hydraulic press at a high pressure of 3000–5000 pounds per square inch. The expressed oil is tanked for a long period, to settle the colouring matter and mucilage. The oil is then filtered and treated with alkali to remove free fatty acids. The oil is bleached with either fuller's earth or charcoal. The refined oil is then chilled to remove the waxy substances.

Characteristics of oil

Colour	Linseed oil is a pale yellow clear liquid
Odour	Characteristic
Taste	Bland taste
Solubility	Slightly soluble in alcohol, soluble in organic solvents and insoluble in water.

Linseed oil gradually thickens on exposure to air forming a thin transparent film. On drying, it forms a hard varnish.

Standards

Acid value	Not more than 4
Iodine value	160 to 200
Saponification value	188 to 195
Refractive index	1.478 to 1.482
Specific gravity	0.927 to 0.931
Unsaponifiable matter	Not more than 1.5
Viscosity	1.47

Chemical constituents   Linseed oil is a drying oil, contains fixed oil (30–40%), mucilage (6%), protein (25%), small amount of enzyme linase, and linamarin, a cyanogenetic glycoside. Linseed oil has high iodine value as it contains considerable quantities of glycerides of unsaturated acids such as 36–50% of linolenic acid, 23–25% of linoleic acid, 10–12% of oleic acid along with some saturated acids like palmitic, stearic and myristic acids, which collectively constitute 5–11%.

Uses

1. Linseed oil is the most important drying oil and is mainly used for external applications like lotions and liniments.

2. It is used in the treatment of scabies, skin diseases, for gouty and rheumatic swellings.

3. Internally it is used for gonorrhoea and irritation of the genito-urinary system.

4. As the oil contains very high iodine value, it is used in the preparation of non-staining iodine ointment, soap, linoleum, greases, polishes, polymers, varnishes, paints, oil cloth, printing inks, artificial rubber, tracing cloth, etc.

5. It is applied to paper and fabrics to render them waterproof and tough.

6. Raw linseed oil alone is used pharmaceutically as an emollient and vehicle in liniments and pastes.

8. COCOA BUTTER

Biological source Theobroma oil is obtained from the seeds of *Theobroma cacao* Linn. of the family Sterculiaceae.

The common names are theobroma oil, cacoa butter.

Geographical distribution Cocoa plant is a small tree, attaining a height of about 10 m. It is native to tropical America (Mexico). The plant is mainly cultivated in Sri Lanka, Philippines, Ecuador, Curacao, Central America, Brazil, Nigeria, Ghana and parts of India. Cocoa butter was prepared as early as 1696 AD. Cocoa is extensively cultivated in Kerala.

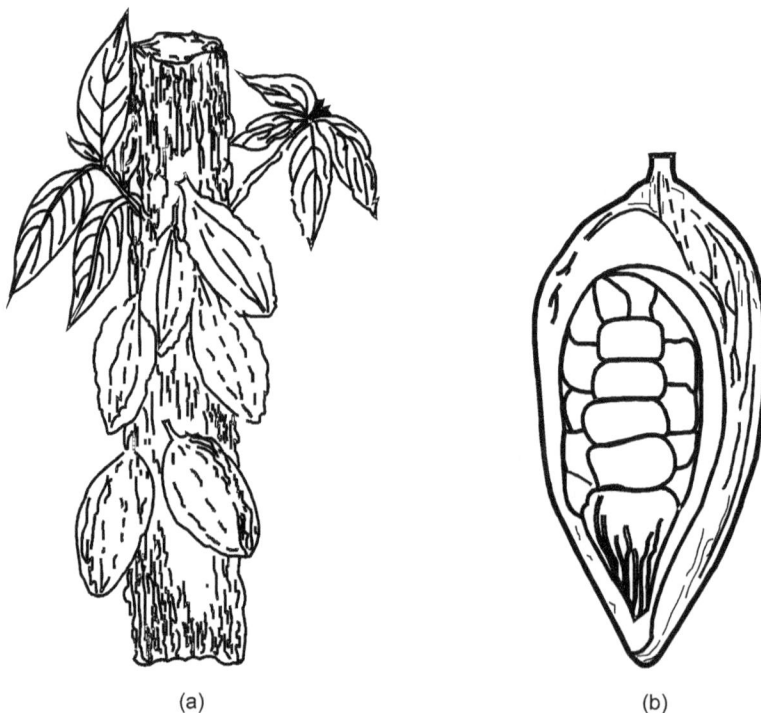

(a)   (b)

Figure 10.7   (a) Cocoa twig   (b) L.S. of cocoa fruit

Collection and preparation Cocoa plant is cultivated up to an elevation of 1000 m above sea level and it needs shade for successful growth. Fruits are formed after three years of planting. As it shows cauliflory, the fruits arise from the older branches or trunk.

As the fruits become ripe, the seeds are separated from the pods and packed in boxes or tubes, and are allowed to undergo fermentation. During the time of fermentation, the colour of the seeds is changed from white to dark reddish brown due to enzymatic reaction. The fermentation process is carried out for 3 to 6 days at 30–40°C. The seeds are dried in the sun. The taste of the seed, which was astringent and bitter at first, now becomes mild and oily. The dried seeds are roasted at 100–140°C, and the seeds lose water and acetic acid, thus facilitating the removal of seed coat. The roasted seeds develop a distinct odour and taste and the seeds are cooled immediately. The cooled seeds are passed through a "nibbling" machine to crack the seed coats, and are removed by winnowing. The kernels are ground between hot rollers to yield a pasty mass containing up to 50% fat called **cocoa butter**. This is further purified to give cocoa butter.

After expressing cocoa butter, the marc, retaining some oil, is powdered, and is called as **prepared cocoa** or **breakfast cocoa**. The cocoa shells are processed to yield an alkaloid.

Characteristics of cocoa butter

Colour	Theobroma oil is yellowish white, solid, brittle below 25°C.
Odour	Pleasant chocolate odour
Taste	Chocolate taste
Solubility	Slightly soluble in alcohol, but soluble in water and boiling alcohol. Very soluble in chloroform, ether and benzene.

Standards

Iodine value	35 to 40
Saponification value	188 to 195
Refractive index	1.454 to 1.458
Specific gravity	0.858 to 0.864
Melting point	35 to 40°C

Chemical constituents Cocoa butter consists of glycerides of stearic (34%), palmitic (25%), and oleic (37%) acids and small amount of arachidic and linoleic acids. These glycerides are present in the form of simple and mixed glycerides.

Uses

1. Theobroma oil is used as lubricant in massage, base for suppositories, ointments, and in manufacturing chocolate, toilet soaps, creams, etc.

2. Breakfast cocoa is a popular beverage.

9. KOKUM BUTTER

Biological source   Kokum butter is the fat obtained from the ripe seeds of *Garcinia indica* Chois. (syn: *Garcinia purpurea* Roxb.) of the family Guttiferae (Clusiaceae).

The common names of the product include kokum fat; Goa butter; kokum oil; mangosteen oil.

Geographical distribution   The plant is a tree. It is indigenous to Thailand, China and Cambodia and also grows wild in India. In India, it is found in Western Ghats, Konkan, Coorg, Malabar and Nilgiri Hills. It is cultivated in Maharashtra.

Figure 10.8   A twig of Kokum butter

Collection and preparation   The mature seeds are separated from the ripe fruits. The dried pericarp consists of fat, which is known as kokum, and its taste is sweet–sour. The outer testa of the seeds are removed and the kernels are crushed and boiled in water. The melted fat floats on hot water and is skimmed off and washed with hot water twice. The separated melted fat is decolorized with animal charcoal or fuller's earth to get solid waxlike fat that constitutes "kokum butter".

Characteristics

Colour	Greyish white solid
Odour	Slight and characteristic

| *Taste* | Oily and characteristic with sweet-sour taste |
| *Shape* | Marketed in the form of egg-shaped lumps. |

Standards

Acid value	Not more than 3
Iodine value	35 to 37
Saponification value	185 to 190
Refractive index	1.47 to 1.458
Unsaponifiable matter	2 to 3%
Melting point	39° to 42°C
Weight per ml	0.895 to 0.899 g

Chemical constituents   Kokum butter consists of triglycerides of stearic acid (55%), oleic acid (40%), palmitic acid (2.5%) and linoleic acid (1.5%).

Uses

1. It is used as nutritive, demulcent, an astringent and emollient.

2. It is used in the preparation of ointments and suppositories and also for the preparation of pomades.

3. It is used in skin diseases, dysentery, mucous diarrhoea and Pthisis pulmonalis.

4. It is employed externally to cure ulcers, fissures of lips and hands, wounds and sores.

5. The dried rind of the fruit is called amsul and is used as a substitute for tamarind.

10. SHARK-LIVER OIL

Biological source   Shark-liver oil is the fixed oil obtained from the fresh or carefully preserved healthy livers of various species of shark, mainly *Hypoprion brevirostris* of the Order Selachii. Species such as *Scoliodon, Carcharias* and *Sphyrna* are abundantly present in India and are mostly used for the extraction of oil.

The common name for this oil is oleum selachoide.

Geographical distribution   Shark is found on the sea coasts of many European countries. In India, the oil is produced on a large scale in the states of Tamil Nadu, Kerala and Maharashtra.

Method of preparation of oil Fresh livers removed from shark are thoroughly cleaned, freed from fatty substances and adhering tissues like gall bladder. The cleaned livers are minced and heated in a boiling pot at a temperature of 80°C. The oil that exudes and floats on the top is separated and treated with dehydrating agent to remove the water. The water-freed oil is chilled to separate stearin. The suspended materials from the oil are removed by centrifugation.

The clear oil is manipulated to adjust vitamin strength as per requirement of IPS. Therefore, the oil is fortified with vitamins A and D.

The oil is sensitive to light and air. Therefore, care is taken to minimize its exposure to sunlight and air. The oil must be stored in amber-coloured containers. According to Indian Pharmacopoeial Specifications, one gram oil should not contain less than 6000 International Units of vitamin A activity.

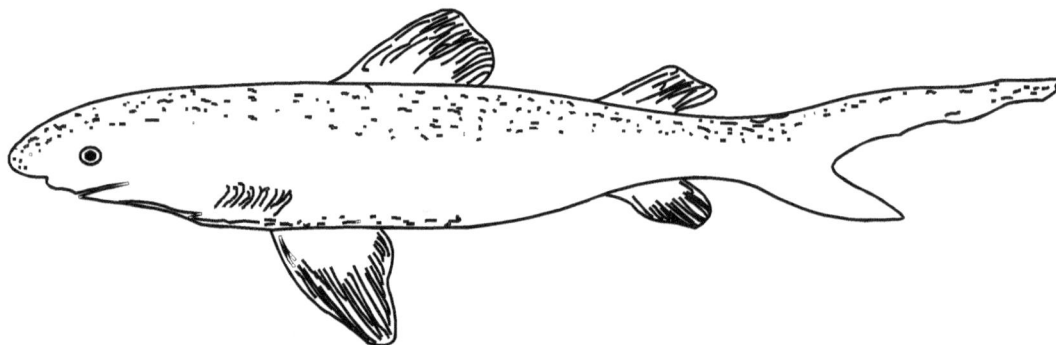

Figure 10.9 Shark fish

Characteristics of shark-liver oil

Colour	Pale yellow to brownish yellow in colour, viscous liquid.
Odour	Characteristic fishy odour
Taste	Bland taste
Solubility	The oil is insoluble in water, slightly soluble in alcohol and miscible with solvents like ether, chloroform and petroleum ether.

Standards The pharmaceutical grade of shark-liver oil should contain the following standards.

Acid value	Not more than 2
Iodine value	Not less than 90
Saponification value	150 to 200

Refractive index	1.459 to 1.477
Specific gravity	0.912 to 0.916

Chemical constituents Shark-liver oil contains the active principle vitamin A. The concentration of vitamin A in the oil varies from 15,000–30,000 International Units per gram. The oil also contains glycerides of saturated and unsaturated fatty acids. Shark-liver oil should contain 16,500 units/g of vitamin A and 40 units/g of vitamin D.

Uses

1. The oil is nutritive and used as tonic.

2. Pharmaceutically, it is used in the preparation of dilute shark-liver oil, shark-liver oil emulsion and shark-liver oil with vitamin D.

3. It is used in burn and sunburn ointments.

4. The oil is a rich source of vitamin A and is used in the treatment of xerophthalmia (abnormal dryness of the surface of conjunctiva due to deficiency of vitamin A).

Chemical test

1. A drop of shark-liver oil in 1 ml of chloroform is treated with one drop of sulphuric acid. A violet colour changes to purple or brown due to the presence of vitamin A.

2. One ml of shark-liver oil is dissolved in 10 ml of chloroform. A few drops of saturated solution of antimony trichloride in chloroform are added to the solution and shaken well. A blue colour is formed due to the presence of vitamin A.

11. COD-LIVER OIL

Biological source Cod-liver oil is obtained from fresh liver of the cod fish, *Gadus morrhua* and other species of *Gadus* of the family Gadidae.

The common name is Cod fish oil.

Geographical distribution The fishes are available in coastal regions of Britain, Norway, Iceland, Germany, Scotland and Denmark.

Collection and extraction of oil The fishes are caught using nets and the livers are removed from the fishes. The fresh and healthy livers are freed from gall bladders and cleaned thoroughly and washed. The livers are then minced and steamed in steam-jacketed containers at a temperature of 85°C for half an hour. It is then cooled and buried in snow for several days using special barrels. This process results in the separation of stearin. The steaming of oil removes the enzyme lipase. Cod fish oil is medicinally used after filtration. The oil is kept in well-closed airtight containers in a cool place away from sunlight so as to avoid loss of vitamins during storage. Cod-liver oil is refined by

i.   removal of impurities

ii. drying

iii. winterization

iv. deodorization

v.   standardization for vitamin content.

The oil contains the growth-promoting and antixerophthalmic vitamin A about 850 units/g, the anti-rachitic vitamin D about 85 units/g and the glyceryl esters of mixed unsaturated fatty acids.

The vitamin A content of the oil is determined spectrophotometrically. The oil is flavoured by adding 1% of a flavouring agent. If needed a small amount of antioxidant like dodecylgallate can be added.

Characteristics

Colour	Pale yellow
Odour	Liquid with slightly fishy odour
Taste	Slightly fishy taste
Solubility	The oil is freely soluble in petroleum ether, chloroform, ether carbon disulphide and slightly soluble in alcohol.

Standards

Acid value	Less than 2
Iodine value	145 to 180
Saponification value	180 to 190
Refractive index	1.471 to 1.475
Specific gravity	0.918 to 0.927

Chemical constituents The oil contains vitamin A about 850 units/g, and vitamin D about 85 units/g. The medicinal value of oil is due to the presence of vitamin A and vitamin D. The oil contains glyceryl esters of oleic, linoleic, gadoleic, myristic, palmitic and other acids. The sterile solution of the sodium salts of the fatty acids from cod-liver oil is known as sodium morrhuate.

Uses

1. Cod-liver oil is used as a source of vitamins, as a nutritive and in the treatment of rickets and tuberculosis.

2. Sodium morrhuate is administered as a sclerosing agent.

WAXES

Waxes are the esters of higher straight-chain fatty acids with long-chain or high molecular weight monohydric alcohol, etc. Waxes may be obtained from animals (beeswax, wool fat), plants (carnauba wax, seasal wax) and minerals (paraffin). Waxes are used in pharmaceuticals for hardening, in cosmetic creams and in ointments and also in the preparation of cerates.

YELLOW BEESWAX

Biological source   Beeswax is obtained by purifying the honeycomb of the bee, *Apis mellifera* L., and other species of *Apis* of the family Apidae.

The common names are beeswax; cera-flava; white beeswax.

Geographical distribution   Beeswax is commercially prepared in Italy, France, West Indies, West Africa, Jamaica, California, Chile, Madagascar and India.

Collection and processing of beeswax   The wax is secreted by the worker bees in the last four segments on the ventral surface of the abdomen. The wax comes out through pores in the chitinous plates. The young worker bee utilizes the wax in the construction of the comb to store the honey.

The honeycomb is separated, broken and melted in water. The heat oozes out the wax and also dissolves the residual honey, and the solid impurities settle down. After straining the water–wax mixture, it is allowed to cool, re-melted and finally strained. The wax is allowed to solidify in suitable moulds to get yellow beeswax.

White beeswax   White wax is prepared from yellow beeswax by bleaching with chemical or by the use of light, air and water. The chemical process of bleaching includes the use of charcoal, potassium permanganate, chromic acid, ozone, chlorine or treatment with hydrogen peroxide.

For slow bleaching, the melted wax is allowed to fall on a revolving moist cylinder. As a result, ribbonlike strips of wax are formed, and are exposed to sunlight and air, till the bleaching is completed.

Characteristics   Beeswax is a solid, non-crystalline material, soft to touch and crumbles under the pressure of the fingers, to a plastic mass.

Colour	Yellowish brown to yellowish white
Odour	Agreeable and honeylike
Taste	Slight balsamic taste
Fracture	Granular
Solubility	It is insoluble in water, but soluble in hot alcohol, ether, chloroform, carbon tetrachloride, fixed and volatile oils. Under molten condition, it can be given any desired shape.

Standards

Acid value	5 to 10
Saponification value	90 to 103
Specific gravity	0.958 to 0.967
Melting point	60°C to 65°C
Ester value	80 to 95

Chemical constituents The chief constituent of beeswax is myricin, i.e., myricyl palmitate (about 80%), free cerotic acid (about 15%). Small quantities of melissic acid and aromatic substance cerolein are the other constituents. The colour of the beeswax is due to the presence of pollen and propolis (bee glue). Indian beeswax is characterized by its low acid value, while European beeswax has an acid value of 17–22.

Uses

1. Yellow beeswax is used as a hardening agent for the preparation of ointments, plasters, cerates, face creams, lipsticks, wax papers and candles, and for moulding artificial fruits and flowers.

2. It is an ingredient of paraffin ointment I.P. and yellow ointment.

3. It is used in engraving, shoe polish, etc.

Tests for identity

1. Heating 0.5 g of wax in 20 ml of aqueous sodium hydroxide for 10 minutes should not produce turbidity. The content is cooled and filtered and treated with HCl to give a clear solution. But, Japan wax produces turbidity due to stearic acid.

2. Beeswax is a true wax and on saponification, the cerotic acid does not form soap. Adulterated wax forms water-soluble soap on treatment with alkali.

REVIEW QUESTIONS

State whether the following sentences are true or false

1. The cuticle layer of paddy grain is called rice bran.
2. Groundnut oil is a semi-drying oil.
3. Cold-drawn oil of castor is used as medicine.
4. *Sesamum indicum* belongs to the family Pedaliaceae.
5. Cocoa butter is taken from linseed.

Choose the correct answer

6. The active principle of shark-liver oil is _____
   a. vitamin C   b. vitamin A   c. vitamin E
7. Beeswax is derived from _____
   a. *Apis mellifera*   b. *Apis lepidra*   c. *Vivera zibitha*
8. Karanja oil is a _____
   a. semi-drying oil   b. non-drying oil   c. essential oil
9. *Garcinia indica* yields _____
   a. cocoa butter   b. fixed oil   c. kokum butter
10. The biological source of cod-liver oil is _____
    a. *Hypoprion brevirostris*   b. *Copernicia terri*   c. *Gadus morrhua*

Answers				
1. True	2. False	3. True	4. True	5. False
6. b	7. a	8. a	9. c	10. c

Answer the following

1. What are lipids? Give an example.
2. Differentiate between fixed oil and fats with suitable examples.
3. Write the medicinal values of the following oils:
   i. Peanut
   ii. Castor

    iii. *Sesame*

    iv. *Pongamia*

    v. Shark

4. Distinguish between the characters of the following drugs given below:

    i. Cod-liver oil and shark-liver oil

    ii. Castor oil and rice bran oil

    iii. Karanja oil and sunflower oil

    iv. Cocoa butter and kokum butter.

5. Write down the biological source of groundnut, sunflower, karanja, linseed and shark.

6. Describe the methods of collection and extraction of oil for groundnut and castor.

7. Give the medicinal values of rice bran oil, linseed oil and shark-liver oil.

8. With suitable examples give the botanical source, distribution, collection and method of preparation, chemical constituents and use of any two semi-drying fixed oils that you have studied.

9. List out the chemical compounds of rice bran oil, castor oil, beeswax and linseed oil.

10. Give the source, family of source plant, collection and extraction of oil, importance and identification of castor oil.

11. How will you test shark-liver oil and beeswax?

11

DRUGS CONTAINING VOLATILE OILS

The volatile principles of plant and animal sources that are odoriferous, constitute volatile oils. When exposed to air at room temperature, they evaporate, hence, they are also called "ethereal oils." As they represent the essence or active constituent of a plant, they are known as "essential oils." The volatile oils are mono- and sesquiterpenes obtained from the sap and tissues of certain plants and animals. Volatile oils are soluble in alcohol, ether and other lipid solvents and insoluble in water. They are colourless, usually lighter than water, and are secreted in special structures such as ducts, cells, schizogenous or lysigenous cavities, trichomes, etc. They are commonly distributed in the members of Umbelliferae, Labiatae, Rutaceae, Zingiberaceae, Myrtaceae, etc.

Volatile oils are extensively used as flavouring and perfuming agents in pharmaceuticals, foods, beverages and cosmetics. In addition, they are also used as medicinal agents, e.g., carminatives, antiseptics, counter-irritants, local anaesthetics, sedatives, etc. Volatile oils are extracted by steam-distillation, or by mechanical methods.

This chapter includes the study of biological origin, geographical source, collection and preparation, characters, active constituents, tests for identity and therapeutic uses of drugs containing volatile oils.

1. CROCUS

Biological source   *Crocus* constitutes the dried stigmas and upper part of styles of the gynoecium of *Crocus sativus* L. of the family Iridaceae.

The common names of the part used are saffron; keshar; kungumappu.

Geographical distribution   The plant is a small perennial herb. The drug is native to South Europe. It is cultivated in Austria, Spain, Italy, France, Greece, China, Germany, Iran and India (Kashmir).

Figure 11.1   *Crocus* plant

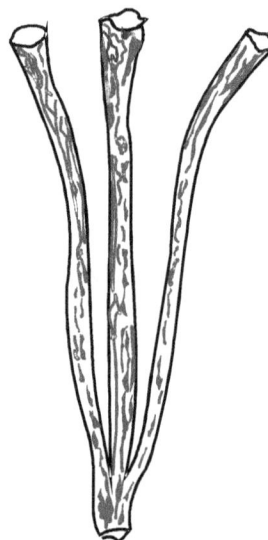

Figure 11.2   *Crocus* stigma

Collection   The flowers are collected in the early morning during the months of November and December.

The stigma of the gynoecium of each flower is separated from the upper part of the style with the nail of the right thumb. The detached stigmas are dried over artificial heat for half an hour, cooled and stored in dry place. About 1 kg of drug is collected from nearly 100,000 flowers.

Characteristics of stigma

Colour	Dark red to reddish brown with some yellowish pieces of tops of styles
Odour	Sweet, pleasant, strong and aromatic
Taste	Characteristic bitter taste
Size	Stigmas are 25 mm long and styles about 10 mm long
Shape	Saffron refers to the flat, tubular, almost threadlike trifid stigmas and cylindrical styles.

Chemical constituents   Saffron contains a number of carotenoid colour compounds such as ester of crocin, a coloured glycoside responsible for colour, picrocrocin, a colourless bitter glycoside, which is responsible for the characteristic bitter taste, aromatic odour and the typical orange-yellow colour of the drug, crocetin, an aromatic compound, gentibiose, α- and γ-carotenes, crocin 2, crocin 3, and 4. Picrocrocin is formed of safranal, an aldehyde and glucose. The drug contains 1.3% volatile oil, fixed oil and wax.

Uses

1. Saffron is used as a colouring dye and flavouring agent.

2. It is used in fevers, cold, melancholia and enlargement of the liver, catarrh, snake-bite, cosmetics and pharmaceutical preparations and as a spice.

3. It is also used as an antispasmodic, stimulant, stomachic, tonic, aphrodisiac, emmenagogue, sedative and as an abortifacient.

Tests for identification

1. When a drop of sulphuric acid is added over the dry stigma, a deep blue colour is produced.

2. Saffron imparts brown colour to water.

3. When kesar is chewed, the saliva becomes orange-yellow colour.

2. VAJ

Biological source   Vaj constitutes the peeled dried or unpeeled rhizomes of *Acorus calamus* L. of the family Araceae.

The common names of the product and source plant are calamus; sweet flag; vasambu.

Geographical distribution   The plant is an aromatic, erect and perennial herb. The drug is native to Eastern Europe. The plant is extensively grown in India, Sri Lanka, North America, Europe, Asia and Holland, in marshes. In India, it is cultivated in Sikkim and found wild in the Himalayas.

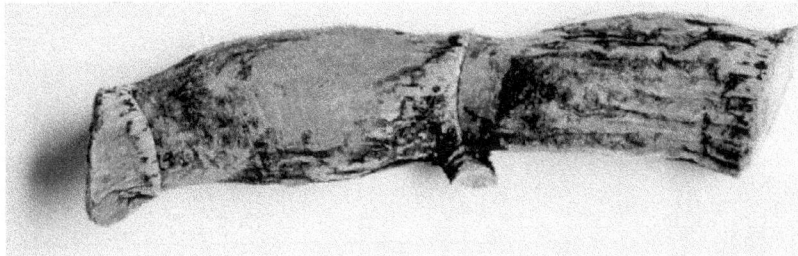

Figure 11.3   *Acorus calamus* (Rhizome)

Collection   The rhizomes of the plant are collected during autumn, when the underground parts are rich in volatile contents. After collection, the rhizomes are freed from roots, mud, and leaf bases and cut into small pieces before drying.

Characteristics of rhizome

Colour	Brown
Odour	Characteristic and aromatic.
Taste	Bitter and disagreeable.
Size	5–15 cm in length and 1–2 cm in thickness.
Shape	The rhizomes are cylindrical and branched, and the pieces are with deep longitudinal wrinkles. They have prominent leaf scars on the upper surface, and lower surface bears small raised circular root scars.
Fracture	Short

Chemical constituents   *Calamus* contains 2–4% volatile oil, starch, resin (2.5%) and tannin (1.5%). Volatile oil contains asaraldehyde. The other contents of the oil are asarone and eugenol. The drug also contains a bitter amorphous principle known as acorine.

Uses

1. *Calamus* is used as a stimulant, stomachic, carminative and vermifuge.
2. Volatile oil is used in perfumery, as insect-repellant and for flavouring gin, beers, etc.
3. It is used in epilepsy and depression and the tranquilizing properties of the drug is due to the presence of asarone.
4. The drug is also used in dyspepsia, colic, remittent fevers, in bronchitis, dysentery of children and snake-bite, as a nervine tonic and also for the preservation of food grains.

Storage   *Calamus* rhizomes must be stored in well-closed containers.

3. ROSEMARY OIL

Biological source   The volatile oil is obtained from the fresh flowering tops of the plant *Rosmarinus officinalis* Linn. of the family Labiatae (Lamiaceae).

Geographical distribution   The plant is an evergreen herb reaching a height of 2 m. It is indigenous to South Europe. It is cultivated in the Mediterranean basin, South Asia and India.

Figure 11.4   Rosemary plant

Collection   After flowering, the plants are cut above the ground and the flowers are separated for the extraction of oil. Commonly effleurage method is used and in this method fresh flowers are mechanically spread on a layer of fatty material and allowed to imbibe and then the exhausted petals are replaced by fresh materials. The process is continued till the fatty layer is saturated with volatile principles, which are then extracted with lipid solvent.

Characteristics of rosemary oil

Colour	Colourless or pale yellow liquid
Odour	Characteristic rosemary flavour
Taste	Camphoraceous

Standards

Specific gravity	0.894 to 0.912
Optical rotation	–5° to 10°
Refractive index	1.464 to 1.476
Solubility	The oil is insoluble in water, but soluble in 10 volumes of 80% alcohol.

Chemical constituents The flowers contain about 1% volatile oil, resin and bitter principle. Volatile oil contains mainly 10–15% borneol, 2.5–3% bornyl acetate, camphor, eucalyptol, pinene, D-camphene, cineol and 45% terpenes.

Uses

1. Rosemary oil is used as a carminative, stimulant and rubefacient.

2. The oil is also used as a flavouring agent in hair lotions, liniments, inhaler, cosmetics and soaps.

Storage The oil should be kept in well-closed containers and protected from light and stored in cool place.

4. TULSI OIL

Biological source The drug constitutes the oil obtained from fresh leaves of *Ocimum sanctum* Linn. of the family Labiatae (Lamiaceae).

The common names of the source plant are sacred basil; holy basil; vishnu–tulsi; thulasi.

Geographical distribution It is an annual herbaceous plant, attaining a height of 40–60 cm. The plant is considered sacred by Hindus, hence it gets the name holy basil. The plant is commonly cultivated in gardens and in Hindu temples and also found wild throughout India.

Figure 11.5 *Ocimum sanctum*

Collection   The leaves and flower tops are collected, cleaned, washed and used for steam distillation. In the steam distillation process, the fresh leaves are subjected to hydrodistillation to extract the oil.

Characteristics of leaves   The leaves are petiolate, simple, elliptical, oblong–obtuse, apex acute, 2–5 cm long, 1.5–3.2 cm wide, margin entire or serrate, base obtuse or acute, surface hairy, pubescent on both sides, minutely gland-dotted, green-coloured and is highly aromatic. The taste of the leaves is slightly pungent.

Chemical constituents   The whole herb contains volatile oil but the leaves contain highest percentage of bright yellow, pleasant volatile oil (0.9%). The prominent constituents of the essential oil are eugenol (72%), methyl eugenol (20%), carvacrol (3%) and caryophyllene (1.8%). The oil content of the drug varies with the type, place of cultivation and season of its collection.

Uses

1. The oil has antibacterial, antiprotozoal, expectorant, antiseptic and insecticidal properties.
2. The fresh leaves are used as stimulant, and are aromatic, anticatarrhal, spasmolytic and diaphoretic.
3. The leaf juice is used as an antiperiodic and also for several skin diseases and to cure ear-ache.
4. The infusion of the leaves is used as stomachic in gastric disorders of children and is also a good immune-modulator agent.
5. The leaves are used as aromatic carminative.
6. The plant is also used in treating snake bite and sting of scorpion.

5. PEPPERMINT OIL

Biological source   Peppermint oil is obtained from the fresh inflorescence of the plant *Mentha piperita* Linn. of the family Labiatae (Lamiaceae)

The common names are mentha oil, colpermin, lamp mint.

Geographical distribution   The plant is a perennial herb. It attains 50–70 cm height. In India, the plant is cultivated in the states of Punjab, Uttar Pradesh, Haryana, Himachal Pradesh, Rajasthan, and Central and South India. It is also found growing wild in Europe, while it is cultivated in England, Japan, France, US, Italy and Bulgaria.

Figure 11.6   A twig of *Mentha piperita*

Collection and preparation   The plants are harvested when they reach the flowering stage. In this stage, the plants are cut, dried for some time in sunlight, then in shade and stored.

To reduce the bulk of the plants, they are dried to ¼ of their weight in the shade. This will save the time of the steam distillation process and also the cost of its production. Fermentation should be avoided during drying of herb to get oil of better quality and quantity. The air-dried material is charged into galvanized iron. Steam under pressure, generated with the help of boiler, is then passed through the drug. It takes about 3 to 4 hours for distillation, and should be completed carefully. The condenser is made up of either aluminium or stainless steel. The mentha oil is collected in a separate can and then decanted and filtered.

Characteristics of peppermint oil   The oil is a clear transparent liquid.

Colour	Colourless to yellow
Odour	Characteristic and pleasant
Taste	Pungent followed by cooling sensation

Solubility   The oil is soluble in 70% alcohol, ether and chloroform, and insoluble in water and lighter than water. The oil darkens and becomes viscous on storage and if cooled, menthol crystals are produced.

Standards

Reaction to litmus paper	Neutral
Refractive index	1.459 to 1.465
Weight per ml	0.9 to 0.912 g
Optical rotation at 25°C	−16° to −30°

Chemical constituents   Peppermint oil contains volatile oil, resin, tannins and gum. The major constituent of the oil is menthol (50–90%) and the other compounds are L- and D-menthone (10%), menthyl acetate, menthyl valerate, jasmone, etc.

Uses

1. Peppermint is used as a carminative and flavouring agent.

2. It is antipruritic and used on the skin or mucous membrane as a counter-irritant, an antiseptic and stimulant.

3. It is used in toothpastes and toothpowders, shaving creams, betel nut, chewing gums, candies, jellies, perfumes and essences.

Storage   Peppermint oil should be stored in well-filled and air-tight containers in cool places and protected from light.

6. CUMIN OIL

Biological source   The drug constitutes the dried ripe fruits of *Cuminum cyminum* L. of the family Umbelliferae (Apiaceae).

The common names of the plant include jira, cheeragum, jiraka, jeeraka, safed jeera.

Geographical distribution   The plant is an annual herb, which is indigenous to the Nile territory. It is found in India, Syria, China, Morocco and Sicily. In India, it is grown in most of the states.

Collection and preparation   The fully ripe seeds are collected, cleaned and dried in the shade. The oil is extracted from the dried seeds by the distillation process.

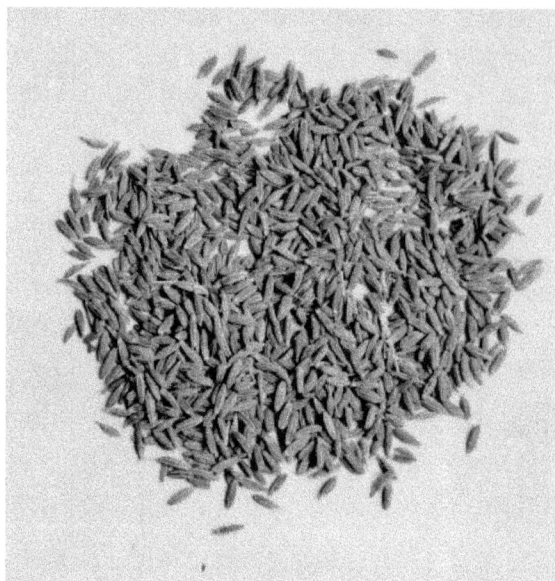

Figure 11.7 *Cuminum cyminum*

Characteristics of fruit and cumin oil

Colour	Fruits brown in colour with light coloured ridges; oil is colourless to pale yellow in colour
Odour	Characteristic and aromatic
Taste	Slightly bitter
Size	4–6 mm in length and about 2 mm thick
Shape	The fruits are elongated and tapering at both ends. Cremocarps are separate and each mericarps is with fine longitudinal ridges. There are secondary ridges alternating with these.

Standards

Ash	Not more than 8.0%
Volatile oil	Not less than 2.5%
Foreign organic matter	Not more than 2.0%
Specific gravity	0.900 to 0.935
Optical rotation	$+4°$ to $+8°$
Refractive index	1.4950 to 1.5090

Chemical constituents The fruits contain 2.5–4.5% volatile oil, 10% of fixed oil and proteins. Volatile oil mainly consists of 30–50% cuminaldehyde, small quantities of α-pinene, β-pinene, phellandrene, cuminic alcohol, hydrated cuminaldehyde and hydrocuminine.

Uses

1. The drug is used as a stimulant and carminative, and also in diarrhoea and dyspepsia.

2. The fruits are used as spice and the powder of the fruits is used for indigestion.

Storage The drug should be kept in well-closed containers.

7. CLOVE OIL

Biological source Clove oil is obtained from the dried flower buds of *Eugenia caryophyllata* Thunb. (Syn. *Syzygium aromaticum* (L.) Merr) of the family Myrtaceae.

The common names of the flower buds from which the oil is extracted include clove buds; lavangappattai; lavangaha; lavang; grambu.

Geographical distribution The plant is an evergreen tree of about 15–20 m height. It is native to Molucca islands, but cultivated chiefly in Penang, Madagascar, Pemba, Mauritius, Sumatra, Zanzibar, West Indies, Brazil, Sri Lanka, South America and South India. In South India, clove is widely grown in the Nilgiris and the Tenkasi hills, in the Kanyakumari district of Tamil Nadu, and in the Kottayam and Quilon districts of Kerala.

Figure 11.8 Flower buds of *Eugenia caryophyllata*

Collection and preparation   The clove plant starts producing the buds after 7 to 8 years and a satisfactory yield is obtained only after 15 to 20 years of growth. The collection operation starts when the base of the clove buds turn from green to crimson red in colour during the dry climate from August to December. The flower buds of clove are handpicked or collected by beating with bamboos. If the trees are tall and the cloves are beyond reach, the natives use platform ladders for collection. The cloves are dried in open air and their stalks are separated. The dried cloves turn crimson red or brownish black in colour. On drying, about 70% water is lost. Dried cloves are graded according to their size, packed into bales and exported.

The separated peduncles (stalks) of cloves are marketed as "clove stalks". If the clove flower buds are allowed to stay too long on the tree, the buds open, the petals fall off and the fruits will develop. These fruits are called "mother cloves". The active principle is maximum only in the flower buds of the clove. Hence, harvesting is done when the cloves are in bud condition. The dried cloves are used for the extraction of clove oil by the distillation process.

Characteristics of clove   Each clove consists of a lower "stalk" and a "head".

Colour	The clove buds are crimson red to dark brown or reddish brown in colour. Clove oil is colourless to pale yellow colour.
Odour	Strong fragrant with spicy odour.
Taste	A pungent aromatic taste.
Size	Clove is about 10 to 17.5 mm in length, 4–5 mm in width and 2–3 mm thick.
Shape	The "stalk" of the clove is called "hypanthium", which is solid, subcylindrical or slightly flattened and tapering below. Its upper part has a bilocular ovary with numerous ovules attached to the axile placenta. The "head" of the clove consists of four calyx teeth, which are slightly projecting outwards. The corolla consists of four membranous petals and numerous incurved stamens surrounding a stiff prominent style of the gynoecium.

Standards

Refractive index	1.527 to 1.535
Specific gravity	1.038 to 1.06
Boiling point	250°C

Chemical constituents Clove contains 14–20% of volatile oil, 10 to 13% of tannin, triterpenic acid, for example, oleanolic acid, esters, glucosides of sitosterol, stigmasterol and campestrol, vanillin, the chromone, eugenin, gum and resin. Eugenol is the important chemical constituent (80–90%) present in the volatile oil. The characteristic aroma of clove is due to the presence of methylamyl ketone, a minor constituent of the oil.

Uses

1. Cloves are used as a carminative and as an aromatic spice and for the production of clove oil.

2. Clove oil is used as a flavouring agent, antiseptic and in perfumery.

3. The oil is commonly employed as a dental analgesic, being applied locally to dental cavities. Eugenol present in the clove oil acts as a local anaesthetic.

4. It is used in throat infection, in the preparation of cigarettes, and also in the manufacture of vanillin and mouthwashes.

Chemical tests for identification

1. Treatment of the hypanthium of clove or clove oil with KOH solution (50%) forms needle-shaped crystals of potassium eugenate.

2. Add $FeCl_3$ solution to an aqueous extract of clove. Blue-black colour is formed due to the presence of tannins.

3. To 1 ml of clove oil in 5 ml of alcohol, add 1 ml of $FeCl_3$. The solution turns blue due to the presence of phenolic hydroxyl group.

Storage Clove, clove oil and its powder should be stored in air-tight containers in cool and dry places.

8. SANDALWOOD OIL

Biological source Sandalwood oil is obtained from the heartwood of *Santalum album* Linn. of the family Santalaceae.

The common names of the oil are chandan oil, sandal oil, yellow sandalwood oil, East-Indian sandalwood oil. The source plant is commonly called safed chandan, chandanam, chandanamaram.

Geographical distribution Sandal is a small to medium-sized evergreen semi-parasitic tree. It is widely distributed in peninsular India and Malaysia. In India, it is extensively cultivated in Mysore and Tamil Nadu. Rajasthan, parts of U.P., M.P. and Orissa are the other states of India that grow sandalwood.

Collection of wood   The trees that are more than 20–25 years old are normally selected for the extraction of both wood and oil. The fully grown plants are uprooted and the barks of the stem and roots are stripped off along with the sapwood. The heartwood part of sandal is cut into small pieces and subjected to steam distillation for oil extraction. The trees grown on stony or gravelly soil produce more highly scented wood, giving a better yield of oil.

Characteristics of sandalwood and oil

Colour	Sandalwood is cream in colour, whereas the sandalwood oil is a pale yellow to colourless viscid liquid.
Odour	Peculiar, heavy, sweet and very lasting odour.
Taste	Unpleasant
Solubility	Oil is very slightly soluble in water and soluble in alcohol and chloroform.

Standards

Specific gravity	0.973 to 0.985
Viscosity	1.5
Acid value	0.5 to 0.8
Optical rotations	$-15°$ to $-20°$
Refractive index	1.500 to 1.510
Esters	Not less than 2% w/w as santolyl acetate
Alcohols	Not less than 90% w/w as santalol.

Chemical constituents   Oil of sandal contains about 95% of two isomeric sesquiterpene alcohols, α-santalol and β-santalol. In addition the oil also contains an aldehyde santalal, santene, santenone, teresantol, santalone and santalene. The oil is present in the elements of the heartwood. The main odorous and medicinal properties of sandalwood is due to the presence of santalol.

Uses

1. Sandalwood oil is used for symptomatic treatment of dysurea and in reducing the frequency of micturition marked in the tuberculosis of the bladder.

2. The oil is highly used in perfumery productions like cosmetics, incense sticks, soaps, face creams, athar, etc.

Storage   Sandalwood oil should be stored in well-filled, well-closed containers, and kept in a cool place, away from light.

9. LEMONGRASS OIL

Biological source Lemongrass oil is obtained from the leaves and aerial parts of the plant of *Cymbopogon citratus* (Dc). Stapf. of the family Gramineae (Poaceae).

The common names of the oil are Indian Melissa oil; East-Indian lemongrass oil; Malabar or Cochin lemongrass; gandha-bina-ka-tel.

Geographical distribution The plant is a perennial herb, indigenous to India and found in the wild. It is also cultivated throughout India. It also occurs in West and East Africa, and Guatemala. In India, the plant is grown in Tamil Nadu, Karnataka, Kerala, Maharashtra, Gujarat, and Punjab.

Figure 11.9 A twig of lemongrass

Collection and production of oil The plant is ready for harvesting at the end of May or early in June. The citral content of the oil is maximum (83%) when it is obtained from grass collected during September. The fresh herbs are collected, washed and cleaned for steam distillation, for the extraction of oil and this process takes about 2–3 hours. The herb should be distilled in the fresh state in the summer season; otherwise an unpleasant odour is developed in the oil. Major production of crop comes from the state of Kerala.

Characteristics of lemongrass oil

Colour	Reddish yellow or brown liquid
Odour	Resembling lemon oil
Taste	Similar to lemon oil
Solubility	Soluble in 70% alcohol, chloroform and in fixed oil, and insoluble in water. The solubility gradually decreases on storage.

Standards

Refractive index	1.4808 to 1.4868
Weight per ml	0.892 to 0.909
Optical rotation	$-3°$ to $+1°$

Chemical constituents   Lemongrass oil chiefly contains citral (75–85%), in addition to methyl heptenol, nerol, citronellal, dipentene and geraniol.

Uses

1. Lemongrass oil is used as a flavouring agent in perfumery, soaps, and cosmetics.

2. β-ionone is prepared from citral, and β-ionone is used in vitamin A synthesis.

3. The oil has bactericidal effect. It is also used as an insect repellent.

The Kerala State Drugs and Pharmaceutical Co. has commissioned at Allepey a plant manufacturing vitamin A from lemongrass oil. The spent grass is a good cattle feed.

Storage   The oil should be kept in well-closed containers in a cool place to protect it from sunlight.

10. CHENOPODIUM OIL

Biological source   Chenopodium oil is obtained by steam distillation from the fresh flowering and fruiting parts (except roots) of the various species of *Chenopodium ambrosioides* L. var. *anthelminticum* of the family Chenopodiaceae.

The common names of the product are oil of American worm seed, oil of Mexican tea, Spanish tea oil, ambrosia oil.

Geographical distribution   It is a small herbaceous plant. It reaches about 30–60 cm height, commonly found in eastern and central U.S.A. It is indigenous to West Indies. It is extensively cultivated in Europe, Canada, Mexico, Maryland, and South India. In India, it is found in West Bengal, Assam, Tamil Nadu, and Karnataka.

The oil is formed in glandular hairs occurring on the leaves, flowers, and fruits, but it is found in abundance in the pericarp and ovary. The oil is obtained by steam distillation method.

Characteristics of chenopodium oil

Colour	Colourless to light yellow-coloured liquid
Odour	Unpleasant and characteristic
Taste	Bitter and burning taste
Solubility	Insoluble in water and soluble in alcohol.

It explodes on heating or on treating with certain acids.

Standards

Weight per ml	0.955 to 0.975 g
Refractive index	1.474 to 1.480
Optical rotation	$-3°$ to $-8°$

Chemical constituents   The herb contains 1–4% volatile oil. The chief active constituent of the oil is ascaridole, which is an unsaturated terpene peroxide (70 to 80%). The other constituents of the oil are limonene, myrcene, pinene, *p*-cymene, ʟ-limonene and ᴅ-camphor.

Uses

1. The oil is a very valuable anthelmintic, particularly for roundworms, hookworms, tapeworms, and intestinal amoebae.

2. It is generally used in veterinary practices.

Tests for identification   Heat one ml of chenopodium oil in a test tube with a piece of porcelain. A deep golden-yellow liquid is produced. (Caution: The test should be carried out very cautiously as the oil is liable to explode).

Storage   Chenopodium oil must be stored in air-tight containers and kept away from sunlight.

11. TURPENTINE OIL

Biological source   Turpentine oil is the volatile oil distilled from oleoresin obtained from *Pinus palustris*, *P. roxburghii* Sarg (*P. longifolia* Roxb.) *P. maritima* and many other species of the genus *Pinus* of the family Pinaceae.

The common names are oleum terbinthae; spirit of turpentine; rectified oil of turpentine.

Geographical distribution   *Pinus* is a perennial tree. Turpentine oil is commercially produced in India, Portugal, China, US, France, Russia and Spain. U.S.A is the world's largest producer of turpentine. The species of the plant are naturally found growing in the Himalayas, Kashmir, Bhutan, Himachal Pradesh, Jammu, Punjab and Uttar Pradesh.

Collection and method of preparation   The trees that are 20 to 25 years old are utilized for the collection of the resin and oil. The production of oleoresin is increased by artificial injury. The resin is tapped by chipping the bark from the tree using a "bark-hack", a long-handled cutting blade. Such tapping is done during spring of every year and the tree can be tapped by this method for 40 years. The exposed surface is sprayed with 50% sulphuric acid. As the oleoresin flows, it is collected in cups fitted just beneath the cut surface of the tree trunk. The thick oleoresin that gets collected in the cups is removed from time to time and brought to the distillery. The crude oleoresin is purified by heating in a steam-jacketed container, after removing the vegetable debris and other impurities. The clarified oleoresin that contains 25% turpentine oil and 70% colophony, is steam-distilled. The distillate collected within 1–2 hours represents turpentine oil, and the residue left behind after distillation is the colophony resin. The volatile oil distilled from the oleoresin is further rectified by treatment with aqueous alkali to remove traces of resin acids, phenols, cresols, etc. and the oil is redistilled. The turpentine oil produced after chemical treatment by alkalies is known as "rectified turpentine oil".

Characteristics of turpentine oil

Colour	Rectified oil of turpentine is a clear, colourless to slightly yellow transparent liquid.
Odour	Strong and characteristic odour
Taste	Pungent and bitter.
Solubility	The oil is insoluble in water but soluble in alcohol, ether, chloroform, fixed oil and glacial acetic acid.

Standards

| Weight per ml | 0.854 to 0.868 g |
| Refractive index | 1.467 to 1.477 |

Chemical constituents   Oil of turpentine is rich in terpene contents and about 40 terpenes are reported in turpentine oil. It mainly contains α-pinene (20 to 30 %), β-pinene (5 to 10%), camphene, β-phellandrene, δ-3-carene, limonene, etc. The turpentine oil produced in Russia and India contains δ-3-carene as the major terpene.

Uses

1. Turpentine oil is chiefly used as an effective counter-irritant and rubefacient in the form of liniments to be used externally.

2. It has antiseptic and disinfectant properties.

3. It is used in the manufacture of synthetic pine oil, insecticides, paints and varnish.

4. It is used in the treatment of chronic bronchitis.

Storage   Turpentine oil should be stored in a cool place in amber-coloured, well-closed containers and kept away from light.

12. OIL OF VETIVER

Biological source   Vetiver oil is obtained from the roots of the plant *Vetiveria zizanioides* Nash of the family Gramineae (Poaceae).

Common names of the product are khus oil, khaj, khas-bene, valo, khaskhas, vetiver oil.

Geographical distribution   The plant is found growing in India, Myanmar, Sri Lanka, East and West Africa, Indonesia, West Indies, Malaysia, Philippines and U.S.A. In India, the plant grows in Punjab, U.P., Rajasthan, Kerala, Karnataka and Tamil Nadu.

Collection   When the plant is about 15 to 18-months old, the roots are collected by uprooting in the dry months of the year. The roots are washed, cleaned of foreign materials and cut into small pieces and used for extraction of oil.

Preparation of oil   The small pieces of roots are chipped, put into the still and are subjected to steam distillation. Since the specific gravity of the oil is 0.990 to 1. 032, it is very difficult to separate the oil from the condensate. Therefore, the condenser should always be kept warm, so that the warm oil after cooling separates very quickly without forming any emulsion. Salt is used to increase the density of water, so as to separate it from the condensate. The traces of water in the oil are removed by treating it with anhydrous sodium sulphate. The yield of oil is 1.2–2% v/w.

Figure 11.10   Plant of vetiver

Characteristics of vetiver oil

Colour	Light brown to deep brown or green
Odour	Characteristic odour
Taste	Characteristic taste
Solubility	The oil is soluble in fixed oil and alcohol, and insoluble in water.

Standards

Specific gravity	0.990 to 1.032
Optical rotation	+15° to +45°
Refractive index	1.512 to 1.523
Acid value	25 to 80

Chemical constituents   Vetiver oil contains 45–60% alcohols such as vetivenol and vetiverol, and 8–35% ketones namely α- and β-vetivones. Indian vetiver oil contains khusal, khusitol and khusinol.

Uses

1. The oil is used as a stimulant, refrigerant, flavouring agent, aromatic and stomachic agent.

2. It is used in the preparations of sharbat, soap, perfumery and toilet preparations and also as a fixative of volatile oils.

3. It is used in the treatment of prickly heat or itches.

Storage   The oil should be stored in well-filled, well-closed containers and kept in a cool place, so as to protect it from direct sunlight.

13. OIL OF PALMAROSA

Biological source   Oil of Palmarosa is obtained from the leaves and tops of *Cymbopogon martini* Wats of the family Gramineae (Poaceae).

The common names are rosha oil; geranium oil.

Geographical distribution   The plant is a tall, perennial and sweet-scented grass, reaching 1.5 to 2.5 m height. The grass is found in the wild in Kashmir, Almora, Punjab and Garwal regions of Uttar Pradesh, Madhya Pradesh, Gujarat, Maharashtra, Andhra Pradesh and Tamil Nadu. In Maharashtra the plant is grown in Jalgaon, Dhule and Nasik districts.

There are two varieties of grass grown for commercial production of oil. The motia variety contains highest percentage of geraniol in oil (about 90%) known as palmarosa or rosha oil or East-Indian geranium oil. Another variety of the grass is called sofia oil, which is poor in geraniol content and is known as ginger-grass oil.

Collection and processing of oil   The crop is harvested when the grass starts flowing in the months of October to December. The grass yields about 7.5 to 23 kg oil per hectare. India is the leading country in the export of oil. The oil is extracted by steam distillation.

Characteristics of oil of palmarosa

Colour	Colourless to light brown-coloured liquid
Odour	Aromatic, resembling that of rose
Taste	Slight
Solubility	Insoluble in water, but soluble in 90% alcohol

Standards

Refractive index	1.4760 to 1.4805
Optical rotation	$+1.4°$ to $-2°$
Specific gravity	0.885 to 0.896

Chemical constituents   The grass contains about 1.25 to 1.5% of oil. The flowers contain maximum amount of oil, while the stems yield minimum. Rosha oil contains about 85–95% of geraniol, citronellal, dipentene, linaliol and terpenes.

Uses

1. Palmarosa oil is used as an insect repellent.

2. It is used in perfumery, cosmetics, soaps, and in the treatment of rheumatism and skin diseases.

Storage   The oil should be stored in well-closed containers in a cool place.

14. EUCALYPTUS OIL

Biological source   Eucalyptus oil is a volatile oil obtained by steam distillation of the fresh leaves of various species of eucalyptus such as *Eucalyptus globulus* Labill, *E. polybractea, E. smithii, E. rostrata, E. australiana, E. citriodora, E. lanceolatus,* etc. that contain high cineol content and belong to the family Myrtaceae.

The common names are gum wood oil; blue-gum tree oil; Australian fever tree oil; Dinkum oil; R.S.Pathy oil.

Figure 11.11   Twig of *Eucalyptus lanceolatus*

Geographical distribution   The plant is a large tree, indigenous to Australia and Tasmania. The tree is largely cultivated in the United States especially in California, Portugal, Spain, Southern France, Angola, South Africa, Brazil, Zaire and India. The Nilgiri district in Tamil Nadu has the pride of being the first place in the Indian subcontinent for the introduction of Eucalyptus species. *Eucalyptus citriodora*, also known as citron-scented gum or lemon-scented gum, a native of Queensland, is grown on a large scale in Kerala, Tamil Nadu and other states of India.

Method of preparation   The trees that are 16 years old are preferred for the collection of leaves for oil distillation. About 700 to 1200 kg of leaves are collected per hectare. The collected leaves are washed, cleaned of foreign matter and dried in shade for about 3 days and then subjected to steam distillation.

The distillation unit consists of a false perforated copper bottom. 350 kg of leaves are used for every charge of distillation. Accordingly sufficient quantity of water is also charged in the still and the steam under pressure is passed through it. It takes about 6 hours for complete distillation. About 0.75 to 1% of eucalyptus oil is produced and collected in the receiver.

The crude oil thus produced is rectified again after treatment with sodium hydroxide. The oil is filtered and filled in suitable containers.

Characteristics of Eucalyptus oil

Colour	Colourless or pale yellow liquid.
Odour	Highly aromatic and camphoraceous
Taste	Pungent and camphorous followed by the sensation of cold
Solubility	The oil is soluble in 90% alcohol, fixed oils, fats, and in paraffin. It is insoluble in water.

Standards

Weight per ml	0.897 to 0.916 g
Refractive index	1.457 to 1.469
Optical rotation	0° to +10°

Chemical constituents   Eucalyptus oil contains 3–6% volatile oil, resins, tannic acid, cinnamic acid, eucalyptic acid, rutin, and *p*-coumaric acid. The oil chiefly contains cineole, also known as eucalyptol (about 80%).

Uses

1. Eucalyptus oil is used as a flavouring agent and also as antiseptic, diaphoretic, disinfectant, diuretic and expectorant.

2. It is used to relieve cough and cold, and in chronic bronchitis in the form of inhalations.

3. The oil is an ingredient of several liniments and ointments.

4. A solution of eucalyptus is used as nasal drops, and as a mosquito repellent and insecticide.

5. It is used in the treatment of malaria, certain skin diseases and asthma.

6. *Eucalyptus* increases the flow of saliva, and gastric and intestinal juices and thus increases appetite and digestion.

7. It increases the rate of heart beat, lowers the arterial tension and quickens respiration.

Storage   The oil should be stored in well-closed containers in cool places and away from light.

15. CIVET

Biological source   Civet is the odorous secretion obtained from the specialized scent glands of the external generative organs of male and female civet cat *Viverra zibetha* Linn. which belongs to the family Viverridae.

The common names are Zibeth, Indian civet.

Geographical distribution   Civet cats are found all over South-East Asia, Madagascar and Southern Europe. Three genera of these cats are also found in India.

Figure 11.12   Indian Civet

Characteristics of the Civet

Colour	Pale yellow semi-liquid
Odour	Obnoxious odour, when diluted becomes very pleasant.
Taste	Bitter
Solubility	It is partly soluble in hot alcohol and ether, and insoluble in water.

Standards

It is fusible and burns without leaving any residue.

Loss in weight at 100°C	Not exceeding 25%
Mineral matter	Not more than 2%
Alcohol extracts	45–65%
Acetone extracts	65–80%
Ether extracts	11–24%

Chemical constituents   Civet mainly contains civetone, civetol, indole, ethylamine, propylamine and few unidentified free acids.

Uses   Civet is used as a flavouring agent and fixative in perfumery.

REVIEW QUESTIONS

State whether the following sentences are true or false

1. Rosemary oil is used as carminative and rubefacient.
2. The binomial of sweet flag is *Glycyrrhiza glabra*.
3. Civet is the odorous secretion of an animal.
4. The gynoecium of *crocus* is used as a drug.
5. *Cymbopogon martini* yields Rosha oil.

Choose the correct answer

6. *Santalum album* contains oil in its _____

   a. sapwood   b. bark   c. heartwood

7. Saffron is the dried _____
   a. stigmas of crocus   b. gynoecium of crocus   c. petals of crocus

8. *Mentha piperita* is a member of the family _____.
   a. Myrtaceae   b. Lamiaceae   c. Poaceae

9. The dried flower buds of *Eugenia caryophyllata* is used as _____
   a. clove   b. khus oil   c. sofia oil

10. The blue-gum tree is _____
    a. *Eugenia jambalona*   b. *Santalum album*   c. *Eucalyptus globulus*

Answers

1.	True	2.	False	3.	True	4.	False
5.	True	6.	c	7.	a	8.	b
9.	a	10.	c				

Answer the following

1. Mention the plant sources of citronellal and pinene.

2. Why are volatile oils called ethereal oils?

3. Discuss the therapeutic importance of the following: vaj, tulsi oil, turpentine oil, saffron, clove and eucalyptus oil.

4. Give the biological source of cumin oil, saffron, lemon grass oil, oil of palmarosa and turpentine oil.

5. How will you test and identify the quality of the following crude drugs? Clove, crocus and chenopodium oil.

6. Write an explanatory note on collection and preparation of peppermint oil, clove oil and turpentine oil.

7. Write the biological source, method of collection and preparation, characters, chemical components and use of any studied by you volatile oils.

12

DRUGS CONTAINING RESINS AND RESIN COMBINATIONS

Resins and resin combinations are the group of phytoconstituents, that may be solid or semi-solid plant exudates. They are produced in schizogenous or schizolysigenous ducts, or cavities as end products of metabolism. Some resins (e.g., benzoin and balsam of Tolu) are formed when the plant is injured. These resins are called pathological resins. They are complex mixtures of compounds like resin acids, resin alcohols, resinotannols, esters and resenes. These are amorphous mixtures of essential oils, and oxygenated products of terpenes and carboxylic acids and contain large number of carbon atoms. They are hard, brittle, translucent, combustible and electrically non-conductive masses.

Chemically resins are classified as resin acids (colophony and myrrh), ester resins (benzoin and storax), resin alcohols (Canada balsam and *Capsicum*) and resenes (asafoetida and gutta percha), and on the basis of their occurrence in combination with other compounds. Resins occurring in association with volatile oils are termed as oleoresins (ginger), those resins that are in association with gums are known as gum resins (asafoetida), those in combination with gum and volatile oil, both called oleogum resins (myrrh), and those found in association with glycosides called glycoresins (*Podophyllum*). Resins that contain high proportions of benzoic or cinnamic acids are called balsams (tolu balsam). Resins and resin combinations are extensively used in the pharmaceutical industry for various purposes. The description of different resins and resin combinations containing drugs are described in this chapter.

1. GINGER

Biological source The rhizomes of *Zingiber officinale* Rosc., of the family Zingiberaceae are used as drug. They are scraped to remove the outer darker skin and dried in the sun. Oleoresin of ginger is obtained by percolating the powdered rhizomes of dried ginger.

The common names of the plant include zingiber and rhizoma zingiberis. Ginger oleoresin is also called gingerin; adrak; adrakam; sukku; ingi.

Geographical distribution Ginger plant is a perennial herb with branching underground rhizomes. It is native of South-East Asia. It is cultivated in China, India, Jamaica, Taiwan, Africa, Australia and West Indies. It is grown in almost all the states of India. Three commercial varieties of ginger namely Jamaica ginger, Cochin ginger and African ginger are available in the world market. Cochin ginger is cultivated in South India.

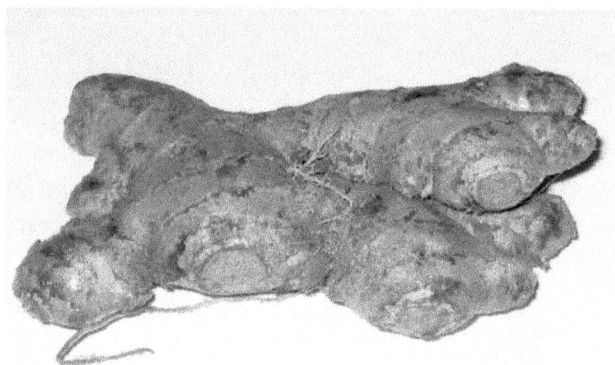

Figure 12.1 Rhizome of *Zingiber officinale*

Collection and preparation of rhizomes Ginger can be harvested in about six months, when the leaves turn yellow. The rhizomes are harvested by digging out, and the soil is removed, and washed thoroughly. The cleaned rhizomes are boiled with water, the skin is carefully scraped and the peeled rhizomes are dried in the sun for five to six days. They may be coated with inert material like calcium sulphate. Green ginger is used as such after harvesting and cleaning.

Method of preparation of powder of rhizome The dried ginger is coarsely powdered and is extracted with acetone or ethyl ether or ethylene dichloride by percolation method. The drug is extracted until exhausted and the extracts are mixed together. The solvent is removed by distillation under reduced pressure. Alcohol gives the maximum yield of oleoresin. The yield of gingerin may vary from 3.5 to 9.0% depending upon the source of the material and the method of preparation. It gives an average yield of 6.5% oleoresin.

Characteristics of rhizome

Colour	Externally buff-coloured
Odour	Agreeable and aromatic
Taste	Agreeable and pungent
Size	Rhizome pieces are about 5–15 cm long, 1.5–6.5 cm wide and 1–1.5 cm thick.
Shape	The dried rhizomes are laterally compressed, bearing short, flattish, ovate, and oblique branches on the upper side, and each with apex-depressed scar.
Fracture	Short and fibrous.
Oleoresin	Dark brown, aromatic and pungent viscous liquid.

Chemical constituents   Ginger contains about 1–3% of volatile oil, which gives a characteristic odour, an acrid resinous matter (5–8%), starch (50%) and mucilage. Zingiberene, bisabolene and zingiberal are the three principal constituents of oil of ginger. The pungency of ginger is due to gingerol, which is a yellow, oily substance. The other pungent principles of ginger are oleoresin such as, zingerone and shogaol. The pungency is lost by boiling with 2% KOH.

Standards

Water-soluble extractive	Not less than 10%
(90%) alcohol-soluble extractive	Not less than 4.5%
Total ash	Not more than 6%
Water-soluble ash	Not less than 1.7%

Uses

1. Ginger is mostly used as a condiment.

2. It is used as a stimulant, carminative, an aromatic and flavouring agent.

3. Ginger oil is used in mouthwashes, ginger beverages and liquors.

4. Limed ginger is available which is ginger coated with lime to improve its colour and quality, and medicinally it is used for various purposes.

5. Oleoresin of ginger is used as a flavour for carbonated beverages in spices and condiments.

6. It is prescribed in dyspepsia, flatulent colic, vomiting spasms, cold, cough, asthma and sore throat. Good relief from sore throat is got by chewing a piece of ginger.

2. ASAFOETIDA

Biological source   It is an oleogum resin obtained from the living roots and rhizomes of *Ferula asafoetida* Regel and other species of *Ferula* of the family Umbelliferae (Apiaceae). The term asafoetida is derived from the latin words *asa* meaning "gum" and *foetida* meaning ill-smelling.

The common names of the oleogum resin are devil's dung; food of the Gods; heeng; perungayam; kayam; hinguka.

Geographical distribution   The plant is a perennial herb about 3 m in height. It is native to Iran and Afghanistan. It is also grown in Turkestan, Persia and India (Kashmir).

Figure 12.2   Oleogum resin of *Ferula asafoetida*

Collection   Asafoetida is obtained from 4–5-year-old plants upon exudation and drying of the milk liquid from the roots. For the collection of the exudates, the stem is removed before flowering during March–April and the upper part of the roots that lies very close to the ground is cut off and the soil is removed. The milky juice oozes out of the cut surface and starts coagulating. The cut surface is covered to avoid contamination with foreign matter and sand. After few days, the coagulated exudate is scraped off and fresh cuts are made to collect more exudates and this process is continued until the plants cease to produce latex. After hardening on the cut surface, it is collected, packed in tin-lined cases and exported. Oil of asafoetida is obtained by steam distillation of the oleogum resin .

Characteristics of asafoetida

Colour Yellowish white or greyish white changing to reddish brown.

Odour Intense, persistent, penetrating and alliaceous.

Taste Bitter, alliaceous and acrid

Size The tears are 0.5 to 3 cm in diameter

Shape Occurs in different forms, i.e., tears and masses. Tears are rounded or flattened. Fresh tears are tough but when dried they become hard and brittle. Tears are internally milky whitish yellow, translucent or opaque, whereas, mass of asafoetida is agglutinated and mixed with root fragment, foreign matter and other impurities.

Chemical constituents Asafoetida contains about 4–20% volatile oil, 40–65% resin and about 25% gum. The oil has a characteristic alliaceous odour, due to the presence of a number of organic disulphides especially isobutyl propanyl disulphide. The resin contains asaresinol ferulate and free ferulic acid. Due to the presence of free ferulic acid, asafoetida gives combined umbelliferone test. The chief constituent of oil of asafoetida is secondary butyl propanyl disulphide and the specific odour of the drug is due to sulphur compounds.

Uses

1. Asafoetida is used as a carminative, expectorant, antispasmodic and laxative.
2. It is also used as a paste externally to prevent bandage chewing by dogs.
3. It is mostly used as flavouring agent for curries, sauces and pickles.
4. It is used in veterinary medicine.

Tests for identification

1. When treated with 50% nitric acid, the asafoetida gives a green colour.
2. The fractured surface of the drug, when treated with sulphuric acid, forms red or reddish brown colour, which when washed with water, changes to violet colour.
3. When the drug is triturated with water, it forms a yellowish orange emulsion.
4. When the drug is boiled with dilute hydrochloric acid, a suspension is formed. When the suspension is filtered into ammonia solution, a blue fluorescence is formed due to the formation of umbelliferone.

3. MYRRH

Biological source   Myrrh is an oleogum resin obtained from the stem of *Commiphora molmol* Engler and from other species of *Commiphora* of the family Burseraceae.

The common names of the oleogum resin are Arabian myrrh; gum myrrh; bol; gum resin myrrh; somali myrrh; vellaipolam; rasgandha.

Geographical distribution   The plant is a small shrub or tree, rarely attaining 3 metres in height. The plant is mostly found in the Arabian Peninsula, Ethiopia, Somalia and North-East Africa.

Commiphora is a Greek word meaning "gum-bearing" and *molmol* means, it is a native of Somalia. Myrrh is derived from the Arabic word "murr" meaning "bitter".

Figure 12.3   Oleogum resin of *Commiphora molmol*

Collection and preparation   Presence of oleogum resin is the characteristic feature of the members of Burseraceae. It is present in the phloem canals and the cavities are formed schizogenously or schizolysigenously in the stems of the plants. The gum resin exudes naturally from the fissures or cracks or may be obtained by giving incisions in the bark of the tree. The oleogum resin is a yellowish white viscous fluid when it exudes, but gradually hardens and becomes dark or reddish brown masses which are collected in goat skins by the natives and sent to the market.

Characteristics of myrrh

Colour	Reddish brown externally and brown or reddish yellow internally
Odour	Aromatic and agreeable
Taste	Bitter and acrid

Size	About 1.5 to 3.0 cm in diameter
Shape	Rounded or irregular masses or tears.
Fracture	Granular and brittle with translucent surface and with whitish spots on broken pieces.

Chemical constituents Myrrh contains 7–17% of yellowish thick volatile oil which gives the characteristic odour of myrrh, 25–40% of resin, and 57–61% of gum. The volatile oil contains terpenes, cuminic aldehyde, eugenol, etc. The gum contains 18% of proteins and 64% of carbohydrate, which is a mixture of galactose, arabinose, glucuronic acid and an oxidase enzyme. The resin contains ether-soluble complex resin acids such as α, β and γ-commiphoric acids, resenes and phenolic compounds.

Uses

1. Myrrh is used as a stimulant and carminative.

2. It is used in incense and perfumes.

3. Due to its disinfectant action, its alcoholic tincture is used in skin applications and in inflammation of mouth and pharynx.

4. It has antiseptic properties and is used as a tooth powder, mouthwash and gargle.

Tests for identification

1. On trituration with water, a yellow-brown emulsion is produced.

2. On treatment with bromine vapours, an ethereal solution of myrrh turns red and the solution becomes purple with nitric acid.

4. TOLU BALSAM

Biological source Balsam of Tolu is obtained by incision from the trunk of *Myroxylon balsamum* Linn. of the family Leguminosae.

The common name of the resin obtained from the source plant is balsam of Tolu.

Geographical distribution The plant is a large tree, indigenous to Columbia. The trees are extensively cultivated in the West Indies, Cuba and Venezuela. Tolu is the name of the place on the north coast of Columbia near Cartagena, hence, the drug is called Tolu.

Collection Balsam of Tolu is a pathological product (resin) secreted in the cortex of young twigs. The resin is collected by making V-shaped incisions in the bark and sapwood. Calabash cups are placed to collect the flow of balsam. The collected balsam is transferred into larger tin containers and exported.

Characteristics of Balsam of Tolu

Colour	Yellowish brown or yellow to brown
Odour	Aromatic and fragrant and slightly pungent
Taste	Aromatic
Solubility	It is soluble in 90% alcohol, chloroform, ether and glacial acetic acid. It is insoluble in water and petroleum ether.

The Tolu balsam occurs as soft, semi-solid or plastic solid transparent in thin layers, brittle when old, dried or kept in cold. The Tolu oil is a yellow liquid with highly fragrant flavour.

Standards

Acid value	100 and 160
Ester value	40 and 95
Saponification value	170 and 230
Alcohol-insoluble matter	Not more than 5%
Loss on drying	Not more than 4%

Chemical constituents   The drug contains about 80% of resin, which is a mixture of resin alcohols combined with cinnamic and benzoic acids. The aromatic acids are also present in free state in proportions of 8–15%. The drug also contains benzyl benzoate, benzyl cinnamate, vanillin, styrene, eugenol, ferulic acid, 1, 2-diphenyl ethane, mono- and sesquiterpene, hydrocarbons and alcohols.

Uses

1. Balsam of Tolu is used as an expectorant, antiseptic and as flavouring agent in medicinal syrups.
2. It is an ingredient of cough mixtures and compound benzoin tincture.
3. It is used in the preparation of confectionery, chewing gums and perfumery.

Tests for identification

1. An alcoholic solution of Balsam of Tolu (1 g) gives green colour with ferric chloride due to toluresinotannols.
2. An alcoholic solution of the drug is acidic to litmus paper.
3. When 1 g of the drug is warmed with 5 ml of potassium permanganate solution, an odour of benzaldehyde is produced.

5. COLOPHONY

Biological source   Colophony is the solid residue left after the distillation of volatile oil from the oleoresin obtained from various species of *Pinus* such as *Pinus palustris* (long leaf pine), *P. longifolia, P. radiata*, etc. of the family Pinaceae.

The common names of the product are rosin; rosina; amber resin.

Geographical distribution   *Pinus* is widely grown in the U.S., Russia, Spain, Portugal, France, China, India, Pakistan, Italy and New Zealand. Colophony is largely produced in U.S.A. contributing to about 80% of the world supply.

Collection and method of preparation   The oleoresin is a normal physiological substance secreted in ducts located in the sapwood. When the tree is about 20–30 years old, it is tapped for oleoresin. Collection of the oleoresin is done by injuring the plant and the injured surface is treated with 50% sulphuric acid to increase the flow. The oleoresin comes out of the injured groove and gets collected in containers fitted below the groove of the trunk. The oleoresin thus collected is taken out periodically and sent for further processing.

The crude oleoresin is steam-distilled and the residue left after distilling the turpentine oil, represents rosin and thus, this is the by-product of turpentine oil production.

Characteristics of colophony

Colour	Commercial grades of rosin vary in colour from pale yellow to yellowish brown or amber colour.
Odour	Faintly turpentine-like odour
Taste	Slight turpentine-like taste
Fracture	Smooth surface with brittle fracture.
Size	Occurs in irregular angular masses of varying size
Appearance	Translucent, hard and shiny.
Solubility	Insoluble in water but soluble in alcohol, ether, benzene, glacial acetic acid and many fixed and volatile oils.

Standards

Acid value	Less than 150
Ash value	Not more than 0.125%
Saponification value	188 and 192

Specific gravity	1.08
Melting point	75°C and 85°C

Chemical constituents   Colophony contains mainly resin acid about 90%, of which two-thirds is abietic acid. The composition of Rosin varies according to the source, storage conditions, etc. Besides this, it contains esters of fatty acids, volatile oils, resenes, sapinic acid, pimaric acid, etc. The commercial abietic acid is prepared by digesting colophony with weak alcohol. Only the light-coloured variety is used for medicinal purpose.

Uses

1. It is used as a stimulant and diuretic.

2. It is used as a stiffening agent in the preparation of plasters, ointments and cerates.

3. It is also used in the manufacture of varnishes, printing inks, cements, soap, sealing wax, wood polishes, floor coverings, paper, fireworks, rosin oil and pressure-sensitive adhesives.

Tests for identification

1. To the solution of 0.1 g powdered colophony in 10 ml of acetic acid, one drop of conc. sulphuric acid is added in a dry clean test tube. A purple colour readily changes to violet, indicating the presence of resin.

2. To a petroleum ether solution of powdered colophony, twice its volume of dilute solution of copper acetate is added and shaken. The colour of the petroleum ether layer changes to emerald-green due to the formation of copper salt of abietic acid.

3. To an alcoholic solution of colophony, water is added. It becomes milky white due to precipitation of chemical compounds.

4. An alcoholic solution of colophony turns blue litmus to red colour, due to the presence of diterpenic acids.

6. PODOPHYLLUM

Biological source   *Podophyllum* consists of the dried rhizomes and roots of *Podophyllum hexandrum* Royle. and *P. emodi* Wall. of the family Berberidaceae.

The common names of the source plant are Indian podophyllum; may apple; mandrake root; Indian apple; papra-bhavan; leghu-patha.

Geographical distribution   The plant is a perennial herb that grows in the high slopes of the Himalayas, Kashmir to Sikkim, Uttar Pradesh, Punjab, Afghanistan and Tibet. It is also found in the eastern parts of Canada and the U.S.A.

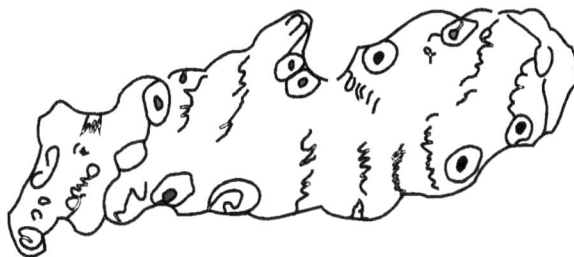

Figure 12.4 Rhizome of *Podophyllum hexandrum*

Collection and preparation   The rhizomes and roots are dug out during spring or autumn season, washed, cleaned and cut into pieces 10–20 cm long, and dried in the sun. The rootlets and other foreign materials are removed. The rhizomes contain a higher percentage of the active constituent (resin) in the month of May. The roots contain more resin than the rhizomes, therefore, the roots are mostly used.

Podophyllum resin is prepared by pouring an alcoholic extract of the dried rhizomes and roots of the plant into acidified water and collecting and drying the precipitate.

Characteristics of rhizome and resin

Colour	*Podophyllum* resin is an amorphous powder with colour varying from light brown to greenish yellow or brownish grey. The rhizomes are sub-cylindrical reddish brown or earthy brown in colour. Roots are irregular and flat.
Odour	Slight and characteristic
Taste	Bitter and acrid taste
Size	Rhizomes are 2.5 cm in length and 1 to 2 cm in thickness. Roots are about 7 cm in length and 2.5 mm in thickness.
Fracture	Both roots and rhizomes are brittle and even short and starchy.
Solubility	The resin is insoluble in cold water, and partly soluble in hot water, solvent ether, chloroform, and in dilute ammonia solution. It is completely soluble in 90% alcohol.

Chemical constituents   Indian podophyllum contains 7–15% resin known as podophyllin. The content of resin varies in roots and rhizomes and also depends on the season of collection. The active principles of resins are lignans. The important lignans are podophyllotoxin (~20%), β-peltatin (~10%) and α-peltatin (~5%) occurring in free state and as glucosides. In addition, the related compounds like dimethyl podophyllotoxin and the glucosides, deoxypodophyllotoxin and podophyllotoxone, are also present in the drug.

Uses

1. *Podophyllum* resin has purgative and anti-cancer properties.

2. It has a cytotoxic action and is used as a paste in the treatment of soft venereal and other warts.

3. The converted semi-synthetic derivative of the resin, named etopsoside, is used as an antineoplastic agent in the treatment of cancer. Etopsoside is marketed as Vepesid.

4. It is used as a bitter tonic.

Tests for identification

1. Macerate 0.5 g of the drug with 10 ml of alcohol and filter. To the filtrate, add strong copper acetate solution (0.5 ml) to form brown precipitate.

2. Take 0.5 g of powdered resin. To it, add 5 ml of 60% alcohol and 5 ml of 1N potassium hydroxide solution. Shake it well. A stiff jelly is produced indicating the presence of resin.

7. GALANGA

Biological source   Galanga is the dried rhizome of *Alpinia officinarum* Hance. and *Alpinia galanga* Sw. of the family Zingiberaceae.

The common names of the source plant are rasna; galangal; colic root; Chinese ginger; Indian root; chithiraththai.

Geographical distribution   The plant is a 1–2 m long perennial herb bearing rhizomes. The plant grows in Thailand, China, Eastern Himalayas, Bengal and South-West India.

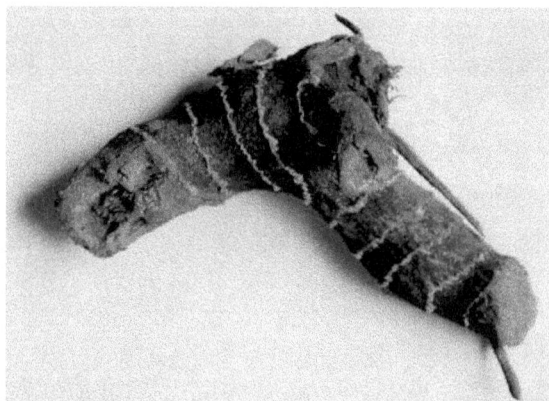

Figure 12.5   Rhizome of *Alpinia galanga*

Collection   The collection of the rhizome is done during autumn season. The plants are dug out and the rhizomes are separated. The rhizomes are washed to remove the mud and cut into small pieces and dried in the sun. The dried rhizomes are graded and marketed.

Characteristics of rhizomes

Colour	The rhizomes are reddish or rusty brown in colour.
Odour	Aromatic, distinct, spicy and agreeable odour.
Taste	Aromatic and pungent
Size	The pieces of rhizome are 5–10 cm long and 2–3 cm thick.
Shape	Irregularly branched and marked with fine annulations.
Fracture	Fibrous and tough.

Standards

Acid–insoluble ash	Not more than 1%
Total ash	Not more than 3.5%
Foreign organic matter	Not more than 2%

Chemical constituents   The drug contains 0.5–1% volatile oil, which contains cineol, α-pinene and eugenol. Resin contents primarily contains alpinol and galangal. In addition, phlobaphene tannins, and an abundance of starch and flavonoids such as kaempferol, galangin and alpinin are also present in the drug.

Uses

1. Galanga is used as a stomachic, stimulant and carminative.
2. The drug is used in the treatment of rheumatism, fever and catarrhal affections especially in bronchial catarrh.
3. It has antifungal and antibacterial properties.
4. It is used as a flavouring agent.

8. TAR

Biological source   Tar is a product obtained by destructive distillation of wood of pine, chiefly *Pinus sylvestris* L. and other species of *Pinus* of the family Pinaceae. Tar differs from the industrially as well as pharmaceutically important coal tar.

The common names of the product are chir tar; pine tar; pix liquida.

Characteristics of tar

Colour	Tar is a viscid, blackish brown liquid. It is translucent in thin layers.
Odour	Peculiar strong aromatic odour resembling naphthalene.
Taste	Bitter pungent taste.
Solubility	Tar is insoluble in water, partly soluble in alcohol and more soluble in chloroform, ether and fixed and volatile oils. It is acidic in reaction.

Standards

Weight per ml	1.07%
Ash content	0.47%

Chemical constituents   Tar consists of a resinous substance admixed with small quantity of turpentine, acetic acid, methyl alcohol and various volatile substances. It contains various phenols and phenolic ethers like phenol, cresols, methyl cresols, catechols, etc. In addition, various aromatic hydrocarbons like benzene, toluene, xylenes, styrene, naphthalene, etc. are also present.

Uses

1. Tar is a local irritant and used as an antibacterial agent.

2. It is used externally as antiseptic and antipruritic in certain skin diseases like chronic dry eczema, scabies and other parasitic skin disease.

3. In the form of syrup, it is used as an expectorant.

Tests for identification

1. To 1 g of the tar, add 20 ml of water. Shake it well and filter. To the filter, add 3 drops of 0.1% solution of ferric chloride. A red colour is produced.

2. Its aqueous solution is acidic to litmus.

REVIEW QUESTIONS

State whether the following sentences are true or false

1. Oleoresins are the resins that occur in association with volatile oils.
2. Asafoetida is an example of latex.
3. The liquid residue formed after distillation of volatile oil is referred to as colophony.
4. Balsam of Tolu is a pathological product.
5. The word "commiphora" means "gum-bearing".

Choose the correct answer

6. The members of Burseraceae are characterized in having _____
   a. volatile oil   b. oleoresins   c. oleogum resins
7. Resins containing high percentage of cinnamic acid are called _____
   a. balsams   b. myrrh   c. gutta percha
8. The by-product of turpentine oil is termed as _____
   a. rosin   b. resin   c. resinol
9. *Podophyllum* is the source of _____
   a. gum resin   b. tar   c. resin
10. *Ferula asafoetida* is placed in the family _____
    a. Zingiberaceae   b. Apiaceae   c. Solanaceae

Answers				
1. True	2. False	3. False	4. True	5. True
6. c	7. a	8. a	9. c	10. b

Answer the following

1. How are the following drugs identified:
   i. Myrrh
   ii. Balsam of Tolu
   iii. Colophony

iv. *Podophyllum*

v. Tar

2. What are balsams? Give an example.

3. Discuss the therapeutic importance of the following: ginger, podophyllum and galanga.

4. Comment on the method of preparation of powder of ginger drug.

5. Write the biological source, distribution, collection, characteristics, chemical constituents and significance of asafoetida.

6. Comment on the characteristics of rhizome of podophyllum and galanga.

7. List out the chemical components of ginger and ferula.

13

DRUGS CONTAINING ALKALOIDS

The term alkaloid is applied to naturally occurring basic compounds. They may be defined as organic nitrogenous substances of plant origin exhibiting well-defined physiological actions. The term is derived from vegetable alkali. Alkaloids have been reported in various plant parts like leaves (*Belladonna*), bark (*Cinchona*), stem (*Withania*), roots and rhizomes (*Rauwolfia*), seeds (nux-vomica), whole aerial plant (*Lobelia*), etc.

The following are the examples of plant drugs containing alkaloids used in various pharmaceutical industries.

1. NUX-VOMICA

Biological source   Nux-vomica constitutes the dried ripe seeds of *Strychnos nux-vomica* Linn. of the family Loganiaceae.

The common names include crow-figs; bachelor's buttons; poison nut; dog buttons; vomit nut; kuchla; yettimaram.

Geographical distribution   The plant is a small tree that grows to a height of 10 to 12 m. It is indigenous to East India. It is found abundantly in the forests of Kerala, Malabar Coast, Orissa, Bihar, Mysore, Konkan, Gorakhpur and west of Tamil Nadu. It is also largely found in Sri Lanka and northern Australia.

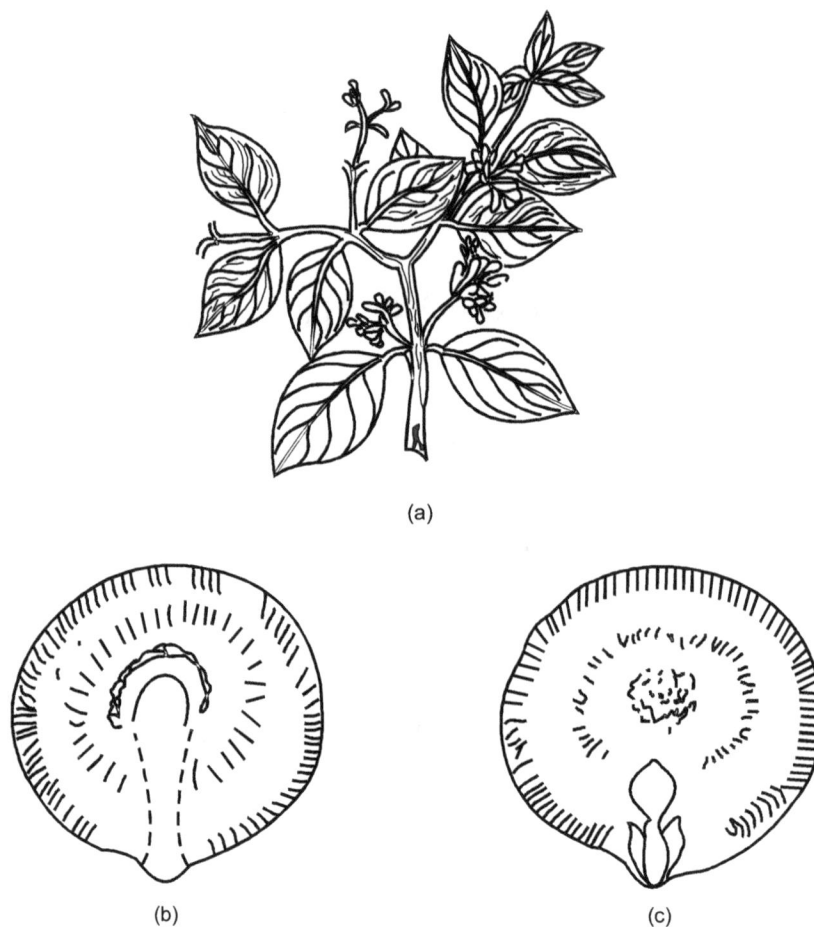

Figure 13.1 Nux-vomica (a) Twig (b) Seed (c) Seed L.S.

Collection In India, the drug is collected from plants grown in the wild by the local tribal community. The collection of fruit is carried out from November to February. The fruit is a berry, which is yellow in colour and about the size of a small orange. Each fruit contains about 4 to 5 seeds, and has heavy bitter pulp. The ripe fruits are collected and the seeds are freed of the pulp. The seeds are washed thoroughly with water. The unripened seeds are removed from the ripened seeds. The mature seeds, which are free from pulp, are dried and packed in gunny bags for marketing. A large amount of seeds and crude alkaloids are exported from India.

Characteristics of seeds

Colour   Seeds are greenish brown or greenish grey.

Odour   Odourless

Taste	Intensely bitter
Size	Seeds 10 to 30 mm in diameter, and 4 to 6 mm in thickness.
Shape	Seeds are disc-shaped, somewhat flat or irregularly bent and concavo-convex, with round or acute edge.
Texture	Extremely hard with silky surface having unicellular hairs.

Chemical constituents   Nux-vomica seeds contain 1.5 to 5% of bitter indole alkaloids. The main alkaloids of therapeutic importance are strychnine (1.25%) and brucine (1.5%). Seeds also contain 3.0% of fixed oil, a glycoside loganin and chlorogenic acid. The other minute alkaloids present in the drugs are vomicine, novacine, α-colubrine, strychnicine; pseudostrychinine and icajine.

Uses

1. The powder of seeds is used as a circulatory stimulant and bitter tonic.

2. Strychnine improves appetite and digestion and is a powerful central nervous system stimulant and also a valuable tool in pharmacological research.

3. Brucine is used as an alcohol denaturant.

4. Nux-vomica increases blood pressure and is recommended in certain forms of cardiac failure.

5. It stimulates respiratory and cardiovascular systems.

Chemical tests   The thin transverse sections of nux-vomica seed are defatted and the following tests are performed:

1. **Potassium dichromate test**   To the cut surface of the seed, a few drops of sulphuric acid and potassium dichromate are added. Formation of violet colour indicates the presence of strychnine.

2. **Strychnine test**   The section is stained with a mixture of ammonium vanadate and sulphuric acid. Purple colour produced shows, the presence of strychnine.

3. **Brucine tests**   A section of the seed is stained with concentrated nitric acid. A yellow colour develops due to the presence of brucine.

2. RAUWOLFIA

Biological source   The drug consists of the dried rhizomes and roots of the plant *Rauwolfia serpentina* Benth. of the family Apocynaceae.

The common names include serpentine root; sarpagandha, Indian snake-root; chhotachand; punernavi.

Geographical distribution   It is a small shrub growing in hot moist regions. The plant is native of India. It is also grown in Sri Lanka, Vietnam, Malaysia, Philippines, Thailand and Indonesia. In India, it is cultivated in Bihar, Orissa, West Bengal, Karnataka, Maharashtra, Gujarat, Uttar Pradesh and Tamil Nadu.

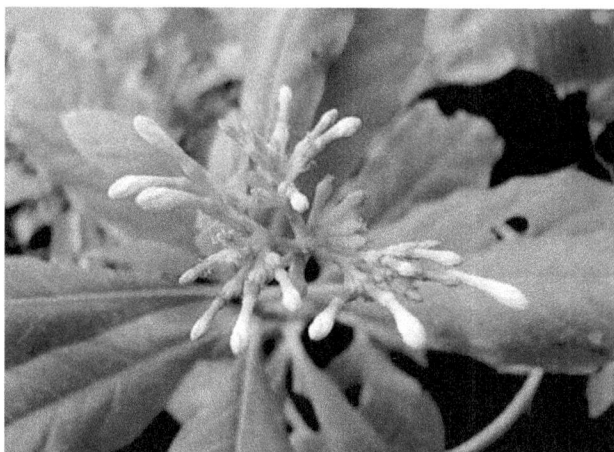

Figure 13.2   Twig of *Rauwolfia serpentina*

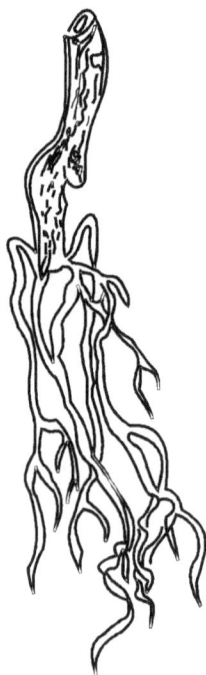

Figure 13.3   Root of *Rauwolfia serpentina*

History   *Rauwolfia* refers to the name of the scientist Dr. Rauwolfia and serpentina refers to snake-like structure of root. The drug is known to Indian system of medicine and found its place as an important drug in the treatment of insanity and snake bite since traditional times. After the isolation of reserpine, an alkaloid, it has received great interest in pharmaceutical industries.

Collection   The drugs are collected mainly from wild plants, when they are about 3 to 4 years old. The rhizomes and roots are dug out during October–November. The roots are washed to remove the mud out properly to get convenient sizes and dried in air.

Characteristics of roots

Colour	Root bark is greyish yellow to brown and wood is pale yellow.
Odour	Slight
Taste	Bitter
Size	About 10 to 15 cm long and 1 to 3 cm in diameter.
Shape	Roots are sub-cylindrical with slightly tapering, tortuous pieces.
Fracture	Short and irregular. The transversely cut surface is white in colour.

Chemical constituents   The drugs contain about 0.7 to 1.4 per cent of alkaloids depending upon the source and are of indole type. Alkaloids are present mostly in the bark of the roots and rhizomes. The important alkaloids isolated from the drug are reserpine, rescinnamine and deserpidine, which have therapeutic significance.

The other alkaloidal components are ajmalinine, ajmalicine, serpentine, serpentinine, reserpinine, rauwolfinine, etc. Other constituents of the drug are fatty acids, phytosterols, unsaturated alcohols, sugars and starch.

Uses

1. *Rauwolfia* is used as a hypotensive in the management of essential hypertension and in certain cases as tranquilizer.

2. The alkaloids reserpine and rescinnamine are used as antihypertensive and sedative drugs.

3. It is specific for insanity, reduces blood pressure and cures pain due to affections of the bowels.

4. Ajmaline is used in the treatment of cardiac arrhythmias.

5. It is also employed in labours to increase uterine contractions and in circulatory diseases.

Chemical tests

1. Reserpine produces violet-red colour, when treated with solution of vanillin in acetic acid.

2. Freshly fractured surface of the root shows red colouration along the medullary rays with concentrated nitric acid.

3. VINCA

Biological source   The drug is the dried whole plant of *Catharanthus roseus* G.Don. (syn, *Vinca rosea* L.) of the family Apocynaceae.

The common names include periwinkle; Madagascar periwinkle; *nithyakalyani; billaganneru; sadabahar;* peenarippu.

Geographical distribution   *Vinca* is indigenous to Madagascar, but now found widely cultivated in the warm regions of the world, such as India, South Africa, USA, Thailand, Taiwan, East Europe, Australia and West Indies as an ornamental plant as well as for its medicinal properties. The use of *Vinca* has been known since BC 50 in European countries as diuretic, antidysenteric, antihaemorrhagic and in wound-healing.

Figure 13.4   Vinca

Collection and preparation   *Vinca* is an erect, everblooming pubescent herb or subshrub, growing to a height of 90 cm to 1m and is woody at the base. The stems and leaves are cut about 7–9 cm above the ground level, after 1–2 years of growth. After cutting the aerial parts, the leaves, stems and seeds are separated, cleaned, washed and dried in air and marketed.

The roots are collected after irrigating the field. The roots are dug out by ploughing, and are then washed to remove earthen materials and dried in shade. The air-dried roots are packed in bales for marketing.

The extraction and separation procedure for alkaloids is based on their separation into insoluble and soluble tartrates in different solvents. The alkaloids vinblastine, vincristine and other weak bases are separated and then fractionated with the help of column chromatography, using alumina as adsorbent. The *Vinca* roots are exported regularly to other countries from India.

Characteristics of the plant

Colour	Leaves are green, roots are pale grey, flowers are violet-pink, white or carmine-red, seeds are black.
Odour	Very characteristic
Taste	Bitter
Size	Leaves are 3–4 cm in length and 3.0–3.5 cm in width.
Shape	Leaves simple, petiolate, ovate or oblong, entire, brittle with acute base and rounded or mucronate apex with glossy appearance and entire margin.

Chemical constituents *Vinca* contains a large number of indole or dihydroindole type alkaloids. Out of them, about 20 dimeric indoledihydroindole alkaloids possess anticancer activity, vincristine and vinblastine being most significant. The other alkaloids present in vinca are ajmalicine, lochnerine, reserpine, serpentine and tetrahydroalstonine. In addition to alkaloids, monoterpenes, sesquiterpene, indole and indoline glycoside have also been reported. Vincristine is clinically more valuable as compared to vinblastine.

Uses

1. *Vinca* is used to cure diabetes and in wasp-sting; its alkaloids are antineoplastic in nature.

2. Vinblastine sulphate is an antitumour alkaloid employed to cure Hodgkin's disease and choriocarcinoma. It is used intravenously.

3. Vincristine sulphate is a cytotoxic compound and is used to treat leukaemia in children.

The two alkaloids (vinblastine and vincristine) are marketed under the trade names "Velban" and "Onconvin" respectively. Another alkaloid of importance is ajmalicine, which is a vasodilator, used as a hypotensive agent.

4. OPIUM

Biological source   Opium is the air-dried milky exudate obtained by incising the unripe capsules of *Papaver somniferum* L. or its variety *P. album* Decand of the family Papaveraceae.

The common names of the product are raw opium, gum opium; crude opium; afim; abin; abhni; naranga. The plant is called poppy.

Geographical distribution   In opium-producing countries, its production is carried out under strict governmental control. Russia, Turkey, China, Iran, Pakistan, Afghanistan, India, Yugoslavia, Thailand and Burma are the main opium-producing countries. In India, opium is grown in Madhya Pradesh (Neemuch) and Uttar Pradesh.

The production of opium is controlled internationally by the international Narcotic Control Board of the United Nations.

Opium has been known to man since centuries due to its narcotic properties. It was first cultivated in the Mediterranean region and probably brought to India by Alexander in BC 327. Narcotine was the first alkaloid to be reported and isolated from opium by Derosne in 1803.

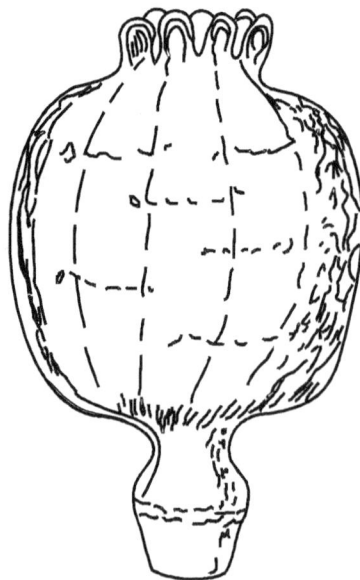

Figure 13.5   Capsule of poppy

Collection and preparation Opium is cultivated under licence from the government. The plant is an erect annual herb, attaining 60–120 cm height, with large solitary flowers, which vary in colour from white to pink or purple. Flowers bloom in the month of April or May and the capsules mature in June or July. Each plant produces 5 to 8 capsules. During maturity, the capsules exude maximum latex. When the capsules mature and are about 4 cm in diameter, the colour changes from green to yellow. Such capsules are incised in the afternoon with a knife about 1 mm deep transversely, without cutting the endocarp, otherwise the latex will be lost. Due to incisions, latex exudes out and thickens due to cold weather in the night. The latex which is white in the beginning, rapidly coagulates and turns brown. The following morning dried latex is removed by scraping with a knife and transferred to a poppy leaf, when sufficient latex is collected. It is then kneaded into balls, wrapped in poppy leaves and dried in the shade.

The incising process is repeated for about 4 times on the same capsule with 2 days interval. Then, capsules are collected and dried in open areas, and the seeds are separated by beating.

The opium collected this way is either exported or some of the part is further processed at government opium alkaloid factory at Ghazipur.

The extraction process of opium is as follows:

1. Add calcium chloride solution to opium powder, shake well and filter it.

2. To the filtrate add 10 per cent solution of sodium hydroxide to get a mixture.

3. Filter the mixture to separate the filtrate from the precipitate.

4. The filtrate is used for the extraction of alkaloids such as, morphine, codeine, and narceine using chloroform.

5. Dissolve the precipitate in alcohol, and acidify it with acetic acid.

6. To this mixture add three volumes of boiling water to get a precipitate and an aqueous solution.

7. Aqueous solution contains thebaine, whereas the precipitate obtained is used for the extraction of narcotine and papaverine, by boiling in 0.3% oxalic acid.

8. The crystals thus formed constitute papaverine oxalate and the aqueous solution contains narcotine oxalate. The commercial varieties of opium are Indian opium, Persian opium, Turkish opium, Chinese opium, and Yugoslavian opium.

Characteristics of opium

Colour	Pale olive grey or chocolate brown
Odour	Characteristic fragrance of the poppy leaves on the upper surface
Taste	Bitter and distinct
Size	Opium mass is 8–15 cm diameter and thickness varies
Texture	Uniform and slightly granular
Appearance	Rounded, flattened masses or cakes

Chemical constituents Opium contains about 25 alkaloids. The most important alkaloids of opium are morphine, codeine, narcotine, papaverine and thebaine. Among these, morphine (10–16%) is the most significant base. Other minor alkaloids are narcein, protopine, laudanine, codamine, etc. These alkaloids are combined with an organic acid called meconic acid. The drug also contains sugar, sulphates, albuminous compounds, colouring matter and water.

Uses

1. Opium acts on the central nervous system, causing its depression.

2. It is used as an analgesic, narcotic and sedative and also used to relieve pain, dysentery and cough.

3. It checks excessive peristalsis and is useful in the control of diarrhoea.

4. It contracts the pupil of the eye.

5. Opium is the source of various useful alkaloids such as morphine, codeine, noscapine and papaverine.

6. Morphine is an analgesic, while codeine is a milder sedative and is used in cough mixtures. Codeine phosphate is used in cough and bronchitis syrups.

7. Noscapine is non-narcotic and also has cough suppressant action acting as a central antitussive agent.

8. Papaverine is a smooth muscle relaxant and is used in the treatment of muscular spasms.

9. Opium, morphine and the diacetyl derivative, heroin, cause drug addiction.

10. Poppy capsules are somniferous, soporific, sedative, and narcotic and used as anodyne and emollient. As astringent, opium checks haemorrhages, lessens bodily secretions and restrain's tissue change.

Chemical tests

1. Morphine when sprinkled on nitric acid gives orange-red colour, whereas codeine does not respond to this test.

2. Aqueous extract of opium treated with $FeCl_3$ solution gives deep reddish purple colour, which persists even on addition of hydrochloric acid, due to the presence of meconic acid.

3. Morphine solution when treated with potassium ferricyanide and ferric chloride solutions gives bluish green colour. Codeine does not respond to this test.

4. Papaverine solution in hydrochloric acid gives a lemon yellow colour with potassium ferricyanide solution.

Storage   Opium is preserved in a well-closed container to prevent loss of morphine.

5. IPECACUANHA

Biological source   Ipecac refers to the dried roots and rhizomes of *Cephaelis ipecacuanha* (Brot.) A. Richard, of the family Rubiaceae, known in commerce as Rio or Brazilian ipecac or of *Cephaelis acuminata* Karsten, known in commerce as Cartagena ipecac, Nicaragua ipecac or Panama ipecac. Ipecac should contain not less than 2% of ether-soluble alkaloids.

The common names include ipecac; Brazilian ipecac; Johore ipecac.

Geographical distribution   The plant *C. ipecacuanha* is indigenous to Brazil and *C. acuminata* is indigenous to northern portions of Columbia and grows in Panama, Nicaragua, Myanmar, Malaysia and also in India. In India, it is cultivated in West Bengal, Darjeeling, Sikkim and in Tamil Nadu in Nilgiris near Kallaar.

Collection and preparation   The plant is a low, straggling shrub with slender rhizome bearing annulated wiry roots. It grows widely in moist, swampy forests. The percentage of all alkaloids is maximum in the third year therefore, harvesting of roots is done after three years of vegetative growth.

The drug is collected in the dry season. The roots are dug out and the soil adhering to the roots is removed and dried in the sun rapidly for 2–3 days.

Figure 13.6 A twig of ipecacuanha

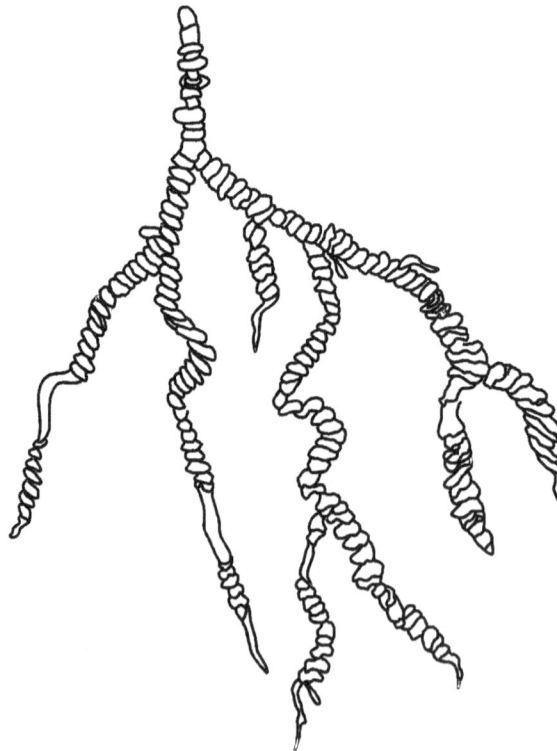

Figure 13.7 Root of ipecacuanha

Characteristics of Rio-ipecac

Colour	Outer surface of the drug is deep brick-red to very dark brown.
Odour	Slight
Taste	Bitter and acrid
Size	150 mm in length with 6 mm diameter.
Shape	Cylindrical, slightly tortuous.
Fracture	Short in the bark and splintery in the wood.
Appearance	Externally annulated broadly with rounded ridges around root.

Chemical constituents The drug ipecac contains about 2–2.5% of quinoline alkaloids. The total alkaloids in Rio-ipecac are up to 2%, Nicaragua ipecac, 2.6–3% and in Panama ipecac, 2.2%. The three main alkaloids are emetine (40–70%) cephaeline and psychotrine contained chiefly in the bark. The drug also contains α-glucosidal tannin, ipecacuanhin and α-glycoside, ipecoside. It also contains starch and calcium oxalate.

Uses

1. Ipecac is used as an expectorant in small doses and emetic in higher doses. Emetine has more expectorant and less emetic action than cephaeline.

2. The drug is also used in the treatment of amoebic dysentery. For this purpose emetine hydrochloride is given parenterally, whereas emetine bismuth iodine is given orally.

3. Ipecac preparations include powdered ipecac, liquid extracts, syrups and tinctures.

4. Ipecac syrup is given to children and used as an expectorant in whooping cough.

5. Ipecac with opium, as in Dover's powder is used as a diaphoretic.

6. Emetine also has anti-tumour properties.

Chemical tests

1. To about 2.5 g of powdered drug, 20 ml of HCl and 5 ml of water are added. The mixture is shaken well and filtered. To the filtrate, 0.5 g of potassium chlorate is added. The yellow colour formed is gradually changed to red, and this is due to emetine.

2. To a small quantity of emetine, sulphuric acid and molybdic acid are added. The presence of bright green colour indicates emetine.

6. BELLADONNA

Biological source The dried leaves and other aerial parts including flowering tops of *Atropa belladonna* Linn. (European belladonna) of the family Solanaceae, are used as drug. The drug should contain not less than 0.35% of alkaloids.

The common names are deadly night shade leaf; banewort; poison black cherry; belladonna herb; sagangur.

Geographical distribution The plant is indigenous to England. It is grown in Germany, US, Canada, UK and India. In India, it is distributed in the Western Himalayas from Simla to Kashmir and Himachal Pradesh.

Figure 13.8 A twig of Belladonna

Collection The plant is an erect, glandular and pubescent perennial herb, about 1.5 m in height. The name Atropa is derived from a Greek word *Atropos* meaning "who cuts the thread of life". It refers to the poisonous nature of the drug. The word

"Belladonna" is derived from Italian words *bella*, meaning "beautiful" and *donna*, meaning "lady". The juice of the fruits when put in eyes makes them more sparkling due to dilatation of the pupil. The leaves and flower tops are collected when the plants are about three years old. Collection is usually made at the end of June or in July, and two to three collections of leaves can be obtained annually. The roots are dug out when the plants are four years old. The collected leaves and roots are dried immediately in sun or in shade. While drying, care is taken to see that the dried leaves retain their green colour. After drying, the unwanted foreign materials are removed, graded, and packed for market. The drug should be stored with great care in a cool place.

Characteristics of aerial parts

Colour	Leaves dull green to yellowish green; flowers, purple to yellowish brown; fruits, green to brown.
Odour	Slight and characteristic.
Taste	Bitter and acrid
Size	Leaves—5 to 25 cm long and 2.5 to 12 cm wide. Flowers—corolla 2.5 cm long and 1.5 cm wide. Fruits—about 10 cm in diameter.
Shape	Leaves—ovate lanceolate to broadly ovate, apex acuminate, margin entire, petiolate, brittle with decurrent lamina; flowers—campanulate, 5 petals; fruits—berries, sub-globular with numerous flat seeds.

Chemical constituents   The total alkaloid content of drug is 0.4 to 1% and varies in different parts of plant—0.6% in roots, 0.05% in stems, 0.4% in leaves, 0.19–0.21% in unripe and ripe berries and 0.33% in seeds.

The main alkaloids of belladonna are L-hyoscyamine (0.3–0.6%), and atropine. The drug also contains belladonine, apoatropine, asparagine, choline, scopoletin, hyoscine, and volatile bases such as pyridine and N-methyl pyrroline.

Uses

1. Belladonna is used as a parasympatholytic drug with anticholinergic properties.

2. It is used for the control of spasms.

3. Belladonna checks excessive perspiration of patients suffering from tuberculosis.

4. The drug is commonly administered as belladonna tincture, and belladonna root is used for making preparations for external use.

5. Belladonna is used as a narcotic sedative, diuretic and as anodyne and to check secretion, sweat and saliva.

Chemical test When ammonia is added to the alcoholic solution of scopoletin, it gives blue fluorescence. This test is useful to detect belladonna poisoning.

7. CINCHONA

Biological source The dried bark of the stem or the root of *Cinchona succirubra* Pav. or other hybrids of the species of *Cinchona ledgeriana, C. officinalis* Hk. and *C. calisaya* Wedd. of the family Rubiaceae are used as drugs. It contains not less than 6.5% of total alkaloids, 30–60% of which are quinine-type alkaloids.

The common names of the drug include Peruvian bark; Jesuit's bark; Calisaya bark; cortex chinae.

Geographical distribution *Cinchona* tree is native to Peruvian Andes, South America, at high altitudes of 1500–2500 m. It is found in Indonesia, Sri Lanka, Columbia, Bolivia, Java, Guatemala, Ecuador, Tanzania and India. In India, the species are cultivated in Anamalai hills of Coimbatore, Nilgiri hills in Nilgiri district (Tamil Nadu), Darjeeling areas of West Bengal, Sikkim, Khasia and Jaintia hills.

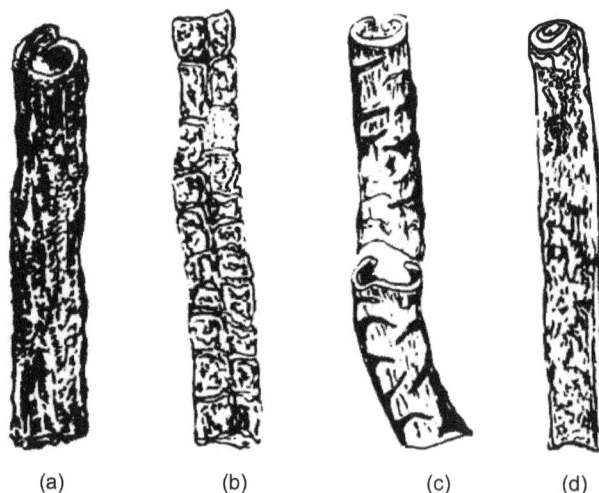

(a)　　(b)　　(c)　　(d)

Figure 13.9 Bark of cinchona (a) *C. succirubra* (b) *C. ledgeriana* (c) *C. officinalis* (d) *C. calisaya*

Collection and preparation *Cinchona* is a tree, which grows profusely, and shade favours the higher production of quinine. Six- to nine-year-old trees possess maximum amount of alkaloids in the bark. The bark is collected by the coppicing method. For this purpose, vertical incisions are made on branches or trunks of trees and these incisions are connected by horizontal circles. The stem bark is then stripped off and dried in sun and further by artificial heat.

During drying, the bark loses up to 70% of its weight. Moulding or fermentation should be avoided during drying of barks. The quills of drugs are packed in gunny bags and marketed. The root bark is collected by uprooting the trees, and the bark is separated manually.

Extraction of quinine For extraction of quinine, the bark is powdered and extracted with benzene or toluene in the presence of alkali. Further, the alkaloids are extracted with dilute sulphuric acid. By bringing the acid extract to neutrality, quinine sulphate separates, as it is sparingly soluble.

Characteristics of bark

Stem bark

Colour	Outer surface dull brown-grey or grey and inner surface pale yellowish brown to deep reddish brown.
Odour	Slight and characteristic.
Taste	Intensely bitter and astringent.
Shape	Quills and curved pieces.
Size	30 cm length and 2 to 8 mm thickness.
Fissure	Bark rough with transverse fissure.
Fracture	Short in external layers and fibrous in the inner portion. The outer surface of the bark bears moss or lichens.

Root bark

Colour	Similar to stem bark
Odour	Similar to stem bark
Taste	Similar to stem bark
Size	2–7 cm length and the diameter and thickness vary with the species.
Shape	Curved, twisted or irregularly channelled or quilled.

Fissure   Broad longitudinal with transverse cracks and number of cracks vary with species.

Powder colour   Cinnamon brown in *C. calisaya* and *C. ledgeriana*, yellow in *C. officinalis*, and reddish brown in *C. succirubra*. The outer surface of root bark is somewhat scaly white, while the inner surface is striated.

Chemical constituents   *Cinchona* bark contains about 25 alkaloids, which belong to the quinoline type. The important alkaloids are quinine (70%), quinidine, cinchonine, and cinchonidine. *C. succirubra* contains 5–7% total alkaloids, of which 30% is quinine. *C. ledgeriana* contains 6–10% and in some cases it may be up to 14% total alkaloids, of which upto 75% is quinine. *C. calisaya* has 6–8% total alkaloids having about 50% quinine.

In addition, cinchona also contains quinic acid, cinchotannic acid, a glycoside quinovin, tannins and bitter essential oil.

Uses

1. *Cinchona* bark is antimalarial in nature. Its main alkaloid quinine is used in the treatment of malarial fever.

2. *Cinchona* is used as a bitter tonic, stomachic, antipyretic and analgesic.

3. Quinidine is used for the prophylaxis of cardiac arrhythmias and atrial fibrillation.

4. *Cinchona* bark and all preparations are valuable in intermittent fever and they have been prescribed as tonic in dyspepsia, gastric catarrh, adynamia and convalescence from fever.

Chemical tests

1. *Cinchona* moistened with glacial acetic acid, on heating in a test tube, gives blood-red drops on the sides of the tube indicating the presence of quinine.

2. *Cinchona* bark moistened with sulphuric acid shows a blue fluorescence in UV light.

3. Thalleoquin test   The powdered drug gives emerald green colour with bromine water and dilute ammonia solution.

4. Quinidine solution gives a white precipitate with silver nitrate solution, which is soluble in nitric acid.

8. GLORIOSA

Biological source   The dried tubers of *Gloriosa superba* Linn. of the family Liliaceae are used as drug.

The common names of the plant include glory lily and *kanvazhippu*.

Geographical distribution   The plant is a herbaceous perennial climber that grows wild in live fences and low jungles. The plant is also grown as an ornamental plant in gardens and also cultivated for its commercial value in Tamil Nadu (India). The drug is indigenous to tropical Asia and Africa.

In recent years *Gloriosa* has become a promising drug in India for the production of colchicine on commercial scale, due to heavy demand for its alkaloid.

Collection and isolation process   The tubers are perennial and fleshy. The tubers are dug out and washed to remove the mud particles and other foreign materials. The tubers are then dried in sun and in shade. The dried tubers are packed for sale.

The extraction of colchicine by conventional methods yields less percentage of alkaloids from *G. superba* whereas the aqueous extraction method yields more than 80% colchicine.

Chemical constituents   It contains proto-alkaloids, the main constituent being colchicine. It also contains different derivatives in minor quantities.

Figure 13.10   Twig of gloriosa

Uses

1. Colchicine is used for the treatment of gout and mammary tumours.

2. It is used as a research tool in plant genetics for producing polyploids.

3. It is used in horticulture to produce novel polyploid plants.

Chemical tests Treating the drug with 60–70% sulphuric acid or with hydrochloric acid, a yellow colour is produced indicating the presence of the alkaloid colchicine.

9. LOBELIA HERB

Biological source It consists of the dried aerial parts of *Lobelia nicotianaefolia* Heyne of the family Lobeliaceae (Campanulaceae). *Lobelia* should contain lobeline in not less than 0.55% of total alkaloids.

The common names of the herb are Indian tobacco; wild tobacco; Asthma weed; nali; kattuppugaiyinai; bibhishna.

Figure 13.11 A twig of lobelia

Geographical distribution It is a large biennial or perennial herb of 1.2–3.6 m height and distributed in many parts of India. It is found in the Western Ghats of Maharashtra up to Kerala at an altitude of 700 to 2300 m. It is also grown in the Nilgiris of Tamil Nadu, and Karnataka.

The name *Lobelia* has been derived to honour a famous botanist Matthias de Lobel. The utility of this plant for asthma was mentioned by Cutler (1813).

Collection During August–September, the inferior ovary of the plant develops into an inflated capsule. Therefore, the drug is collected in this season and the aerial parts are cut carefully, dried in shade to preserve the green colour and packed into bales or packets.

Characteristics of aerial parts

Fruit	Inflated capsule with longitudinal ribs
Seeds	Small, ellipsoidal and compressed
Colour	Stems are green with purplish tinge and cylindrical or somewhat angular. Leaves are green. Flowers are white.
Odour	Similar to tobacco
Taste	Extremely acrid and irritating
Size	Leaves are 5 to 25 cm in length and 1 to 5 cm in breadth.
Shape	Leaves are oblong, lanceolate, sub-sessile, and margin finely serrulate. The upper leaves are shorter, and lower leaves are bigger in size, apex acute to acuminate and the base decurrent and tapering.

Chemical constituents Indian lobelia contains about 0.5 to 1.2% total alkaloids and the most important of them is lobeline. The alkaloid content of *L. nicotianaefolia* shows seasonal variation. That is, the alkaloid content is found to be in the highest concentration in the month of October and November. Lobelidine, lobelanine, isolobelanine and lobelanidine are the other alkaloids isolated from Indian lobelia. In addition, the drug also contains gum, volatile oil, resin and fixed oil.

Uses

1. *Lobelia* is mainly used in the treatment of asthma and as a respiratory stimulant. It is also used for bronchial asthma and chronic bronchitis.

2. An infusion of leaves is used as an antiseptic.

3. It is also given in the resuscitation of newborn infants through umbilical vein. It also helps smokers to get rid of their smoking habit.

Chemical tests

1. Lobeline solution treated with sulphuric acid gives red colour with formaldehyde.

2. Lobeline solution on boiling produces acetophenone with a characteristic smell.

10. COFFEE

Biological source The dried ripe seeds of the species of *Coffea* of the family Rubiaceae are used as drug. The important species of *Coffea* are

   i.   *Coffea arabica* Linn. (Arabian coffee)

   ii. *Coffea canephora* Pierre. (Robusta or Congo coffee) and

   iii. *Coffea liberica* Bull. (Liberian coffee).

   The common names are coffee beans and coffee seeds.

Figure 13.12 *Coffea arabica*

Geographical distribution The plant is an evergreen shrub, which grows well in shady places. It is distributed in Sri Lanka, Indonesia, India, Brazil, Europe, Java, Sumatra, Malaya, Philippines, Colombia, Mexico, Uganda, US, Germany, France, Italy, Canada and the Netherlands.

It is known by the name *qahuah* in Middle-East countries. The word has its origin from Turkish and Arabic languages. In South India, the plant is grown largely in the hills of Nilgiris, and Kodaikanal regions.

Harvesting and processing The fully ripe coffee fruits are picked either by hand or by shaking the coffee bushes and collecting the fallen berries. Ripe red fruits yield the best quality coffee. The coffee berries are processed by two methods—i. Dry method and ii. Wet or wash process. The first method yields coffee of superior grade.

i. **Dry method**   The berries are dried in the open sun or in hot-air driers. The fruits are agitated to permit uniform and thorough drying. They are bagged and stored in warehouses or separated immediately from the dried skin and pulp either by hand pounding or by using hulling machines.

ii. **Wet method**   In this method, the berries are placed in large tanks filled with water. The well-developed fruits sink to the bottom, while the defective ones and the leaves, twigs, dried fruits and trash float on the surface and are removed. Thereafter, the ripe fruits are subjected to pulping, fermenting, drying and curing.

The fleshy outer parts of the fruit are removed with the help of pulping machines. The remainder pulp adhering to the parchment coat of the beans is separated by controlled fermentation in water-filled concrete tanks for 12–24 hours, during which the gummy or pectic matter is loosened and freed by repeated washings in running water. The depulped berries are dried to moisture content of about 12% either by exposing to sun or by hot-air driers. Hulling removes the seed parchment and the seed coat, exposing the coffee seeds. After polishing, coffee beans are graded, bagged and shipped for export.

Polished coffee beans are then roasted in coffee roasting machines for five minutes at a temperature of 260°C. The seeds become a deep rich brown in colour and their texture becomes porous and crumbly under pressure. They also develop the characteristic coffee aroma and flavour. The flavour of coffee is due to the presence of a volatile oil, caffeol. The roasted beans are rapidly cooled in a vat and are ready for grinding. Powdered coffee should be immediately put into sealed containers, otherwise it loses its aromatic quality and turns rancid.

Extraction of caffeine   Caffeine is prepared from coffee beans which is one of its major sources. For extraction, coffee roasters are used in which caffeine sublimed during roasting is recovered. It is the commercial method for extraction of caffeine.

Chemical constituents   The main constituents of coffee bean are caffeine (2–3%), tannins (3–5%), fixed oil (10–15%), proteins (13%), chlorogenic or caffeotannic acid and sugars in the form of dextrin, glucose, etc. In the seeds, caffeine is present as a salt of chlorogenic acid.

During the process of roasting, the agreeable smell of coffee is developed, which is due to an oil called caffeol, which mainly contains furfural along with minor quantities of phenol, pyridine and valerianic acid.

Uses

1. Caffeine, an alkaloid, is used as a stimulant.
2. Sometimes, it is used to combat the toxic effects due to CNS-depressant drugs.

11. ASHWAGANDHA

Biological source   The dried roots and stem bases of *Withania somniferum* Dunal. of the family Solanaceae are used as drug.

The common names include withania root; asgandh; ashvagandi; ashwagandha.

Geographical distribution   The plant is a perennial shrub with tuberous roots that grow wild in drier parts of India, especially in Madhya Pradesh, Uttar Pradesh, Punjab plains and north-western parts of India like Gujarat and Rajasthan. It is also found in Congo, South Africa, Egypt, Jordan, Europe, Pakistan and Afghanistan.

Figure 13.13   Plant of ashwagandha

Collection and preparation   The plants from different sources vary in their morphological and therapeutic properties. Collection is done during January which may extend up to March. The roots are collected by uprooting the plant and washed thoroughly. The entire tuberous roots or the pieces of roots are dried immediately and packed for marketing.

Characteristics of roots

Colour	Roots are buff to grey-yellow on the outer surface.
Odour	Not characteristic
Taste	Bitter and mucilaginous
Size	5–12 mm thick
Shape	Straight conical with longitudinally wrinkled outer surface.
Fracture	Smooth and powdery

Chemical constituents Alkaloids and steroidal lactones are the main constituents of Ashwagandha. Among the various alkaloids, withanine is the main constituent. The steroidal lactones include withaferins and withanolides. The other alkaloids isolated include tropine, pseudotropine, hygrine, cuscohygrine and withasomnine.

Uses

1. Ashwagandha is used in a large number of ayurvedic formulations, and has sedative and hypnotic effects.

2. It is used as respiratory stimulant, and immuno-modulatory agent.

3. Traditionally, it has been used in the treatment of rheumatism, gout, hypertension, nervine and skin diseases.

4. It is also used as a sex-stimulant and rejuvenator, and is considered as a strength- and vigour-promoting drug, especially in geriatric cases.

5. The leaves have antitumour activity.

12. VASAKA

Biological source Vasaka refers to the fresh and dried leaves of the plant *Adhatoda vasica* Nees. of the family Acanthaceae.

The common names include adhathoda; arusha; adatodai; vasa; vasaka.

Geographical distribution Vasaka is a perennial shrub, reaching a height of 2–3 m. It is indigenous to India, where it is found in the sub-Himalayan tracts up to an altitude of 1300 m, and in the Konkan region of Maharashtra. Besides India, the plant is found in Sri Lanka, Myanmar and Malaya.

Figure 13.14 Twig of vasaka

Collection   The plant is used medicinally since olden times and is included in different formulations in ayurveda. The collection can be done in all seasons of the year. Fresh leaves are collected, washed and either used as such or dried in the shade, powdered and packed in containers for sale.

Characteristics of leaves

Colour	Fresh leaves are green but on drying, colour changes from dull brown to grey.
Odour	Characteristic
Taste	Bitter
Size	10–30 cm length and 4–10 cm width
Shape	Lanceolate to ovate, slightly petiolate, margin crenate to entire, apex acuminate and base tapering. Surface glabrous or slightly pubescent.

Chemical constituents   Vasaka contains a number of alkaloids. Vasicine and vasicinone are the two major alkaloids isolated from the drug. The drug also contains volatile oil, betain, vasakin and adhatodic acid.

Uses

1. Vasaka is used as expectorant and bronchodilator.

2. It is available in liquid extract, syrup and powder form and has abortifacient effect and also has action similar to oxytocin.

3. It is also used in the treatment of asthma and phthisis.

REVIEW QUESTIONS

State whether the following sentences are true or false

1. The powder of nux-vomica seeds is used as bitter tonic and circulatory stimulant.
2. The alkaloid vincristine is used as anticancer drug.
3. The bark of *Cinchona* yields quinine.
4. The botanical name of Indian tobacco is *Nicotianum tabacum*.
5. The dried seeds of *Coffea canephora* is called Robusta coffee.

Choose the correct answer

6. _____ is a bronchodilator.
   a. vasaka   b. *Annona*   c. cumin

7. *Withania somniferum* is a member of _____
   a. Acanthaceae   b. Solanaceae   c. Loganiaceae

8. Formation of violet-red colour with a solution of vanillin in acetic acid indicates the presence of _____
   a. papaverin   b. vincristine   c. reserpine

9. The alkaloidal drug vinblastine is marketed under the trade name _____
   a. Onconvin   b. Velban   c. Ominovin

10. The unripe capsules of *Papaver somniferum* yield an alkaloid called _____
    a. Opium   b. Atropine   c. Colchicines

Answers				
1. True	2. True	3. True	4. False	5. True
6. a	7. b	8. c	9. b	10. a

Answer the following

1. What are alkaloids? Cite an example.
2. Discuss the method of collection and preparation and medical application of vinca drug.
3. Write an essay on any two drugs containing alkaloids.

4. Critically comment on the characters and chemical constituents of Ipecac and *Cinchona*.

5. Write a short note on the following:

    i.   *Gloriosa*

    ii. Ashwagandha

6. Give the biological source, distribution, harvesting and processing, extraction, chemical components and importance of coffee.

7. -List out the importance of *Adhatoda* and *Rauwolfia*.

14

DRUGS CONTAINING TANNINS

Tannins are derivatives of polyhydroxybenzoic acids, which are widely distributed in the plant kingdom, as secondary metabolites. These are present in the cell sap as solution and also in distinct vacuoles. They have been used since long as astringent, and this depends upon their ability to precipitate proteins. They are therefore used in medicines for allied purposes. They are used as antiseptics, in arresting diarrhoea, as antidotes and to check haemorrhages. Commercially tannins are used in the leather industry to prevent putrefaction of animal hides. Tannins are classified into hydrolysable tannins, (myrobalan, clove, chestnut) condensed tannins, (cinchona, cinnamon, green tea) and pseudotannins (nux-vomica, coffee, cocoa). The following are the examples of tannin-containing crude drugs.

1. MYROBALAN

Biological source   Myrobalan refers to the dried, ripe fruits of an Indian tree called *Terminalia chebula* Retz. of the family Combretaceae.

The common names are chebulic myrobalan; kadukkai; harana; haritaki.

Geographical distribution   The tree is cultivated and also found abundantly in wild state, in all deciduous forests of India, especially in West Bengal, Bihar, Assam, Madhya Pradesh, Maharashtra, Gujarat and Kerala.

Figure 14.1   Fruit of myrobalan

Characteristics of fruits

Colour	Fruits are yellowish brown.
Odour	Slight
Taste	Astringent, slightly bitter and sweetish at the end.
Size	2–4 cm long and 1.5–2.5 cm broad
Shape	Ellipsoidal to ovate hard and stony drupe with longitudinal wrinkles.

Chemical constituents   Myrobalan fruits are an important source of tannin. The fruit consists of 30–40% of hydrolysable tannins (pyrogallol type), which on hydrolysis yield chebulic acid and D-galloyl glucose. In addition, it contains chebulagic acid, chebulinic acid, ellagic acid, gallic acid, and glucose sorbitol. The drug also contains anthraquinones and fixed oil containing mainly esters of palmitic, oleic and linoleic acids.

Uses

1. It is mainly used as an astringent, laxative, stomachic and tonic.

2. It is also used as an anthelmintic and the pulp of the fruit is used to cure bleeding.

3. It is used for external application in the treatment of piles and ulcers, and finely powdered drug is used in carious teeth.

4. Myrobalan in combination with *Emblica officinalis* (Amla), and *Terminalia bellerica* (Bahera) forms the ingredient of the well-known ayurvedic preparation called "triphala" which is used as a laxative and in stomach disorders.

2. BAHERA

Biological source It refers to the dried ripe fruits of the plant *Terminalia bellerica* Roxb. of the family Combretaceae.

The common names include beleric myrobalan; bahira; akkam; thannikkai.

Geographical distribution It is a large deciduous tree measuring 30 to 40 m in height and 2–3 m in girth. It is commonly found wild in all deciduous forests of India up to an altitude of 1000 m. It grows abundantly in Madhya Pradesh, Uttar Pradesh, Punjab, Maharashtra and also in Sri Lanka and Malaya.

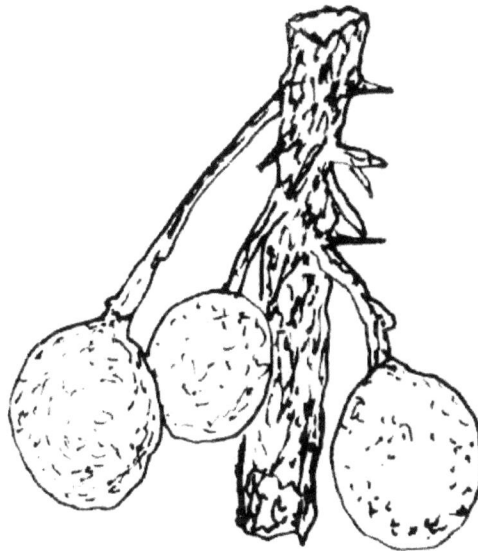

Figure 14.2 Fruit of bahera

Collection The collected ripe fruits are dried in shade and packed. The packed fruits are marketed.

Characteristics of fruits

Colour	Fruits are dark brown to black.
Odour	Nil
Taste	Astringent
Size	1.3–2 cm in length and 1.3–2.5 cm diameter
Shape	Fruits are ovoid globular drupes and 5-angled.

Appearance Velvetty outer side with irregularly wrinkled and fine longitudinal ridges. A prominent round scar of pedicel at one end of the fruit.

Chemical constituents Bahera fruits contain 20–30% tannins. In addition, it contains gallic acid, ellagic acid, phyllemblin, ethylgallate, chebulaginic acid and galloyl glucose. It also contains colouring matter, 30–40% of fixed oil, esters of palmitic, stearic, oleic and linoleic acids.

Uses

1. Bahera is used as an astringent, tonic, laxative, and shows antipyretic activity.
2. It is used in the treatment of dyspepsia, dysentery, diarrhoea, leprosy and headache.
3. Bahera is one of the ingredients of the ayurvedic purgative preparation "Triphala".
4. The half-ripe fruit contains fixed oil and has purgative property.
5. The gum is used as a demulcent and purgative, and the oil is used in the manufacture of soap.

3. ASHOKA

Biological source The drug is constituted by the dried stem bark of the plant *Saraca indica* Linn. of the family Caesalpiniaceae (Leguminosae).

The common names are ashoka bark, *Ashogam*.

Figure 14.3 Bark of ashoka

Geographical distribution   It is one of the most sacred trees of the Hindus and is commonly grown as an ornamental and evergreen avenue tree in India. It reaches a height of 9 m. It is distributed in south Asia, i.e., in Malaysia, Indonesia, Myanmar, Sri Lanka and India. In India, it is found in Central and Eastern Himalayas, Eastern Bengal, Western peninsula, etc.

Collection of bark   Bark of the tree is collected by making transverse and longitudinal incisions. The collected bark is dried in shade, cleaned and packed after removing the unwanted matter from the bark.

Characteristics of bark of the stem

Colour	Externally the bark is yellowish to grey and black, with internally reddish brown colour.
Odour	Nil
Taste	Astringent and bitter
Size	Bark is 40 cm long, 4–6 mm wide and 5–8 mm thick.
Shape	Various sizes and shapes
Fracture	Short and fibrous
Appearance	Bark is warty outside with smooth longitudinal striations inside. The bark pieces may be channelled or flat.

Chemical constituents   The drug contains about 6% tannin. In addition, it also contains catechol, sterol, ketosterol, phlobaphenes, saponin, organic calcium and iron compounds. The bark contains a powerful oxytocic principle, known as phenolic glycoside, P_2.

Uses

1. It is used in the traditional Indian system of medicine as a domestic remedy for the treatment of various disorders of the uterus, particularly for menorrhagia (excessive menstruation) and leucorrhoea (white discharge from the vagina).

2. It is also used as a sedative.

4. BLACK CATECHU

Biological source   It is the drug obtained from the dried aqueous extract prepared by the boiling of heartwood chips of *Acacia catechu* W. & A. of the family Mimosaceae (Leguminosae).

The common names are kattha, cutch, catechu, karun-kali; khadira.

Geographical distribution The tree is native of India and is found in the wild as well as cultivated in Myanmar.

Collection and manufacture of catechu The red heartwood of the tree is removed by felling the trees. The bark as well as the sapwood is removed and the heartwood is cut into small chips mechanically and boiled with water in earthenware or steel vessels to avoid any reaction with iron. The decoction is filtered and concentrated to the syrupy consistency. The syrup is poured into wooden frames and allowed to cool or cooled by refrigeration. It is then centrifuged to isolate the cake of kattha. The cake is moulded in desired sizes and dried in proper conditions. After drying, the cakes are packed and marketed. The mother liquor left after centrifugation is concentrated, and cooled to give cutch.

Characteristics of the cutch

Colour	Light brown to black
Odour	Nil
Taste	Bitter in the beginning and very astringent afterwards.
Size	2.5 to 5 cm.
Shape	Cubes or irregular fragments of broken cubes or brick-shaped pieces.
Fracture	Hard and brittle and the broken surface is dark brown with dull gloss and porous.
Appearance	Irregular mass or cubes depending upon the mode of solidification and the outer surface is rough and dull and rarely glossy.

Chemical constituents It contains 10% of acacatechin (acaciacatechin). In addition, the drug contains catechins, catechutannic acid and flavonoids like quercetin, gum and quercitrin. It is partially soluble in cold water and alcohol, and completely soluble in hot water.

Uses

1. The drug possesses cooling and digestive properties.
2. Kattha used as an astringent externally for boils, skin eruptions and ulcers.
3. It is used in cough and diarrhoea.
4. It is also used as a mouthwash.
5. Cutch is used for dyeing and colouring, water softening, reducing the viscosity of drill mud, protective agent for fishing nets and also in preparation of resins.
6. Black catechu is mainly used as an ingredient of "paan masala," and betel leaf.

Chemical tests

1. The drug gives pink or red colour with vanillin and hydrochloric acid, due to the presence of catechin.

2. It gives positive result to catechin test (matchstick).

3. The aqueous extract of black catechu gives brown colour, when treated with lime water, which turns to red precipitate on standing for some time.

4. A dilute solution of the drug gives green colour with ferric ammonium sulphate. On addition of sodium hydroxide, the green colour turns purple.

REVIEW QUESTIONS

State whether the following sentences are true or false

1. The biological source of bahera is *Terminalia arjuna*.

2. Appearance of green colour with ferric ammonium sulphate indicates the presence of black catechu.

3. Myrobalan is mainly used as laxative, astringent and tonic.

4. Leather industry uses tannins to prevent putrefaction of animal hides.

5. Green tea is a condensed tannin.

Choose the correct answer

6. Triphala is used as a _____
   a. sedative   b. cooling   c. laxative

7. Tannins are the derivatives of polyhydroxy _____
   a. citric acid   b. benzoic acid   c. flavonoid

8. The medicinally valued part of *Saraca indica* is the _____
   a. leaves   b. root bark   c. stem bark

9. The fruit of *Terminalia chebula* is referred to as _____
   a. nutmeg   b. clove   c. myrobalan

10. The taste of tannin is generally _____
    a. pungent and acid   b. sweet and chocolate   c. astringent and bitter

Answers				
1. False	2. True	3. True	4. True	5. True
6. c	7. b	8. c	9. c	10. c

Answer the following

1. What are tannins? Mention the types of tannins.
2. Write the herbal combinations of triphala.
3. Give the pharmaceutical importance of myrobalan, black catechu, and bahera.
4. Give an account of any three tannin-containing drugs you have studied. Add a note on their medicinal values.

15

ANALYTICAL PHARMACOGNOSY

DRUG ADULTERATION AND TYPES

Adulteration is a practice of substituting the original crude drug with other similar-looking inferior, useless and harmful substances.

Adulteration in simple terms is debasement of an article.

The main purpose of adulteration is a commercial one and it is done purposely to cheat and to earn more money. Adulteration is practised when the drug is scarce or when the price of a drug is high in the market. Adulteration is normally done deliberately, but it may also occur due to faulty collection, imperfect preparation, transportation and identification, and incorrect storage.

If adulteration occurs accidentally, this type of unintended adulteration is called undeliberate adulteration.

Undeliberate Adulteration

Undeliberate adulteration may occur due to the following reasons.

Faulty collection The collection of crude drugs during correct seasons or age is very important to avoid any adulteration. This is because, in some plant species, the presence of medicinally active principles reaches a maximum at a particular age, season or stage of development. Solanaceous leaves should be collected at the time of flowering during summer. If the leaves are collected during spring season, it results in adulteration by faulty collection of less valuable part of a genuine plant. For example, coriander seeds should be collected when fully grown and ripe. If it is collected in immature stage, the quality of the seeds becomes inferior; also, the seeds may get mixed up with other small flower buds to adulterate it.

Imperfect preparation Sometimes collection of unwanted parts of the crude drug may also end in adulteration. For example, along with flower buds of clove, flower stalks and leaves are also collected.

Confusion of common vernacular names Confusion of common vernacular names of plants also leads to adulteration of drugs. For example, the plants *Evolvulus alsinoides* and *Clitoria ternatea* are called by the same name "*Shankhupushpi*". In this case, it is very difficult to get the original specimen for the preparation of crude drugs.

Deliberate Adulteration

There are a number of adulterants, which are deliberately substituted by the adulterators for commercial purpose and this type of adulteration is called deliberate adulteration.

Adulteration involves different conditions such as deterioration, admixture, sophistication, substitution, inferiority and spoilage.

* Deterioration is an impairment in the quality of drug.
* Admixture is addition of one article to another due to ignorance or carelessness, or by accident.
* Sophistication is the intentional or deliberate type of adulteration.
* Substitution occurs when some totally different substance is added in place of the original crude drug.
* Inferiority refers to any substandard drug
* Spoilage is due to the attack of microorganisms.

Types of Adulterants

Generally drugs are adulterated with substandard commercial varieties, inferior drugs, artificially manufactured substance, exhausted drugs, synthetic chemicals, vegetative matter from the same plant, and non–plant materials. The different types of adulterants found in the market are given below.

Substitution with substandard commercial varieties In this type, the substituted products are very similar in appearance to the original drug. Hence, the adulterants used here may resemble the original crude drug by morphological, chemical and medicinal characters, but are substandard in nature, therefore, cheaper in cost. This type is a rather very common practice of adulteration. Some examples are as follows:

Drug	Materials substituted
Strychnos nux-vomica	*Strychnos potatorum,* *Strychnos nux-blanda*
Indian senna	Arabian senna, dog senna
Ginger	African, Japanese and Cochin ginger
Ashoka bark (*Saraca indica*)	*Trema orientalis, Polyalthia longifolia*

Substitution with superficially similar inferior drugs   In this type, inferior drugs are substituted and they may or may not have any chemical or medicinal value as that of the original drug. Due to their morphological resemblance to the original drug, they are marketed as adulterants. Some of the examples are as follows:

Drug	Substituted materials without medicinal values
Black pepper	Papaya seeds
Coffee powder	Tamarind seed powder
Black mustard seeds	Argemone seeds
Saffron	*Carthamus tinctorius*
Clove buds	Clove stalk and dried leaves
Coriander powder	Faecal matter of donkey
Cinnamon bark	*Cassia*
Beeswax	Japan wax
Senna leaves—*Cassia angustifolia*	Leaves of *Cassia auriculata*
Paddy grains	Small white stones

Substitution with artificially manufactured substances   In this type, substances that are artificially prepared to resemble original drugs are used as substitutes. This method is practised for much costlier drugs. Some examples are as follows:

Drug	Substituted materials without medicinal values
Coffee powder	Compressed chicory powder
Beeswax	Yellow-coloured paraffin wax

Substitution of exhausted drugs In this type, the same drug is admixed, but it is without any medicinal value, as they are already extracted out. Some examples are as follows:

Drug	Exhausted drugs
Tea leaves, ginger	Used tea leaves
Pepper seeds, liquorice	Used pepper seeds
Cumin seeds	Used cumin seeds
Fenugreek	Used fenugreek seeds
Fennel	Used fennel seeds

Used tea leaves collected from hotels are dried, sometimes dyed, if needed coloured, and mixed with fresh tea leaves again and packeted for marketing.

In the same way, cloves, fennel, caraway, pepper, cumin, coriander, etc. are used to obtain volatile oils by steam distillation. After extracting the volatile oil from them, they are dried, if needed, coloured and mixed with the original drugs for sale. Steam distillation does not change the external physical appearance of these seeds. Sometimes natural tastes of these seeds are also manipulated by adding other additives to adulterate it. For example, exhausted gentian made bitter with aloes, and artificial colouring of exhausted saffron is done before marketing.

Substitution with synthetic chemicals Sometimes, synthetic chemicals are added to enhance the natural character as in the case of addition of benzyl benzoate to balsam of Peru, citral to citrus oils like lemon oil and orange oil, etc. Similarly drugs like oleo gum-resin, myrrh, gum kino, balsam of Tolu, etc. are also adulterated with different products of similar nature as a matter of routine.

Substitution of vegetative matter from the same plant Sometimes, the other plants growing along with medicinal plant are mixed with the drug due to their resemblance in colour, odour and in some cases in similar constituents. Lower plants like moss, liverworts and epiphytes growing on bark portion are mixed with *Cascara* or *Cinchona* and Lichens (Kalpasi).

The stem portions are mixed along with leaf drugs like *Stramonium*, *Lobelia* and *senna*.

Substitution with harmful adulterants Several times, wastes from the market are collected and admixed with authentic drugs. This is particularly noticed for liquids or unorganized drugs. Some examples are as follows:

* In colophony, pieces of amber-coloured glass are mixed.
* In asafoetida, limestone is mixed.

- In opium, lead shot is mixed.
- In coconut oil, cocoa butter is adulterated with stearin or paraffin. These adulterations are very common practice in the market.
- Rodent faecal matter is added to cardamom seed and is a very harmful adulterant.

Substitution of powders with non-plant materials Adulteration with non-plant materials or powder to a particular original crude drug is very common nowadays in the market. For example,

- Nux-vomica powder is mixed with olive stone powder.
- Asafoetida is adulterated with gum arabic, red clay, gum resins, rosin, etc.
- Nutmeg is mixed with broken kernels moulded with clay and shaped pieces of wood.

The powdered forms are also frequently found to be adulterated.

Original powdered drug	Adulterated powder
Ipecacuanha	Dextrin
Liquorice, gelatin	Powdered olive stones
Colocynth, ginger	Exhausted ginger powder
Chilli powder	Red sandalwood, brick powder
Coriander powder	Powdered faecal matter of donkey

METHODS OF DRUG EVALUATION

Evaluation of drugs deals with the correct identification and determination of quality and purity of the crude drugs.

Identity denotes the correct identification of the origin or biological source of the drug.

Quality denotes the quantity of the active constituents present in the drug.

Purity denotes the amount of foreign materials present in the drug.

The evaluation of a crude drug is necessary to find out the following:

i. biochemical variation in the drugs.
ii. deterioration due to treatment and storage.
iii. substitution and adulteration, as a result of carelessness, ignorance or fraud.

Before any botanical drug is used, it is very essential that it should be properly identified. Initially, crude drugs were identified by comparison only with the standard description of the plant or animal available. Due to advancement in the chemical knowledge of crude drugs, at present, evaluation also includes method of estimating the active constituent present in the crude drug, in addition to its morphological and microscopic analysis.

Therefore, evaluation of crude drugs is done on the basis of their morphological or organoleptical, microscopical, physical, chemical and biological properties. The different techniques involved in standardization of crude drugs are discussed in the following section.

MORPHOLOGICAL (OR) ORGANOLEPTIC METHODS

Organoleptic evaluation is done by means of our organs of sense, which includes external morphology and other sensory characters like odour, colour, taste, texture and nature of fracture.

Study of Morphological Characters

Proper authentication of a drug depends almost entirely on morphological characters. Most of the crude drugs are derived from a part of a plant. The morphological or macroscopical study of the respective part is done by observing it with the naked eye or with the aid of magnifying lens.

The plant parts that are used as crude drugs are the stem, bark, underground structures, leaves, flowers, fruits, seeds and herbs. For each group, a particular systematic examination should be carried out.

i. Stem

The characteristics used for the correct identification of stem are as follows: Dimensions, form, colour, position (erect or prostate), consistency (woody or herbaceous), presence or absence of structures like glands, thorns, spines, prickles, hairs, etc.

ii. Barks

Barks are the tissues in a woody stem present outside the vascular cambium or in roots. Barks may be broken or complete around the stem or roots. Barks are collected by stripping from the trunk or branches of the trees and are thus obtained as relatively narrow strips, e.g. *Cinchona, Quillaia*, cinnamon, ashoka, kurchi bark, etc.

The external features of the bark include colour, smoothness, rugged or scaly, etc. The fracture type includes short, fibrous, glandular, etc. On drying, the bark exhibits different shapes such as flat, curved, recurved, channelled, quill, double quill and compound quill as shown in Figure 15.1.

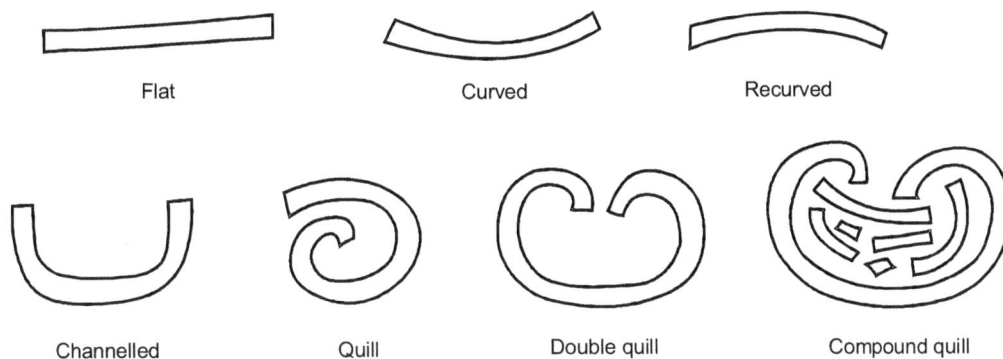

Figure 15.1   Different shapes of barks of drugs

iii. Underground Structures

Underground structures include the stems, such as bulbs, tubercles and rhizomes, which have buds, scale leaves and scars (Figure 15.2). They have a central pith surrounded by a ring of xylem tissue and the drugs include ginger, turmeric, dioscorea, etc. The root includes the main taproot and adventitious roots and also the tubercles, e.g., roots of *Podophyllum*, liquorice, jatamansi, *Rauwolfia* and *Hemidesmus*. They are often swollen due to storage of carbohydrates and other chemicals, and exhibit a wide variety of forms. To facilitate drying in preparation for the market, these are sliced or cut into small pieces. Some drugs are scraped to remove the dark cork, to give a light-coloured product.

Underground structures are identified by the following characters:

Direction of growth	Horizontal, oblique, vertical
Shape	Cylindrical, conical, fusiform, napiform, straight, tortuous
Fracture	Brittle, hard, horny, mealy, splintery.

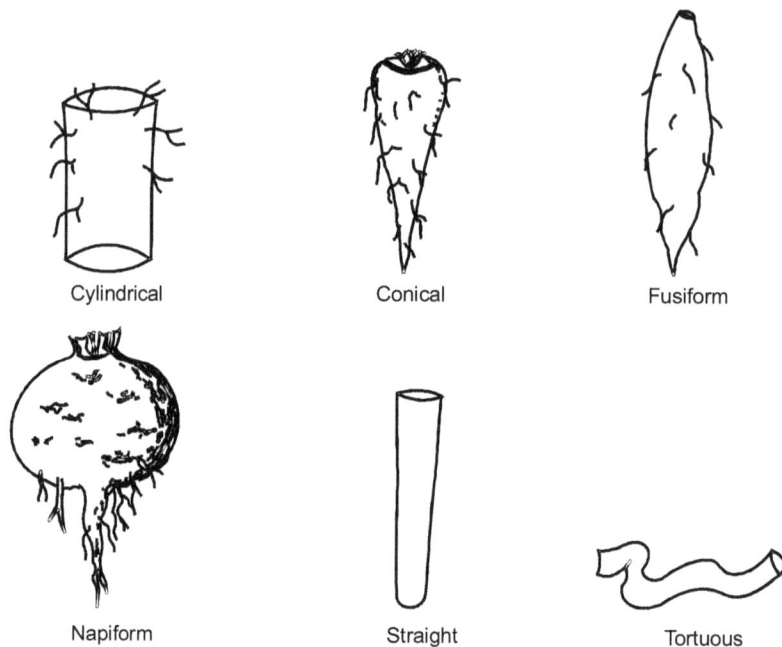

Figure 15.2 Different shapes of underground drugs

iv. Leaves

These are photosynthetic organs arising from a node on a stem. A leaf may be simple (one leaf at a node), e.g., *Hibiscus* (Figure 15.3a), and Vasaka or may consist of many leaflets, in which case, it is termed as a compound leaf (paripinnate or imparipinnate). e.g., *Murraya, Azadirachta*, etc. (Figure 15.3b).

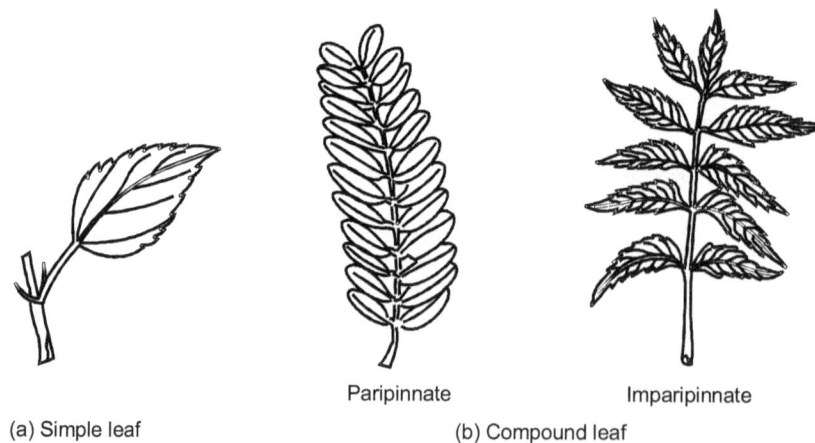

Figure 15.3 Types of leaves

The following are the morphological characters used in the identification of leaf drugs:

Shape   Acicular, linear, oblong, ovate, lanceolate, sagittate, cordate, hastate, reniform, spathulate, cuneate, elliptical, rounded, obovate, obcordate.

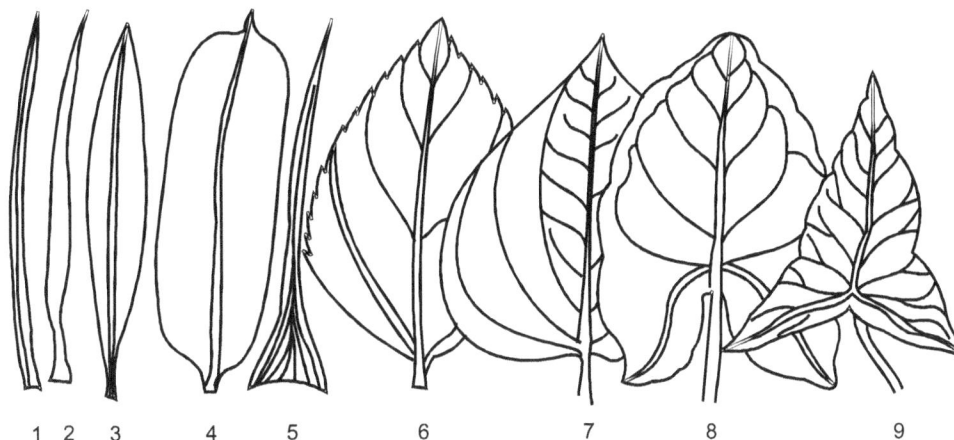

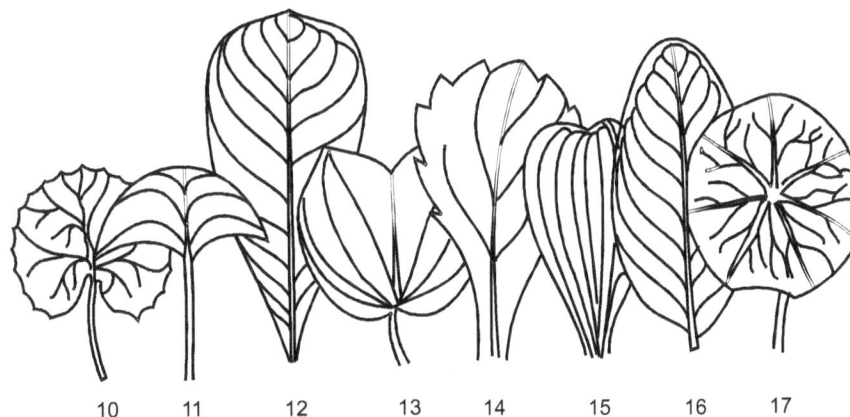

1.	Acicular	2.	Linear	3.	Lanceolate	4.	Oblong
5.	Subulate	6.	Ovate	7.	Cordate	8.	Sagittate
9.	Hastate	10.	Reniform	11.	Lunate	12.	Obovate
13.	Obcordate	14.	Spathulate	15.	Cuneate	16.	Elliptical
17.	Rotund						

Figure 15.4 Different shapes of leafy drugs

Margin   Entire, serrate, serrulate, dentate, crenate, spiny, sinuous, etc.

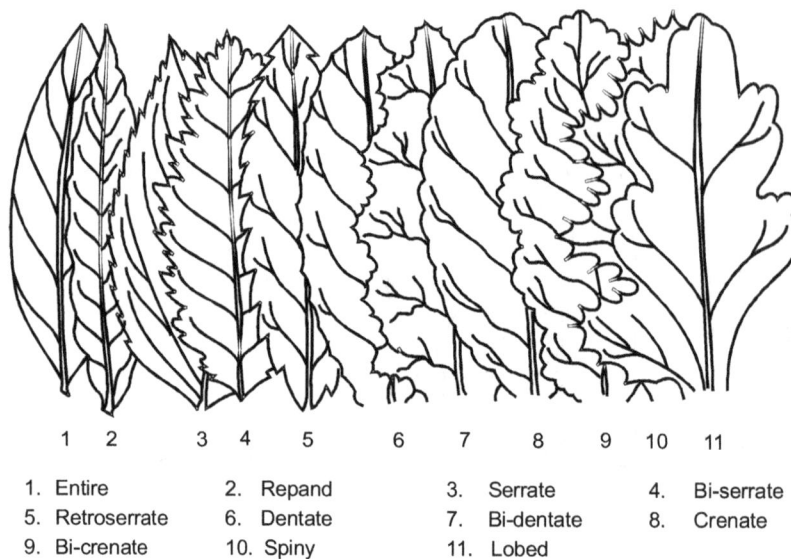

1. Entire	2. Repand	3. Serrate	4. Bi-serrate
5. Retroserrate	6. Dentate	7. Bi-dentate	8. Crenate
9. Bi-crenate	10. Spiny	11. Lobed	

Figure 15.5   Types of margins of leafy drugs

Apex   Acute, acuminate, obtuse, cuspidate, mucronate, truncate, retuse, emarginate, recurved, obtuse.

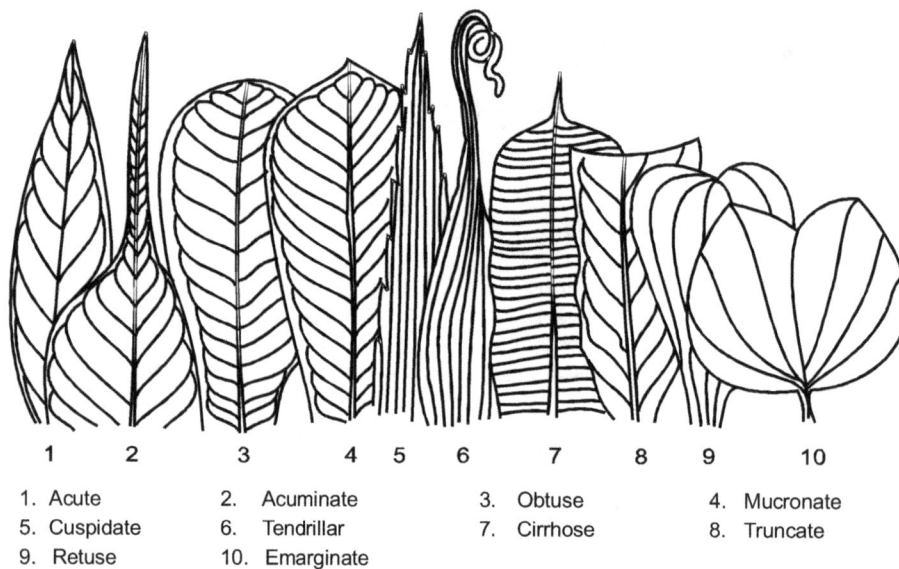

1. Acute	2. Acuminate	3. Obtuse	4. Mucronate
5. Cuspidate	6. Tendrillar	7. Cirrhose	8. Truncate
9. Retuse	10. Emarginate		

Figure 15.6   Apex types of leafy drugs

Base   Asymmetric, cordate, reniform, sagittate, hastate, decurrent.

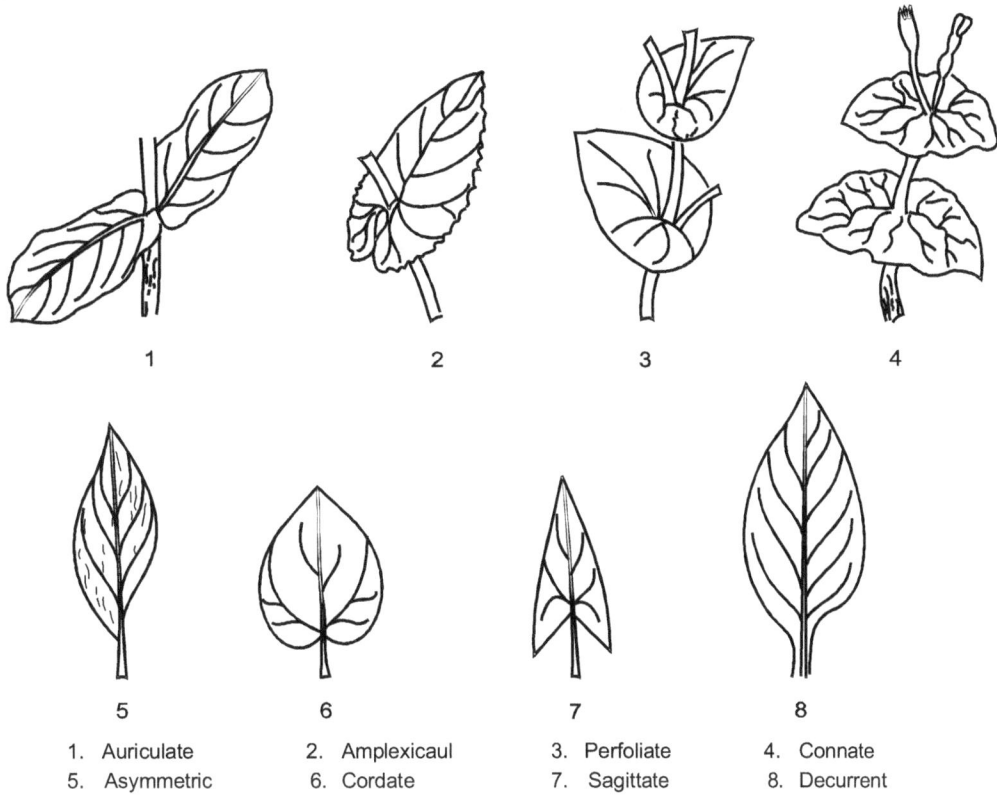

1. Auriculate   2. Amplexicaul   3. Perfoliate   4. Connate
5. Asymmetric   6. Cordate   7. Sagittate   8. Decurrent

Figure 15.7   Types of bases of leaves

Venation   Reticulate (pinnate or palmate), parallel (divergent or convergent).

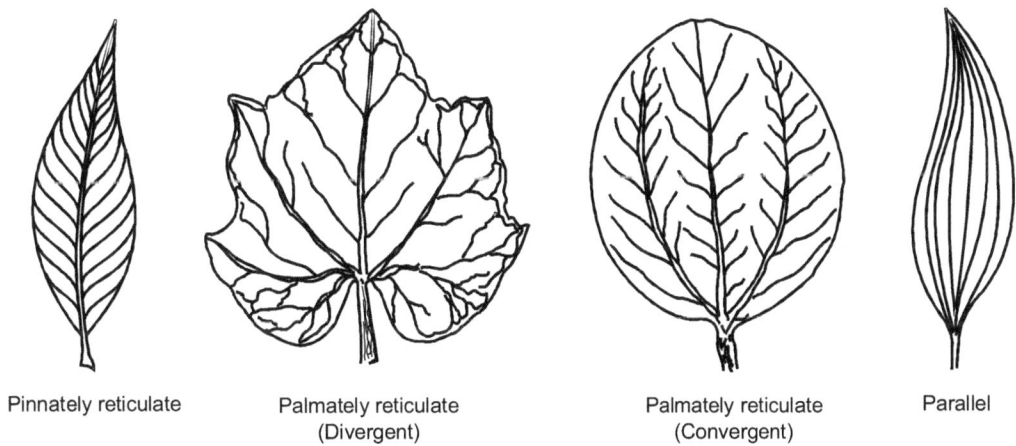

Pinnately reticulate   Palmately reticulate   Palmately reticulate   Parallel
                            (Divergent)   (Convergent)

Figure 15.8   Types of leaf venation

v. Inflorescence

These are the reproductive organs of a plant and consist of four basic parts, the calyx, corolla, androecium and gynoecium. There is a great diversity in the form of flowers and the way in which they are borne on the plant. Many plants have a solitary flower or may consist of many individual flowers forming an inflorescence.

The term inflorescence refers to a cluster of flowers together with their axes, and with bracts and bracteoles. An inflorescence may be simple or compound, it may be small or large. The number of flowers of an inflorescence also varies from few to numerous. The position of the inflorescence may be "terminal", i.e., if it terminates the main shoots or axillary shoots (e.g., mango, poppy) or "axillary", i.e., if it is formed in the axil of foliage leaves as in *Hibiscus*.

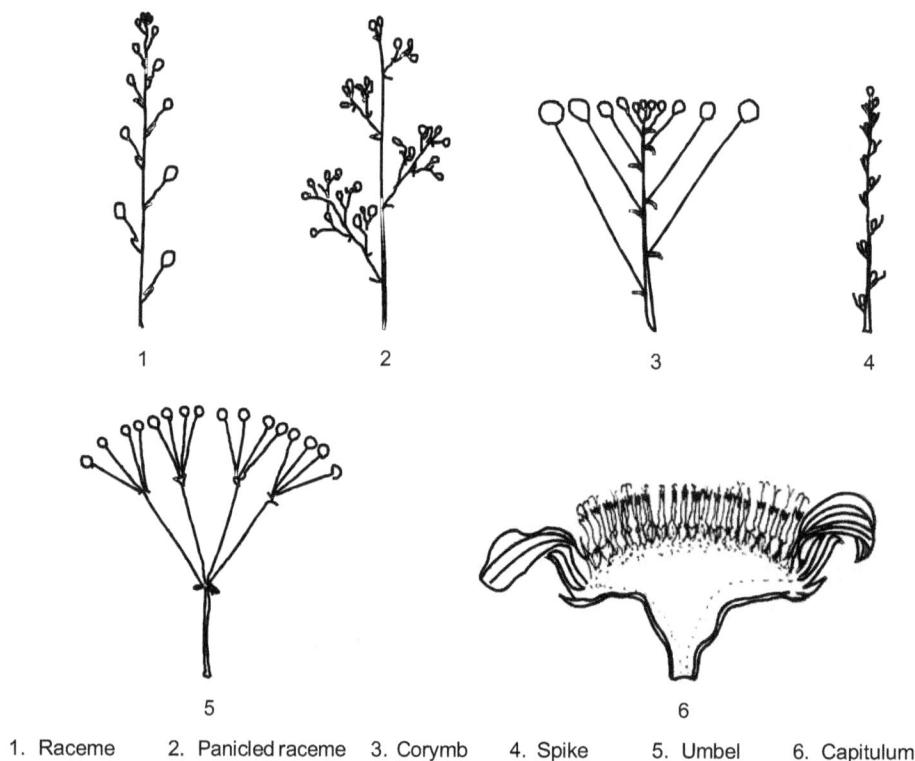

1. Raceme   2. Panicled raceme   3. Corymb   4. Spike   5. Umbel   6. Capitulum

Figure 15.9   Types of racemose inflorescence

Types of inflorescence   Based on the nature of growth and arrangment of flowers, the inflorescence may be classified into:

1. Racemose type
2. Cymose type

1. *Racemose type*   It is otherwise called indeterminate, i.e., the growth of the apex of the inflorescence axis continues to form lateral flowers. Hence, the youngest flower is situated apically and oldest basally, i.e., in acropetal succession. When the flowers form a cluster, the youngest is towards the centre, i.e., the development is "centripetal".

Racemose inflorescence may be classified into the following types:

i. Raceme   It consists of an elongated axis, the peduncle, with pedicellate or stalked flowers, oldest at the base and younger flowers borne apically (e.g., *Crotalaria*).

In some plants, e.g., Dracaena, the main peduncle gives rise to secondary penducles in an acropetal order, to produce flowers and this is called panicled raceme.

ii. *Spike*   It is similar to raceme but the flowers are sessile or without stalk (e.g., *Achyranthes aspera*). Very small spikes of secondary order are the spiklets which may be grouped as units of inflorescence (e.g., paddy) to form various arrangements like panicle, racemes or spikes (e.g., *Triticum aestivum*).

iii. *Catkin*   It is a spike or spikelike inflorescence which is usually pendulous, and is generally composed of apetalous (without petals) small unisexual flowers subtended by a scaly bract (e.g., *Acalypha hispida*).

iv. *Spadix*   It is a spike with thick, fleshy axis bearing usually minute and unisexual flowers. The entire inflorescence is generally covered over by a large enveloping, green, white or brightly coloured bract called a "spathe" (e.g., *Colocasia*). If the spadix is branched and is covered by a stiff boat-shaped spathe as in the family Palmae (e.g., *Coconut*), it is called compound spadix.

v. *Corymb*   It is a modification of raceme. It has a very short/central peduncle and the lower pedicels of the flowers are much elongated, thus forming a flat-topped inflorescence. In mature condition all the buds, flowers and fruits come to the same level. The opening of flower starts from periphery towards the centre (e.g., *Caesalpinia* spp.).

vi. *Umbel*   It is a modified raceme, whose peduncle is very much reduced and the flowers have stalks of nearly equal length, so they seem to arise from the summit of the peduncle reaching the same level and forming a sort of an umbrella. The blooming sequence is from outside towards centre. An umbel may either be simple or compound. It is usually subtended by an involucre of green bracts (e.g., *Coriandrum sativum*).

vii. *Capitulum or head*   It is a modification of racemose inflorescence, in which the peduncle is reduced and modifed into a "disc" or "receptacle" which may assume various shapes—flat, concave, convex, columnar, etc. Numerous sessile flowers arise from the receptacle. The opening of flower is from periphery towards centre. A "head" may either be solitary (e.g., sunflower) or it may be aggregated into arrangements forming panicles, racemes, corymbs, cymes (e.g., *Vernonia*). A head is usually subtended by an involucre of green bracts as in sunflower. The capitulum may be heterogamous type if it contains both ray florets and disc florets (e.g., sunflower) or homogamous head if it contains either ray florets (e.g., *Chrysanthemum*) or disc florets (e.g., *Vernonia*).

2. *Cymose type*   It is otherwise called definite or determinate. In cymose inflorescence, the apical growth of the primary axis (main axis) and of the lower axes are terminated by the formation of flowers. The oldest flower is produced at the top and the youngest flower is formed basally, i.e., the progression of blooming is basipetal (e.g., jasmine).

Cymose inflorescence may be classified into the following types:

i. *Solitary cyme*   It is a cyme of single flower which is produced by the apex of the axillary shoot that is formed at the axil of a leaf. The apical growth of the primary axis of the inflorescence terminates its growth by the formation of a single flower, (e.g., *Hibiscus* and *Thespesia*).

ii. *Monochasial cyme*   It is a type of cymose inflorescence in which, the peduncle (main axis) bears a terminal flower and below it a lateral branch which too terminates into a single flower and so on. The terminal flower is the oldest. It is further differentiated into the following two types:

a. Helicoid cyme   When the successive lateral axes develop only towards one side of the main axis, thus forming a spiral coil unilaterally, the cyme is called helicoid cyme (e.g., *Hamelia patens*).

b. Scorpioid cyme   When the successive lateral axes develop alternatively on both sides of the main axis (peduncle), the cyme is called scorpioid cyme. The whole inflorescence coils downwards (e.g., *Heliotropium*).

iii. Dichasial cyme   It is a cyme where the peduncle bearing the terminal flower develops below it a pair of lateral branches, which too, terminates into a flower, the central flower being the oldest (e.g., *Clerodendron*).

iv. Polychasial cyme It is a cyme where more than two lateral axes terminating into a flower, are formed below the main axis which bears a terminal and the oldest flower of the inflorescence (e.g., *Calotropis*).

Special types of inflorescence   Special types of inflorescence usally have condensed axes due to which the flowers become densely crowded on it. The special types of inflorescence may be of the following types:

a. *Hypanthodium*   It is the characteristic condensed cymose inflorescence of *Ficus* spp. It has a hollow flask-shaped fleshy axis developed by the involution of a receptacle. It has a small opening, the ostiole, at the apex and numerous small and greatly reduced flowers which are closely arranged all along its inner surface. The female flowers that are produed towards the centre of the receptacle mature first; the male flowers are crowded near the periphery. At the base of the ficus fruit are present three scaly structures. One of these is a bract in the axil of which is borne the main shoot and the other two are the bracteoles.

Thus hypanthodium is a greatly condensed cymose inflorescence in which pedicels and peduncles have all coalesced to become fleshy, pulpy and edible fruit.

b. *Verticillaster*   It is a condensed cyme characteristic of the family Lamiaceae (Labiatae). It consists of whorls of flowers around the axis (the main stem). These whorls of flowers become congested and continuous around the axis as in *Leucas*. Usually the flowers are sessile with reduced axis. The whorls at each node are placed opposite to each other in the axis of opposite bracts or leaves. Each whorl is a dichasium of scorpioid cymes, i.e., a dichasial cyme to start with, passes on either sides into monochasial type by suppression of one branch alternately at each branching.

c. *Cyathium*   It is a cymose inflorescence reduced to such an extent as to stimulate a single flower. It is the characteristic inflorescence of the genus *Euphorbia*, the cyathia may either be solitary or are arranged in cymose manner.

An involucre of usually five bracts forms a cup-shaped structure, the margins of which bear cresent-shaped nectary glands. A naked female flower is present in the centre of the inflorescence. The gynoecium is borne on a long stalk. The gynoecium terminates with tricarpellary ovary with trifid stigma. In the axil of each bract is present a group of male flowers arranged in scorpioid cymes. Each male flower is represented by a single jointed stamen with bicelled anther at the top of the filament. Hairs are generally present at the joint between the pedicel and filament. The oldest male flower matures first towards the centre.

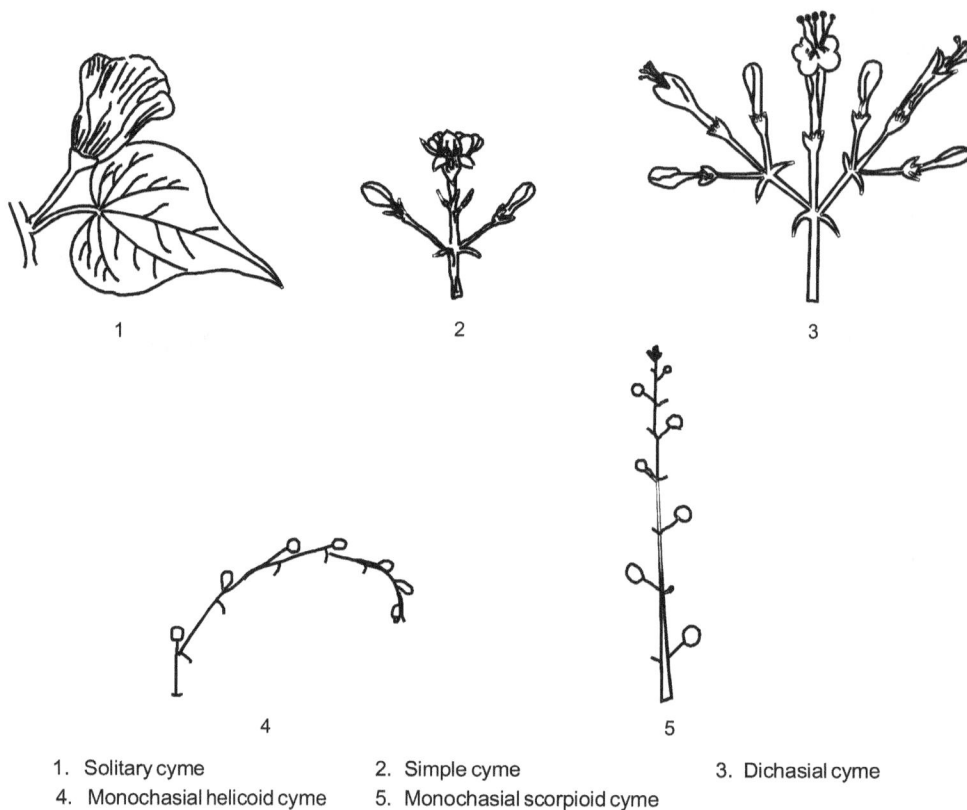

1. Solitary cyme
4. Monochasial helicoid cyme
2. Simple cyme
5. Monochasial scorpioid cyme
3. Dichasial cyme

Figure 15.10 Types of cymose inflorescence

Special types   Hypanthodium, verticillaster and cyathium.

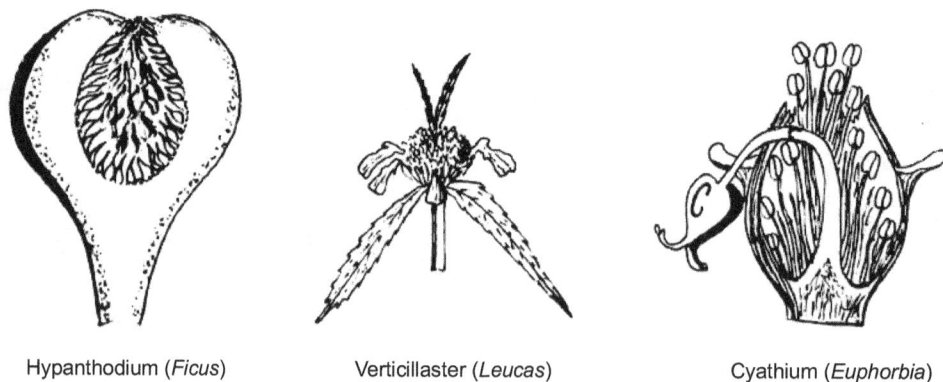

Hypanthodium (*Ficus*)   Verticillaster (*Leucas*)   Cyathium (*Euphorbia*)

Figure 15.11 Special types of inflorescence

Flower The solitary flower may be bisexual, unisexual, sessile, sub-sessile, pedicellate, regular, irregular, actinomorphic, zygomorphic, trimerous, tetramerous, pentamerous, cyclic, and hemicyclic.

Arrangement of receptacle Hypogynous, perigynous, and epigynous.

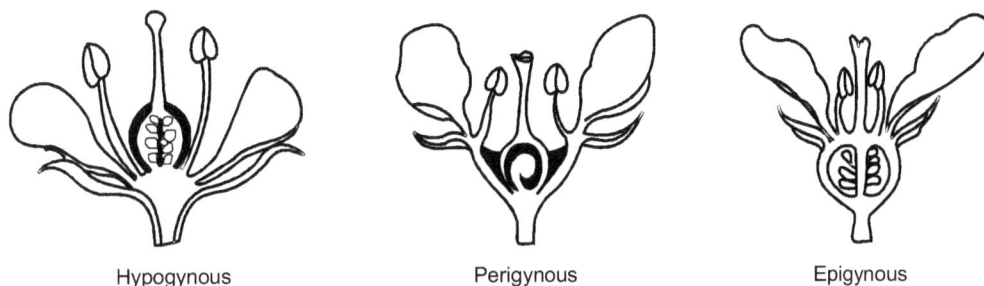

Hypogynous   Perigynous   Epigynous

Figure 15.12   Types of receptacles

Calyx and corolla Number of sepals—polysepalous, gamosepalous, aestivation; number of petals—polypetalous, gamopetalous, aestivation, form, colour and special features.

Androecium Number of stamens, polyandrous, or united, monoadelphous or diadelphous, syngenesious, didynamous, epipetalous, epiphyllous, gynandrous, basifixed, adnate, dorsifixed, versatile, anthers monothecous or dithecous, line of dehiscence apical, transverse, or longitudinal.

Gynoecium Number of carpels; position—superior, inferior or half-inferior; union—apocarpous, or syncarpous; placentation—axile, marginal free-central, parietal, or basal; ovule—orthotropous, anatropous, amphitropous, or campylotropous; number of locules and position of ovules, and type of stigma.

vi. Fruits

Fruits are the organs of the plant that arise from the ovary. They contain seeds, the mature ovules and the pericarp, the wall of the fruit. The shapes may be globular, oblong or ellipsoidal, etc. The types of fruits also vary with the species.

Types of fruits

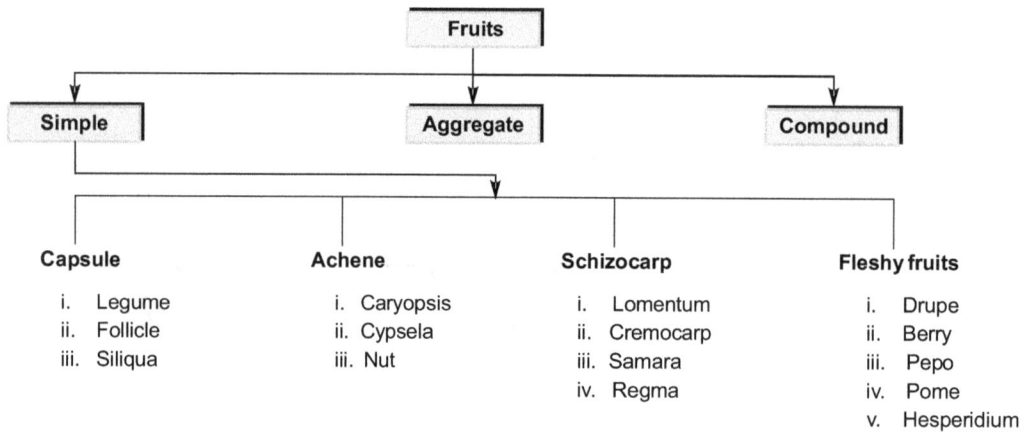

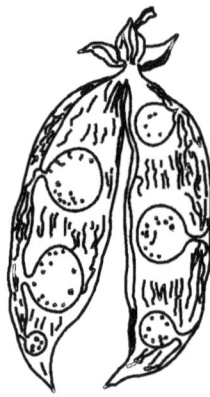

```
                    ┌──────────┐
                    │  Fruits  │
                    └──────────┘
         ┌──────────────┼──────────────┐
    ┌─────────┐   ┌───────────┐   ┌───────────┐
    │ Simple  │   │ Aggregate │   │ Compound  │
    └─────────┘   └───────────┘   └───────────┘
```

Capsule	Achene	Schizocarp	Fleshy fruits
i. Legume	i. Caryopsis	i. Lomentum	i. Drupe
ii. Follicle	ii. Cypsela	ii. Cremocarp	ii. Berry
iii. Siliqua	iii. Nut	iii. Samara	iii. Pepo
		iv. Regma	iv. Pome
			v. Hesperidium

Legume (Pea)   Follicle (Calotropis)   Siliqua (Mustard)

(a) Capsule

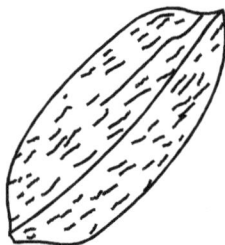

Caryopsis (Paddy)   Cypsela (Sunflower)   Nut (Cashew nut)

(b) Achene

Figure 15.13 Types of Fruits (*Continues*)

| Lomentum (Acacia) | Cremocarp (Coriander) | Samara (Pterocarpus) | Regma (Castor) |

(c) Schizocarp

| Drupe (Mango) | Berry (Tomato) | Pepo (Cucumber) |

| Pome (Apple) | Hesperidium (Orange) |

(d) Fleshy fruits

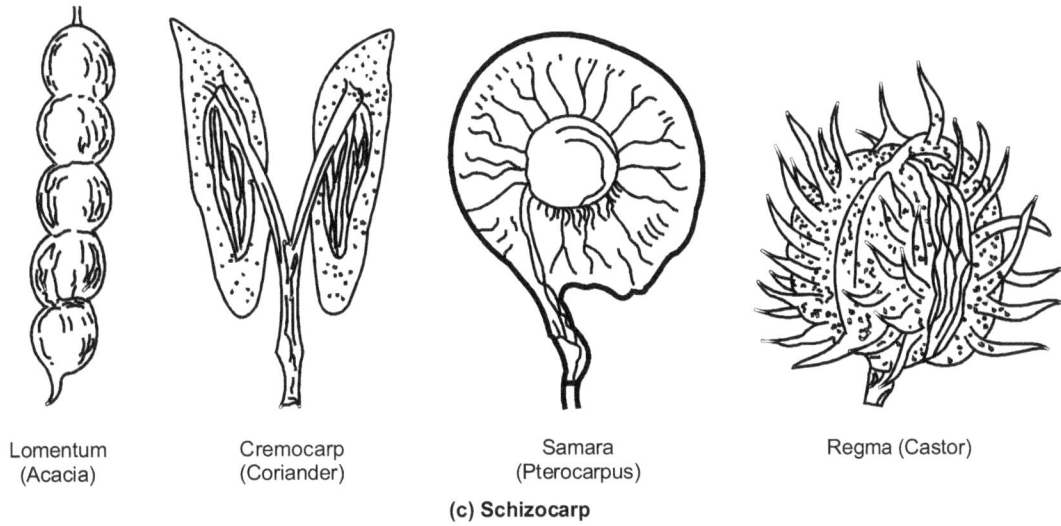

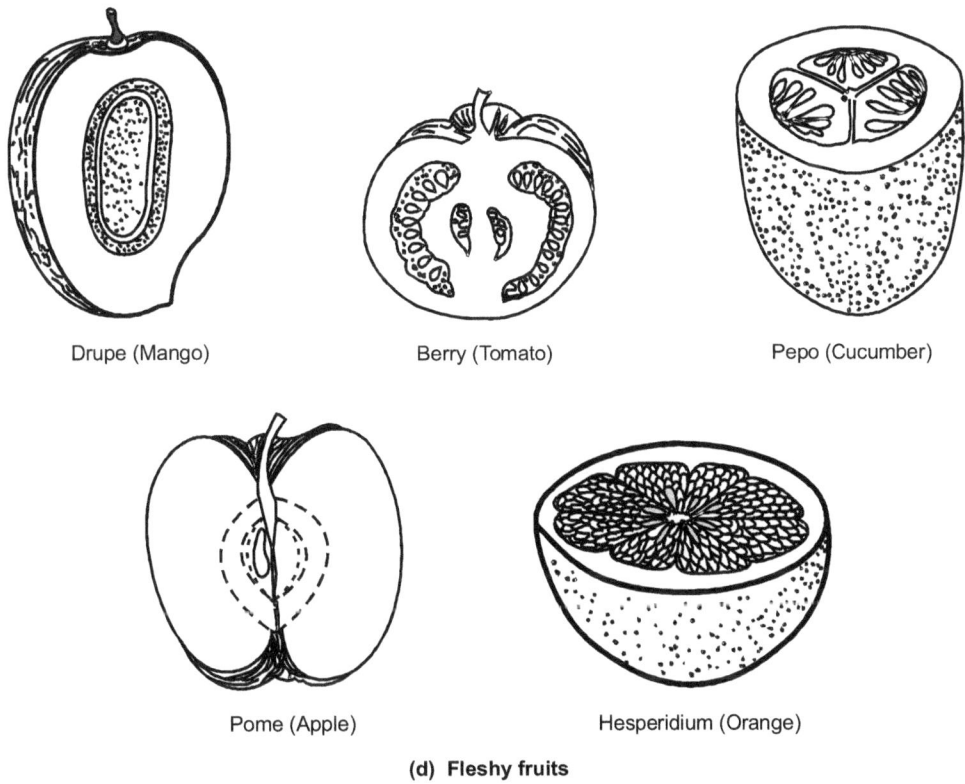

Figure 15.13   Types of fruits

vii. Seeds

Seeds are the propagules developed from the ovules in carpels of the flowers and characterized by the hilum, micropyle and sometimes raphe. The seed drugs are linseed, *Azadirachta*, nux-vomica, *Psoralea*, etc. The shape of seeds include globular, oblong, oval, reniform, planoconvex, spherical, angularly ovate, etc.

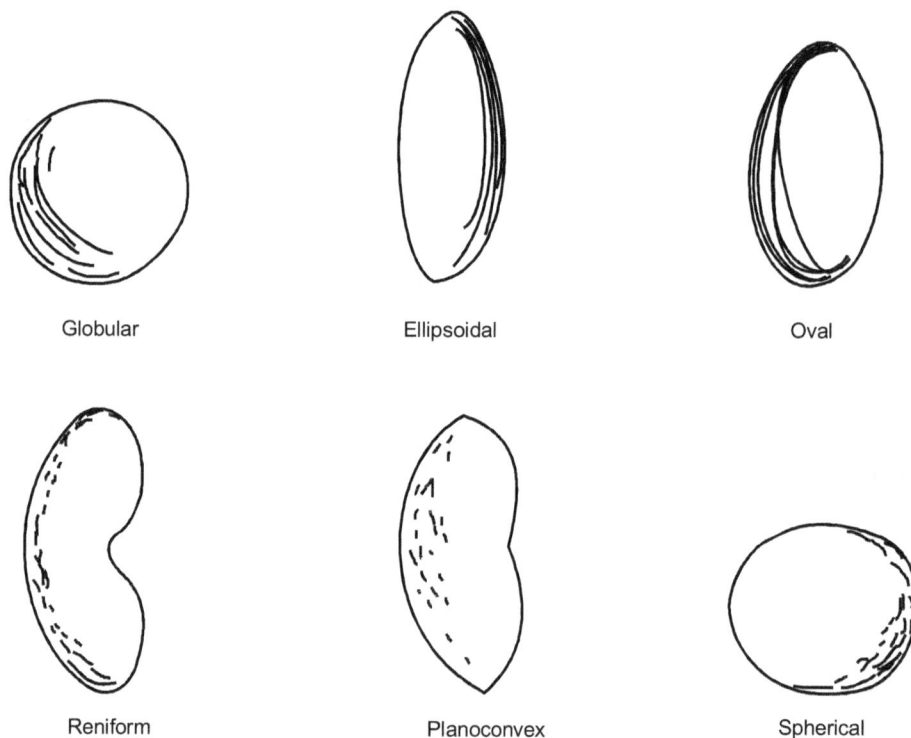

Globular   Ellipsoidal   Oval

Reniform   Planoconvex   Spherical

Figure 15.14 Shapes of seeds

viii. Herbs

The whole plant is sometimes used as a drug, e.g., brahmi, chirata, kalmegh, *Phyllanthus*, pudina, tulsi, shankupushpi, etc. A full description of a herb must include a systematic description of the stem, leaves, flowers and fruit.

Study of Sensory Characters

The sensory characters include making observation on the colour, smell, taste and texture of the drug under evaluation. This method is especially applicable to drugs containing volatile oils or pungent principles (e.g., *Capsicum*, garlic, onion), and to the detection of the effects of inadequate drying or damp storage. These characters are felt by our sensory organs.

Colour

The external colour varies from white to yellowish grey, brown, orange or brownish black. The colour of some drugs change if they are dried in sunlight in place of shade. Colour of drugs are standardized and determined by the Inter–Society Colour Council National Bureau of Standard Method. For example, reserpine is described as a white or pale buff to slightly yellowish colour, odourless and crystalline powder.

Odour

Some drugs have a characteristic smell which helps in their easy identification, e.g., cardamom, cinnamon, eucalyptus and clove, which contain volatile oils. Also clove and exhausted clove can be distinguished by odour. The smell of a drug may be distinct or indistinct. The terms used to define odour are aromatic, balsamic, spicy, alliaceous (garliclike), camphoraceous (camphorlike), terebinthinate (turpentinelike) and others. Leaves of different species of *Mentha* can be distinguished by smell. Similarly, deterioration of some drugs like ergot or cantharides due to improper storage can emit undesirable ammoniacal odour.

Taste

Drugs can be evaluated by taste also. The taste may be sour (acidic), salty (saline), sweet (saccharine), bitter, alkaline and metallic. The various tastes due to characteristic odours can be grouped as aromatic, spicy, balsamic, alliaceous, camphoraceous and terebinthinate. For example, *Glycyrrhiza* and honey are sweet in taste. Gentian and chirata are bitter, ginger and capsicum have a pungent taste. Fixed oils have bland taste, calcium oxide is astringent, *Podophyllum*, *Ipomoea* and Jalap are acrid to taste.

Texture

Sometimes drugs can be examined by their consistency, texture and nature of fracture. The texture of the plant part used as drug can be observed by breaking a piece of the drug under examination. For example, colocynth, can be compressed easily since its parenchyma is loosely present. The fracture would depend upon the nature of the tissues and the mode of drying the drug. For example, in *Glycyrrhiza* the fracture is hard and fibrous due to the presence of fibrous and woody tissues, whereas in nux-vomica and Aconite, horny fracture is present due to gelatinization of starch which breaks at high temperature on drying.

MICROSCOPICAL METHODS

These methods allow more detailed examination of a drug and can be used to identify the organized drugs by their known histological characters. Schleiden (1847) used the microscope for the examination of drugs.

Prior to examination through a microscope, the material must be suitably prepared. This can be done by powdering, cutting thin sections of the drug or preparing a macerate. After this, mounting and staining techniques can be followed to differentiate particular tissues from adulterants.

In the identification of whole crude drugs, it is often necessary to examine the anatomical structure and this involves the microscopical examination of sections of the material as appropriate for the particular morphological group. In some drugs, transverse and longitudinal sections are taken and the arrangements of tissues are observed. The arrangement of tissues is best ascertained by observation of transverse sections. The basic layout of tissues in stems, roots, leaves, etc. is fairly constant, but there is a considerable variation in the amount or extent of different regions, in the extent of lignifications, in the detailed appearance of the cells, etc. between different plants while in some cases, definite anomalous arrangement of the tissues occur. The cells, which are most useful for the purpose of identification are those least affected by drying, e.g., fibres, sclereids, tracheids, vessels and cork cells. For example, lignified trichomes in nux-vomica, warty trichomes of senna, wavy medullary rays of cascara bark, glandular trichomes of mint, etc.

For effective results, various reagents or stains can be used to distinguish cellular structure. Microscopic evaluation also covers the study of the constituents by application of chemical methods to small quantities of the drug in powdered form or to histological sections of the drug (micrometry or chemomicroscopy). A drop of phloroglucinol and concentrated hydrochloric acid give red stain with lignin. Mucilage is stained pink with rhuthenium red and also when treated with corallin soda and few drops of sodium carbonate solution. Cellulose swells and dissolves in cuoxam, while N 50 iodine solution stains starch and hemicelluloses blue. Microscopic linear measurements and quantitative microscopy are also covered under this technique of evaluation.

In the identification of powdered crude drugs, a microscopical examination is essential and a detailed study of the fragments present must be made. In powdered drugs, the fragments may consist of parts of cells or groups of cells; so it is essential to recognize the cells and tissues under these conditions. Cell contents will be either in free state or present within the cells. The cells of significant diagnostic value are those which are least affected by communication, for example, thick-walled cells, and cells with lignified walls. During microscopic examination, careful drawings should be made of the important features present.

If it is the establishment of authenticity that is required, (to determine whether the material is correctly labelled or not) the illustrated report would be compared with a description of the drug in question from any reference book.

Powdered cloves do not contain sclereids or calcium oxalate crystals, but both of them are present in powdered clove stalks. Powdered clove fruits show presence of starch, while it is absent in cloves. Presence of non-lignified vessels in powders of rhubarb and ginger indicate adulteration. The presence or absence of crystals of aloin indicates different varieties of aloes.

Quantitative Microscopy

Walles (1916) published the use of quantitative microscopy in pharmacognosy, which helps to determine the percentage of extraneous organic matter in powdered drugs, in addition to the measurements of the size of cells, tissues and cell contents.

The other important histological aspect is quantitative microscopy and linear measurements. The various parameters studied here are stomatal number and index, palisade ratio, vein-islet number, size of starch grains, length of fibres, etc. Senna varieties are distinguished by differing stomatal number and palisade ratio. The diameter of starch grains in *Cinnamomum* and *Cassia* is 10 microns and hence useful for detecting adulterants. The number of sclerenchymatous cells per square mm of cardamom is useful for detecting different varieties of cardamom seeds.

Leaf constants   The average number of palisade cells beneath each epidermal cell is called palisade ratio. It can be determined with powdered drugs with the help of camera lucida. Vein-islet number is defined as the number of vein-islets per sq. mm of the leaf surface midway between the midrib and the margin. Levin in 1929 determined the vein-islet numbers of several dicot leaves. Vein-termination number is defined as the number of veinlet terminations per sq. mm of the leaf surface midway between midrib and margin. Stomatal number is the average number of stomata per sq mm of the epidermis of the leaf. Stomatal index (SI) is the percentage proportion of the number of stomata forms to the total number of ordinary epidermal cells of a leaf. It is calculated by using the following formula:

$$SI = \frac{S}{E+S}$$

where, SI = Stomatal index

S = Number of stomata per unit area

E = Number of ordinary epidermal cells in the same unit area

The technique of determination of leaf constants can be used for microscopic evaluation of several leaf drugs such as senna, datura, digitalis, buchu, coca, belladonna, etc.

Water pores Water pores and stomata resemble each other to some extent in structure and functions. Water pores are immovable unlike stomata and they are present on the teeth of the margin.

Trichomes There are other important diagnostic characters helpful in the identification of drugs and detection of adulterants. Trichomes are the tubular elongated or glandular outgrowths of the epidermal cells. Trichomes are also called plant hairs. Trichomes consist of two parts viz., root (in the epidermis) and body (outside the epidermis). Trichomes are present in most of the parts of the plant such as leaves (senna and digitalis), seeds (nux-vomica and *strophanthus*), fruits (*Helicteris isora* and Lady's finger, etc.) and are absent in roots. Trichomes are as such functionless, but sometimes, perform secretory function and in xerophytes, hairs act as a screen against light to cut short the light intensity. The trichomes excrete, at times, volatile oil as in case of peppermint. They are absent in coca and hemlock, and are rarely present in buchu, and henna. Depending upon the structure and the number of cells present in trichomes, they are classified into the following types:

i. Covering tichomes or non-glandular trichomes or clothing trichomes

ii. Glandular trichomes

iii. Hydathodes and emergences

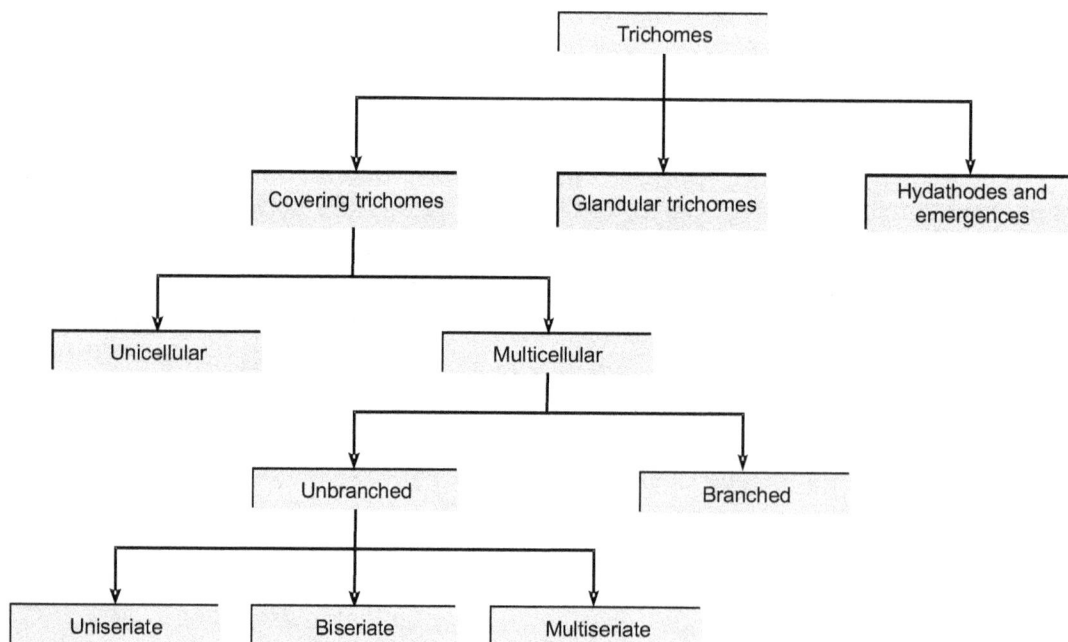

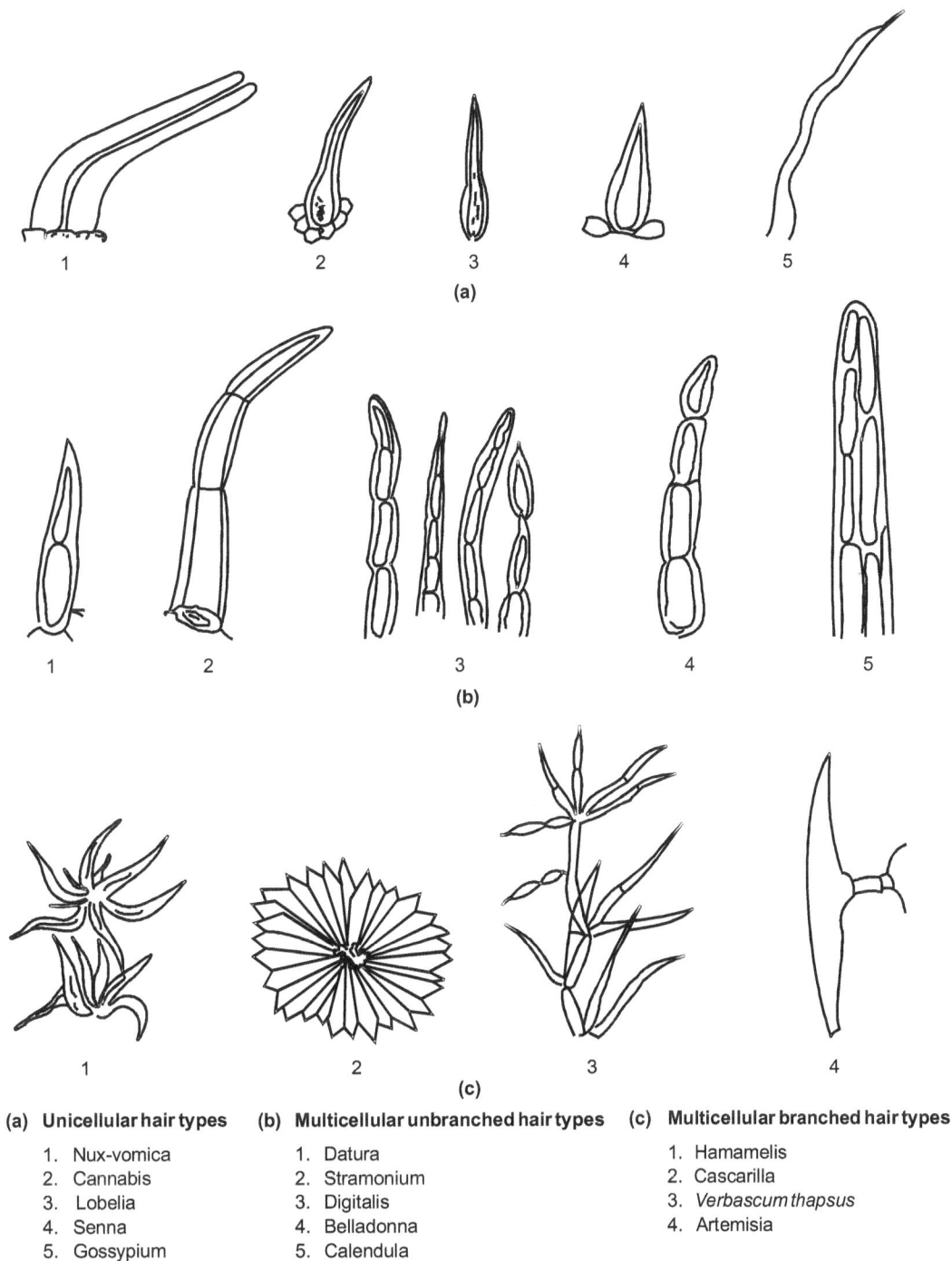

(a) Unicellular hair types	(b) Multicellular unbranched hair types	(c) Multicellular branched hair types
1. Nux-vomica	1. Datura	1. Hamamelis
2. Cannabis	2. Stramonium	2. Cascarilla
3. Lobelia	3. Digitalis	3. *Verbascum thapsus*
4. Senna	4. Belladonna	4. Artemisia
5. Gossypium	5. Calendula	

Figure 15.15   Types of covering hairs

1. Covering trichomes
    a. *Unicellular trichomes*
        * Lignified trichomes, e.g., nux-vomica, *Strophanthus*
        * Short, sharply, pointed, curved, e.g., *Cannabis*
        * Large, conical, strongly shrunken, e.g., *Lobelia*
        * Short, conical, unicellular, e.g., tea, buchu
        * Strongly waved, thick-walled, e.g., *Yerba santa*
    b. *Multicellular, unbranched trichomes*
        i. *Uniseriate*
            * Bi-cellular, conical, e.g., *Datura*
            * Three-celled, long, e.g., Stramonium
            * Three- to four-celled long, e.g., *Digitalis*
            * Four- to five-celled long, e.g., Belladonna
        ii. *Biseriate*
            * *Calendula officinalis*
        iii. *Multiseriate branched trichomes*
            * Stellate, e.g., *Hamamelis, Helicteris isora*
            * Peltate (platelike arrangement of surrounded cells), e.g., *Humulus*
            * Candelabra (Uniseriate, branched axis), e.g., *Verbascum thapus*
            * T-shaped trichomes, e.g., *Artemisia*
2. Glandular trichomes   These are characterized by the presence of glandular (spherical) cells at the top of the trichome. These are sub-classified as follows:
    a. *Unicellular glandular trichomes*
        The stalk is absent, e.g. *Piper betel*, Vasaka
    b. *Multicellular glandular trichomes*
        * Unicellular head and unicellular stalk, e.g., *Digitalis purpurea.*
        * Unicellular head and uniseriate multicellular stalk, e.g., *Digitalis thapsi, Atropa*, etc.
        * Multicellular head, multicellular, biseriate stalk, e.g., sunflower
        * Unicellular stalk and biseriate head, e.g., *Digitalis purpurea*
        * Short stalk with secreting head formed of rosette or club-shaped cells, e.g., *Mentha* species.

❋ Trichomes with multicellular, multiseriate, cylindrical stalk and rosette of secretory cells, e.g., *Cannabis sativa*.

❋ Multicellular multiseriate head and multicellular uniseriate stalk, e.g., Indian hemp and tobacco.

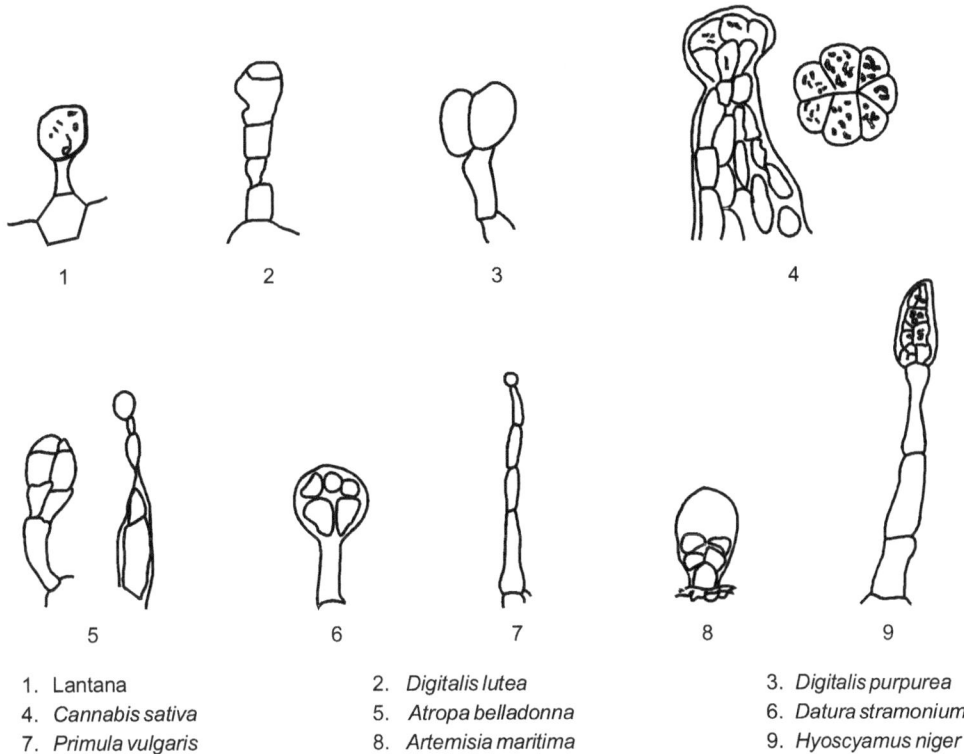

1. Lantana	2. *Digitalis lutea*	3. *Digitalis purpurea*
4. *Cannabis sativa*	5. *Atropa belladonna*	6. *Datura stramonium*
7. *Primula vulgaris*	8. *Artemisia maritima*	9. *Hyoscyamus niger*

Figure 15.16 Types of glandular trichomes

3. Hydathode and Emergences

These are the organs of absorption or secretion of water developed in certain plants, e.g., *Piper betel*, London pride, etc.

Emergences are otherwise called prickles and are small outgrowths on the epidermal walls of the aerial parts of the plant. They are epidermal and sub-epidermal in origin. They may be present on the stem or fruits. Emergences are not microscopic structures. They are very hard, stout in nature and meant for plant protection, e.g., Rose.

Figure 15.17 Emergence of rose

Observation of Powdered Drugs

The preliminary tests for the observation:

* Colour observation and its description.

* Mixing one quantity of powder with few drops of water and then leaving it for softening for some time. The water extract and concentrated liquid are dissolved and the gummy and mucilaginous nature may be manifested.

* *Presence of calcium* A small portion of powder is mixed with dilute sulphuric acid; this mixture will give an effervescence if calcium is present.

* *Observation of fixed oils* One quantity of powder is pressed on a piece of filter paper, to form an oily spot, which is extended and persists when the paper is heated.

* *Presence of saponins* One quantity of powder is stirred in a test tube with water; if abundant foam appears, the result is positive; boil softly and a smell is emitted, then it is divided in two portions and each portion may be tested for tannins and a derived anthraquinonine.

Lycopodium Spore Method

It is an important analytical technique for powdered drugs, especially when chemical and other methods of evaluation of crude drugs fail as accurate measures of quality.

Wallis demonstrated that the spores of *Lycopodium* are exceptionally uniform in size (25 microns) and that 1 mg of *Lycopodium* contains 94,000 spores. The number of spores per milligram is determined by direct counting and by calculation based on specific gravity and dimensions of the spores. With this valuable technique, it is possible to evaluate powdered drugs with the following characteristics:

* Well-defined fragments/particles which may be counted (pollen or starch).

* Single-layered tissues or cells whose area may be traced under suitable magnification and the actual area calculated or the objects of uniform thickness, the length of which can be measured at a definite magnification and their real size calculated. Mounts containing a definite proportion of the powder and lycopodium are used and the lycopodium spores counted in each of the fields in which the number or area of the particles in the powder is determined.

In this method the moisture content of the powdered material is determined. A mixture of weighed quantities of the powder and lycopodium spores are suspended in a suitable viscous liquid. A drop of this suspension in mounted and examined with a 4 mm objective. The number of lycopodium spores and the number of characteristic particles are counted in 25 various fields. The same experiment is repeated with a second similar suspension. From the mean of these results and a knowledge of the weights of *Lycopodium* and powder in the mixture, the number of characteristic particles in 1 mg of the powder may be determined.

The percentage of an authentic powdered ginger is calculated using the following equation:

$$\frac{N \times W \times 94,000 \times 100}{S \times M \times P}$$

where,

N = Number of characteristic structures (e.g., starch grains) in 25 fields

W = Weight in mg of *Lycopodium* taken

S = Number of *Lycopodium* spores in the same 25 fields

M = Weight in mg of the sample calculated on the basis of sample dried at 105°C

P = 2,86,000 in case of starch grains in ginger powder.

Lycopodium spore method can be used for evaluation of powdered clove, ginger, cardamom, nutmeg, umbelliferous fruits, number of pollen grains in pyrethrum powder, etc.

PHYSICAL METHODS

Physical standards are to be carried out to determine the quality and purity of drugs. Physical evaluation of drugs include moisture content, specific gravity, density, optical rotation, solubility, melting point, refractive index, viscosity, foreign matter, ash values and extractives. Some of the methods are described below:

i. Moisture content   The percentage of active chemical constituents in crude drugs is described on air-dried basis. Presence of moisture in a crude drug can lead to deterioration either due to chemical change or growth of microbes. Hence, the moisture content of a drug should be minimized.

Moisture content can be determined by heating a drug at 105°C in an oven and calculating the loss of weight. For drugs containing volatile active constituents, the toluene or xylene distillation method is used. Karl Fischer method is a standard procedure for determining moisture content.

Crude drugs with limit of moisture content

Drugs	Moisture content % (W/W)
Acacia	Not more than 15
Aloe	Not more than 10
Digitalis	Not more than 05

ii. Viscosity   Viscosity of a liquid is constant at a given temperature and is an index of its composition. Hence, it can be used as a means of standardizing liquid drugs.

❋ Liquid paraffin—Kinematic viscosity not less than 64 centistokes at 37.8°C.

❋ Pyroxylin—Kinematic viscosity not less than 1100–2450 centistokes.

iii. Volatile content   The efficacy of volatile-oil-containing drugs like clove, fennel, cinnamon, etc. depends upon the amount of the volatile content present in the drug. A weighed quantity of the drug is boiled with water in a round-bottomed flask fitted with Clevenger apparatus. The distillate collected in the graduated receiving tube separates into volatile oils, which being lighter than water remains on the top and the oil is thus collected from the tube.

The pharmaceutical significance of aromatic drugs is due to their odourous principles, i.e., volatile oils. Such crude drugs are standardized on the basis of their volatile oil content.

Volatile oil content of few crude drugs

Drugs	Volatile oil content (% w/w)	
Clove	Not less than	15.0
Fennel	"	1.4
Dill	"	2.5
Caraway	"	2.5
Fresh lemon peel	"	2.5
Cardamom seed	"	4.0

iv. Melting point In phytochemicals, melting points are very sharp and constant. Since the crude drugs of animal or plant origin contain mixed chemicals, they are described with certain range of melting points. The purity of the following crude drugs can be ascertained by determining their melting points in the range shown against each of them.

Melting point range for few crude drugs

Drugs	Melting point (°C)
Colophony	75–85
Kokum butter	39–42
Cocoa butter	30–33
Beeswax	62–65
Wool fat	34–44

v. Solubility The presence of adulterant in a drug could be indicated by solubility studies.

✤ Castor oil is soluble only in 3 volumes of 90 per cent alcohol, while the adulterated form may show good solubility in alcohol.

✤ Balsam of Peru is soluble in chloral hydrate solution and colophony is freely soluble in light petroleum.

✤ Asafoetida is soluble in carbon disulphide.

✤ Alkaloidal bases are soluble in chloroform.

✤ Alkaloidal salts are soluble in polar solvent.

✤ The glycosides are extractable with alcohol and water, while their aglycone moieties are soluble in non-polar solvents like benzene or solvent ether.

vi. Optical rotation

✤ Many substances of biological origin have the property of rotating the plane of polarized light in the pure state or in solution. Plane of polarized light may be rotated either to the right (dextrorotatory) or to the left side (laevorotatory).

✤ Such compounds are said to be optically active and this property is known as optical rotation.

✤ Normally, optical rotation is determined at 25°C using sodium lamp as the source of light.

✤ The extent of rotation is expressed in degrees, plus (+) indicating rotation to the right and minus (-) indicating that to the left.

✤ A polarimeter is used to measure optical rotation.

✤ Most volatile oils contain optically active components, and the direction of the rotation as well as its magnitude, are useful criteria for purity of drugs.

Optical rotation values for few crude drugs

Drugs	Angles of optical rotation
Castor oil	3.5° to +6.0°
Honey	+3° to −15°
Caraway oil	+75° to +80°
Chenopodium oil	−30° to −8°
Eucalyptus oil	0° to +10°
Clove oil	0° to −1.5°

vii. Refractive index For the evaluation of volatile and fixed oils, measurement of refractive index is of significance.

When a ray of light passes from one medium to another of different density, it is bent from its original path. Thus, the ratio of the velocity of light in vacuum to its velocity in the substance is termed as refractive index of the second medium. Depending upon purity, it is constant for a liquid, and it is one of the criteria for standardization of drugs. Refractive index of a compound varies with the wavelength of the incident light, temperature and pressure.

Refractive index is measured by means of a refractometer. Refractive indices of the following compounds are for sodium light at a temperature of 25°C.

Refractive indices of some phytoconstituents

Drugs	Refractive index
Caraway oil	1.4838 to 1.4858
Castor oil	1.4758 to 1.527
Clove oil	1.527 to 1.535
Arachis oil	1.4678 to 1.470

viii. *Ash values and extractives* To prove the purity of drugs, the following test can be carried out.

a. Ash content   The total ash content is determined by incinerating a known quantity of the air-dried crude drug in a silica crucible. The residual ash is weighed to calculate the percentage of ash and is done on the basis of initial dry weight of the drug sample.

* Total ash of unpeeled drug is more than the peeled drug.

* Most of the drugs are admixed with various mineral substances like sand, soil, calcium oxalate, chalk powder or with different organic contents.

* For determining ash, powdered drug is incinerated so as to burn out all organic matter. The residue remaining after incineration is the ash content of the drug, which represents inorganic salts naturally present in the drug or deliberately added to it as adulterants.

* Ash value is a criterion to judge the purity of crude drugs.

* Total ash usually consists of carbonates, phosphates, silicates and silica.

* Acid-insoluble ash, which is a part of total ash insoluble in dilute hydrochloric acid, is also recommended for certain drugs. Adhering dirt and sand may be determined by acid-insoluble ash content. Higher content of acid-insoluble ash indicates possibility of sand or other materials mixed with the drug.

Crude drugs with their ash contents

Drugs	Total ash (% w/w)	Acid-insoluble ash (% w/w)
Ginger	6.0	1.7 (Water-soluble ash)
Valerian	12.0	–
Aloe	5.0	–
Ashoka	11.0	–
Belladonna	–	3.0
Cannabis	15.0	5.0

b. Extractives The extracts obtained by exhausting crude drugs indicate the approximate measures of their chemical constituents.

❋ Based on the chemical nature and properties of contents of drugs, various solvents are used for determination of extractives.

❋ The solvent used for extraction is in a position to dissolve appreciable quantities of substances desired.

i. *Water-soluble extractives* This method is practised for drugs that contain water-soluble active constituents such as tannins, mucilage, sugars, glycosides, etc.

Water-soluble extractive values of crude drugs

Drugs	Water-soluble extractive (% w/w)
Senna leaves	Not less than 30.0
Linseed	Not less than 15.0
Ginger	Not less than 10.0
Aloe	Not less than 25.0
Glycyrrhiza	Not less than 20.0

ii. *Alcohol-soluble extractives* Alcohol is an ideal solvent for extraction of various chemicals like resins, tannins, etc. Hence this method is very commonly employed to determine the approximate content of resin present in the drug.

❋ This method is also used for myrrh and asafoetida.

❋ In general, 95% ethyl alcohol is used for determination of alcohol-soluble extractive.

Alcohol-soluble extractive values of some crude drugs

Drugs	Alcohol-soluble extractive (% w/w)
Myrrh	Not less than 70.0
Ginger (90% alcohol)	Not less than 4.5
Aloe	Not less than 10.0
Asafoetida (90% alcohol)	Not less than 50.0
Benzoin	Not less than 24.0

iii. *Ether-soluble extractives* It is used for crude drugs containing volatile oils and resins or fixed oils. The volatile ether-soluble extractive represents volatile oil content of the drug, while the non-volatile ether-soluble extractives represent resins, fixed oils or colouring substances present in drugs.

Non-volatile ether-soluble extractive values of some drugs

Drugs	Limit for non-volatile ether-soluble extractive (% w/w)
Male fern	Not less than 01.5
Linseed	Not less than 25.0
Capsicum	Not less than 12.0
Nutmeg	Not less than 25.0

ix. Spectroscopic analysis Specific groups of phytoconstituents are reported to be present in crude drugs and thus assay of a particular group of phytochemicals help in the evaluation of drugs.

* For example, glycosides of cardiotonic drug or alkaloids of a solanaceous drug help in the evaluation of drugs.

* In spectroscopic analysis, the capacity of certain molecules of a drug to absorb vibration at specific wavelength is the basis for drug evaluation.

* UV, IR, NMR, mass and radioimmunoassays are applied more frequently in spectroscopic analysis of phytoconstituents.

* Chromatographic techniques such as paper, column, thin-layer, gas-liquid (GLC) and high-performance liquid chromatography (HPLC) provide information about the chemical constituents present in the drug.

✤ Nowadays most groups of active phytochemicals of drugs are determined by spectrometry including colorimetry and fluorescence analysis.

x. Rf values Thin layer chromatography (TLC) has become increasingly popular for both qualitative and quantitative evaluation of drugs.

✤ Rf values refer to the ratio of distance moved by the solute to the distance moved by the solvent on a thin layer of an adsorbent.

✤ Rf value of a compound is a characteristic and can be used to identify the components by comparison with the reference standard.

✤ The intensity of the colour of spot of the compound under test can be used for quantitative estimation of the principle in the drug.

$$Rf = \frac{\text{Distance travelled by the component (Solute)}}{\text{Distance travelled by the solvent}}$$

xi. Foreign organic matter The parts other than the drug are considered as foreign organic matter.

✤ The maximum limit for the foreign organic matter is defined in the monograph of crude drugs, and they should be in pharmacoepial limits.

✤ If it exceeds the limits, it shows deterioration in quality of the drug due to adulteration.

✤ The foreign organic matter is determined by sedimentation or floatation method.

✤ Sometimes crude drugs get contaminated with certain microbes like fungi, bacteria and moulds. Under such conditions the drugs may have to be subjected to sterilization.

CHEMICAL METHODS

The determination of active constituents in a drug by chemical methods is called chemical evaluation.

It consists of different chemical tests and chemical assays. The isolation, purification and identification of active constituents are chemical methods of evaluation. Quantitative chemical tests such as titrimetric assay, acid value, iodine value, ester value, saponification value, ash values, etc. are also employed. Some of these tests are useful in evaluation of resins (sulphated ash, acid value), balsams (acid, saponification and ester values); volatile oils (acetyl and ester values), and gums (determination of methoxyl value and volatile acidity).

Preliminary photochemical screening is a part of chemical evaluation. Qualitative chemical tests are useful in the detection of adulteration. For example, colophony admixed as an adulterant for resins, balsams and waxes is detected by copper acetate. The following are the specific chemical tests used for determining various substances present in drugs.

✤ Halphen's test for cottonseed oil

✤ Van Urk's reagent for ergot

✤ Vitali's test for propane alkaloids

✤ Potassium chlorate and hydrochloric acid test to estimate emetine in Ipecac

✤ Ammonium vanadate and sulphuric acid test for detecting strychnine in nux-vomica

✤ Borntrager's tests for detecting anthraquinone glycosides present in senna, cascara, aloe and rhubarb

✤ Acid-base titration for determining total alkaloid contents

✤ The ash value for determining authenticity and purity of drugs

Assay of Drugs

The purity of crude drugs is ascertained by quantitative estimation of active constituents present in them. This can be done by one or more of the following methods as may be applicable.

✤ Physical assay method

✤ Chemical assay method

✤ Biological assay method

The physical method may be useful in determining single active constituent or the group of related constituents present in the same drug. Spectroscopic analysis using the specific absorption of the constituents in the ultraviolet, visible or infrared regions, and fluorimetric analysis are examples of physical assay methods. The technique of thin layer chromatography is becoming very popular for the qualitative and quantitative evaluation of phytoconstituents. Comparison of Rf values with standards ascertains the qualitative evaluation of constituents.

The conventional titrimetric estimations as applicable to the estimation of alkaloids from crude drugs, ester and aldehyde contents of volatile oils, gravimetric methods, etc. are the techniques of chemical assay. Alkaloidal drugs are regularly analysed for their total alkaloidal content; the main alkaloids are opium (morphine), belladonna (atropine), nux-vomica (strychnine), cinchona (quinine), rauwolfia (reserpine), and ergot (ergometrine).

Chemical assays are carried out for cineole in eucalyptus oil, aldehydes in lemon oil, carvone in caraway oil and dill oil. The chemical assays are also extended to anthraquinone and cardiac glycosides, content of balsamic acids like benzoic acid and cinnamic acid in balsam of Tolu, balsam of Peru and prepared storax. Thus chemical evaluation of crude drugs is more appropriate because, this identifies as well as estimates the phytoconstituents present in a drug.

BIOLOGICAL METHODS

The drugs that cannot be assayed satisfactorily by chemical or physical methods are evaluated by biological methods. The tests carried out on intact animals, animal preparations, isolated living tissues, or microorganisms are called "biological assays".

Biological evaluation is also done in cases where chemical evaluation does not justify therapeutic efficacy of a drug, for example, the chemical assay method for some plant (digitalis) in mixtures does not always distinguish between therapeutically active glycosides and those which are less active or even therapeutically or physiologically inert.

Thus, in biological evaluation, the drug is administered to animals under identical conditions and the response on a particular organ, tissue or even the whole animal is measured and compared with the standard drug under the same set of conditions.

Drug containing cardiac glycosides can be biologically evaluated on cats, frogs or pigeons. The bitter drugs like quassia, gentian and chirata can be evaluated by the bitter value whereas, ginger and capsicum can be evaluated for their pungency. Anthelmintic drugs like male fern can be evaluated by their action on earthworms.

The accepted procedure for biological evaluation is to carry out a series of several determinations on the same lot of material and then to average the result. But biological standardization methods are more time-consuming and more expensive to conduct than chemical assays. Therefore, they are generally used only when the drug cannot be evaluated by any other method.

Indications for Biological Evaluation

✤ When chemical methods are insensitive or not available.

✤ When the quantity of the drug is small, and cannot be evaluated chemically.

✤ When the chemical nature of the drug is not known but it has a specific biological action.

✤ Drugs, which have different chemical composition with same biological activity.

Biological assay methods are of three types—toxic, symptomatic, and tissue methods.

In toxic and symptomatic techniques, animals are used, whereas in tissue method, the effect of a drug is observed on an isolated organ or tissue. Among the drugs that are subjected to bioassay are cardiac glycosides, natural pesticides and antibiotics.

A biological assay measures the actual biological activity of a given sample. In any one test, the animals of only one strain are used. For some assays, a specific sex must be used. The male rat has faster growth rate than the female. Therefore, use of both sexes in a growth test should be avoided.

Bacteria such as *Salmonella typhi* and *Staphylococcus aureus* are used to determine antiseptic value of certain drugs. Yeasts and moulds are used for assaying vitamins and for determining the activity of antibiotic drugs. Human beings are also used to note the activity of drugs in clinical trial.

There are some disadvantages of bioassays. Techniques and interpretations involved vary with different operators. The effect measured in the test animals is different from that observed in patients who are treated.

REVIEW QUESTIONS

State whether the following sentences are true or false

1. Undeliberate adulteration occurs accidentally.
2. Quality refers to quantity of the active principles present in the drug.
3. Study of sensory characters are useful to drugs containing lipids.
4. Lignin stains blue with phloroglucinol and concentrated HCl.
5. Silica crucible is used for determining the total ash content of a drug.
6. The measurement of refractive index is very much useful for volatile and fixed oils.
7. Sedimentation method is used to determine the presence of foreign organic matter.
8. Ash value of crude drugs is useful for identifying the presence of active principle
9. The drugs in capsicum and ginger can be biologically evaluated for their pungency.
10. Black pepper is commonly adultered with papaya seeds.

Choose the correct answer

11. Symptomatic method is a _____
    a. physical method   b. biological assay method   c. chemical analysis
12. The distance travelled by component divided by the distance travelled by the solvent gives _____
    a. R.Q. values   b. R.F. values   c. R index values

13. High content of acid-insoluble ash in a drug indicates the presence of _____
    a. sand   b. wax   c. minerals
14. The uniform size of *Lycopodium* spore was identified by _____
    a. Gupta   b. Levin   c. Wallis
15. The organs that absorb or secrete water are called _____
    a. hydathodes   b. stomata   c. lenticels
16. Sclereids are absent in powdered _____
    a. cloves   b. pea seeds   c. cardamom seeds
17. The use of quantitative microscopy in pharmacognosy was given by _____
    a. Waller   b. Walles   c. Vavilo
18. A substance staining pink with ruthenium indicates the presence of _____
    a. lignin   b. lipid   c. mucilage
19. Tamarind seed powder is mixed with _____
    a. argemone powder   b. coffee powder   c. mustard powder
20. Chilli powder is adulterated with _____
    a. brick powder   b. dextrin   c. ginger powder

Answers				
1. True	2. True	3. False	4. False	5. True
6. True	7. True	8. False	9. True	10. True
11. b	12. b	13. a	14. c	15. a
16. a	17. b	18. c	19. b	20. a

Answer the following

1. Define the terms adulteration, palisade ratio, vein-islet number, vein-termination number.
2. Why is it necessary to evaluate crude drugs?
3. Distinguish between deliberate and undeliberate adulterations.
4. Explain any five types of adulteration with suitable examples.
5. "Original crude drugs in the market are adulterated with different adulterants"—Justify the statement.
6. How are drugs adulterated by faulty collection?

7. Comment on adulterants.

8. Write an essay on organoleptic methods of drug evaluation.

9. Write a critical note on chemical methods of evaluation of herbal drugs.

10. How do sensory characters help in drug evaluation?

11. Explain the role of quantitative microscopy in the evaluation of drugs.

12. How are physical standards used to determine the quality and purity of crude drugs?

13. Comment on the assay of crude drugs.

14. List out the common adulterants for the following:

    i. Mustard seeds

    ii. Paddy grains

    iii. Coriander power

    iv. Beeswax

16
GENERAL ANATOMICAL FEATURES AND GROSS ANATOMY OF SELECTED PLANTS

INTRODUCTION

The cell is the basic structural and functional unit of a living organism. Each cell contains an outer cell wall and the inner living material, the protoplast. The protoplast consists of cytoplasm, nucleus and all other organelles.

The discovery of the electron microscope by Knoll and Ruska in 1932 aided in obtaining a clear picture of cells. A group of cells that perform a common function and have a common origin is known as tissue. Plant tissues are basically categorized into permanent and meristematic tissues.

The permanent tissues include simple, complex, and secretory tissues. Meristematic tissues include different meristems such as primary and secondary, apical, intercalary and lateral meristem, which are categorized, based on their mode of origin and positions.

The following sections give an essential and relevant explanation about the various tissues and their functions.

CLASSIFICATION OF PLANT TISSUES (SACHS, 1875)

* Dermal tissue system
* Fundamental tissue system
* Vascular tissue system
* Secretory tissues

DERMAL TISSUE SYSTEM

The dermal tissue forms the outermost covering of various organs of plants and can be studied under the following headings.

    i. Epidermis and periderm

   ii. Stomata

  iii. Trichomes

Epidermis

Epidermis is the outermost single layer of cells in various plant organs. The term epiblema or rhizodermis is generally used for the root epidermis. Epidermis gives mechanical protection to the internal tissues, checks loss of water through transpiration in aerial parts, helps in gaseous exchange through stomata and also helps in storage of water (ice plant) and metabolic products. The epidermal cells are parenchymatous, with mostly colourless cell sap.

Depending upon the multiple functions of the epidermis, it contains a wide variety of cell types, with different shapes, sizes and arrangement. The nature of cell wall of epidermis varies from plant to plant. The outer walls of the epidermis are variously cutinized and the surface of the cuticle may be smooth, rough, ridged or furrowed. The cuticle layer is absent in root epidermis and it stains red with Sudan IV. Some specialized epidermal cells in the upper epidermis of the leaf of the members of Poaceae and in other monocotyledons are called **bulliform cells**, that are present at the grooved regions. Epidermis is generally single-layered but, in some cases the epidermis is composed of several layers of cells when it is called multiple epidermis, as in the leaves of Nerium, Ficus, Begonia and in epiphytic roots of *Vanda*.

The characteristic features of epidermal cells are helpful in the identification of leaves of different plant species. Epidermal cell walls are polygonal in coca (*Erythroxylon coca*) and senna leaves, wavy-walled in *Stramonium*, *Hyoscyamus* and belladonna, beaded walls in *Lobelia* and *Digitalis* species, and Papillate in the leaf of *Euphorbia*.

In addition to ordinary epidermal cells, guard cells, various types of outgrowths such as unicellular and multicellular hairs, glands, etc. are also present.

Stomata

In leaves and aerial parts of the plant, the epidermis possesses numerous minute openings called stomata. The continuity of the epidermis is interrupted by these stomata. Each of which is limited by two specialized cells, that may be kidney-shaped or dumb-bell shaped cells called guard cells. The guard cells, together with the opening between them, constitute the stoma, through which gaseous exchange takes place. In some plants, for example, *Ixora, Petunia*, etc., the guard cells are bounded by two or more cells that are different in shape and arrangement from epidermal cells and called subsidiary cells. Stomata are absent in roots and in plants, and cells that lack chlorophyll. They are present in a few submerged water plants, e.g., *Vallisneria*.

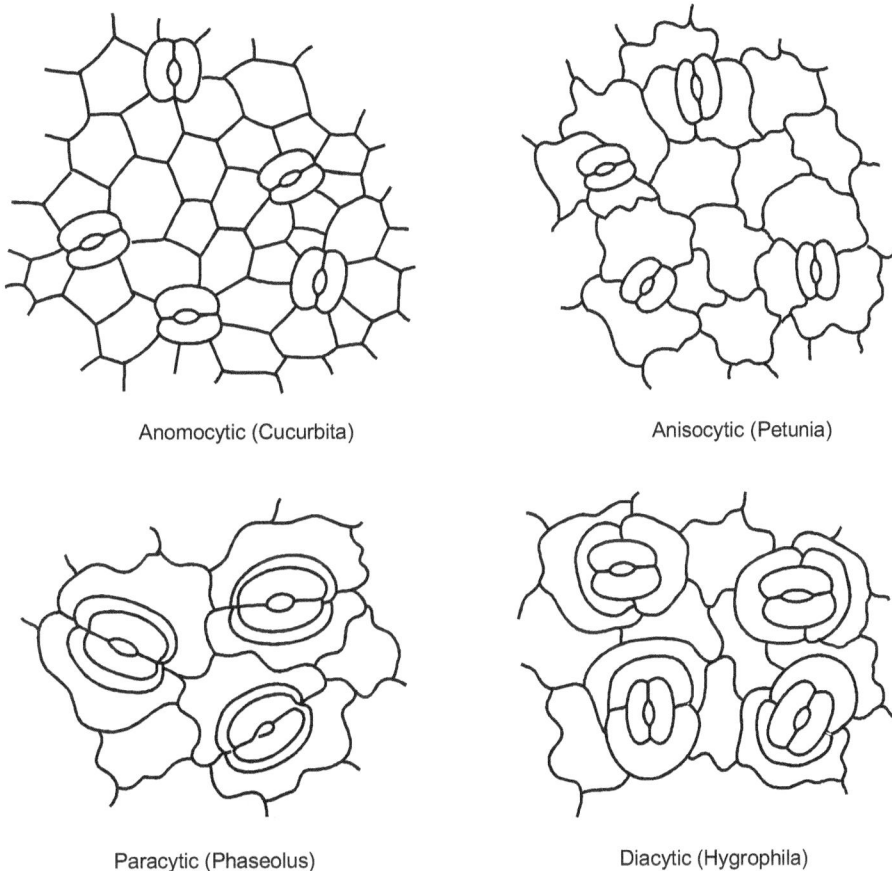

Anomocytic (Cucurbita)

Anisocytic (Petunia)

Paracytic (Phaseolus)

Diacytic (Hygrophila)

Figure 16.1 Stomatal types of dicotyledons

On the basis of the arrangement of epidermal cells with subsidiary cells, the following types of stomata are distinguished in dicotyledons and monocotyledons.

Ranunculaceous type (Anomocytic) In this type of stomata, the guard cells are surrounded by a limited number of cells that do not differ in size and shape from the other epidermal cells. They have no subsidiary cells, e.g., members of Ranunculaceae, Cucurbitaceae, Scrophulariaceae, Papaveraceae, etc.

Cruciferous type (Anisocytic)   The guard cells are surrounded by three subsidiary cells of unequal size. Two are large and one is smaller, e.g., members of Cruciferae, *Petunia*, etc.

Rubiaceous type (Paracytic)   The guard cells are surrounded with two subsidiary cells with their long axis parallel to the longitudinal axis of guard cells, e.g., Members of Rubiaceae, Mimosae, Convolvulacae, etc.

Caryophyllaceous type (Diacytic)   Two subsidiary cells occur with their long axis at right angles to the guard cells of the stomata, e.g., members of Caryophyllaceae, Lamiaceae, etc.

Actinocytic stomata   The stomatal pore is surrounded by a circle of radiating subsidiary cells.

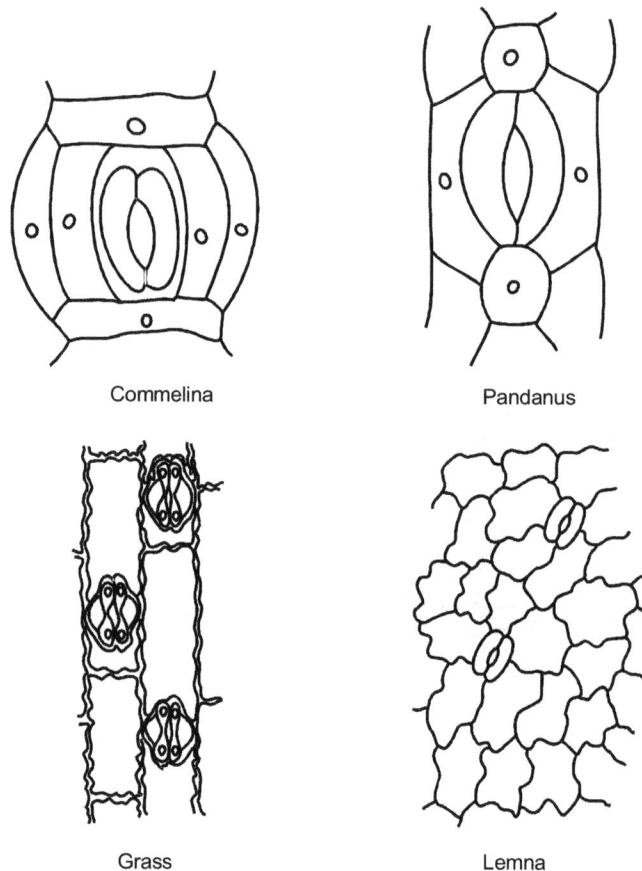

Commelina

Pandanus

Grass

Lemna

Figure 16.2   Stomatal types of monocotyledons

The following types of stomatal complexes have been noticed in monocotyledons:

* In Araceae, Commelinaceae, Musaceae and Zingiberaceae, the guard cells are surrounded by 4-6 subsidiary cells.

* In Palmae, Pandanaceae and Cyclanthaceae, the guard cells are surrounded by 4-6 subsidiary cells of which two are roundish, smaller than the rest and are situated at the ends of guard cells.

* In Flagellariaceae, the guard cells are accompanied laterally by two subsidiary cells, one on each side.

* In Dioscoreaceae, the guard cells are not associated with any subsidiary cells.

* In Poaceae the guard cells are dumb-bell-shaped and surrounded by two subsidiary cells that lie parallel to the guard cells and this type of stomata are called graminaceous type.

Stomata are characteristic features of the plant species and therefore, help in their diagnosis.

Epidermal Trichomes

Epidermal appendages (both unicellular and multicellular) present in the plant body are called trichomes. Trichomes may persist throughout the life of an organ or may soon fall off. A trichome may be differentiated into a base (stalk), embedded in the epidermal cell and a tubelike projecting body. The walls are thick due to cellulose deposition or may be highly lignified. They are highly variable in form, structure and function. Trichomes play an important role in protecting the plant organs from external heat and thus reduce the rate of transpiration.

On the basis of structure, trichomes are classified into non-glandular and glandular trichomes.

Non-glandular trichomes These are also called covering trichomes and are of the following types:

Unicellular hair It contains a single cell, short or long (*Gossypium*) and may be branched or unbranched and occur on the stem, leaves, flowers and roots.

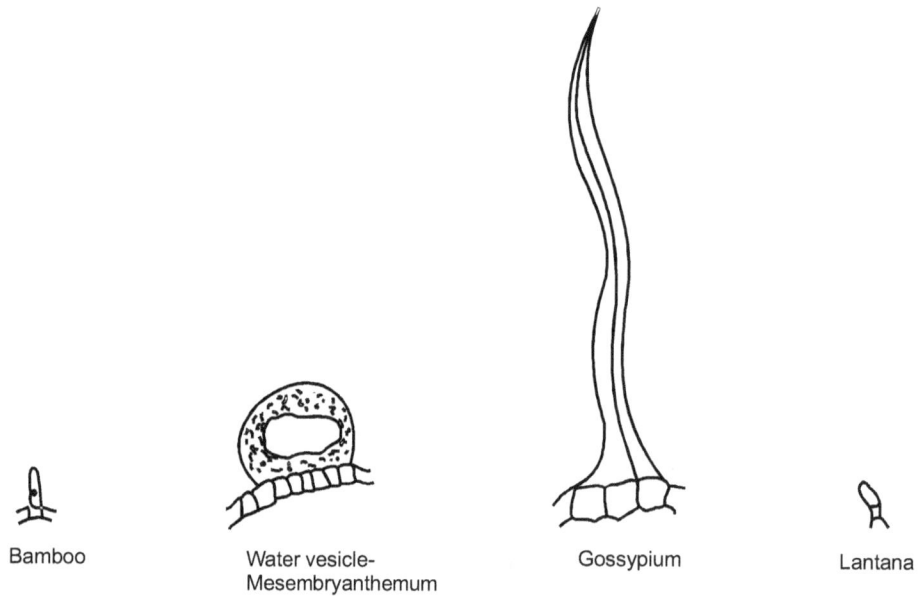

Figure 16.3   Unicellular hair types

Multicellular hairs   These consist of more than one cell and may be short, long, unbranched and branched. They occur both on the stem and the leaves. The sepals, petals, stamens and the carpels also possess such hairs. Their shape varies with species or even in different organs of the same species. Usually they are long and tubular, but may be peltate, stellate or vesicular. Multicellular hairs can be further classified as uniseriate (*Tridax*), biseriate (*Lantana*) and multiseriate types (*Portulaca*).

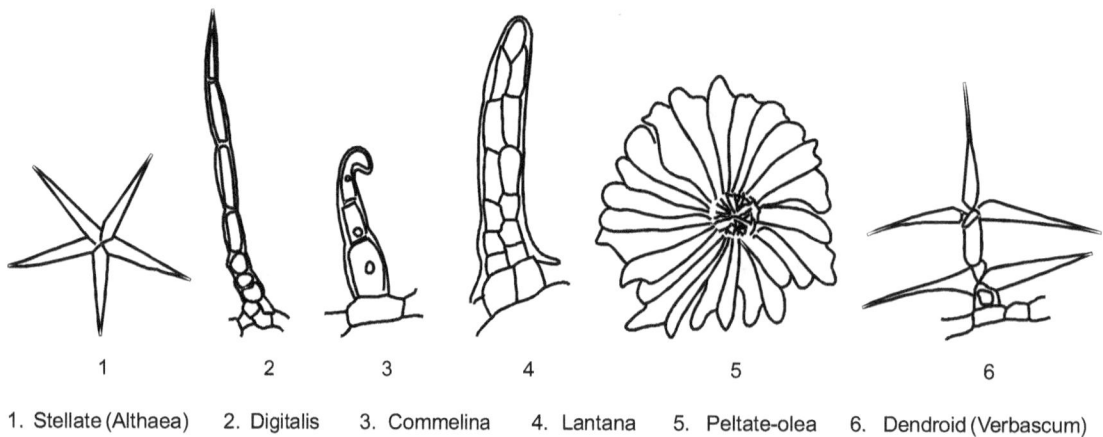

1. Stellate (Althaea)   2. Digitalis   3. Commelina   4. Lantana   5. Peltate-olea   6. Dendroid (Verbascum)

Figure 16.4   Multicellular hair types

Glandular trichomes These are involved in the secretion of various substances like salt solution, sugar solution, essential oils, resins and gums. The secreting trichomes are often called glands. These are of the following types:

Hydathode These are found in the tip of young leaves and stems of grasses and secrete an aqueous solution containing some organic acids.

Salt-secreting trichomes In these glands, the cytoplasm is dense, rich in mitochondria, endoplasmic reticulum, Golgi bodies, and has many vesicular structures. The salt solution is secreted onto the surface of the secretory cells.

Nectar-secreting trichomes The vesicles mainly of endoplasmic reticulum origin are involved in the secretion of nectar.

Mucilage-secreting glands These glands secrete mucilage, mainly a polysaccharide. The extruded mucilage accumulates in spaces between the cell wall and cuticle.

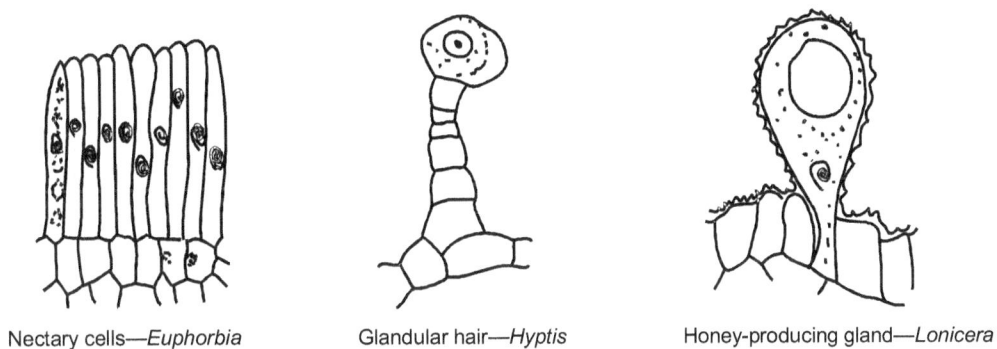

Nectary cells—*Euphorbia*   Glandular hair—*Hyptis*   Honey-producing gland—*Lonicera*

Figure 16.5 Glandular trichome types

Periderm

In plants having secondary growth, the epidermis is replaced by periderm, the secondary protective layer. It is formed due to the meristematic activity of cork cambium or phellogen, the secondary meristem found near the epidermis. The cork cambium may either arise from the epidermis or cortex, or even from endodermis. The periderm is made up of three different layers. They are,

* outer phellum or cork layers,
* middle phellogen or cork cambium and
* inner phelloderm or secondary cortex.

The cork cambium cuts off tissues on the outer side called cork layers and on the inner side called secondary cortex. The cork cells thicken to form bark cells. The cork layers, cork cambium, and the secondary cortex together constitute the periderm, the secondary protective layer, e.g., dicots and gymnosperms. In periderm, lenticels, which are pores that help in gaseous exchange, are also present. In lenticels, there are no guard cells and they remain always open.

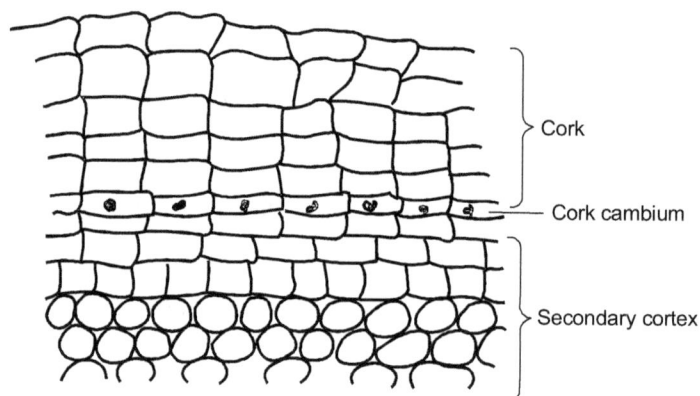

Figure 16.6   Structure of periderm

FUNDAMENTAL TISSUE SYSTEM

Fundamental tissue includes hypodermis, cortex, pith of stem, mesophyll and midrib regions of leaves. It constitutes the main body of the plant and consists of different kinds of tissues and has many functions. The tissues that constitute the fundamental tissue system are as follows:

Parenchyma

The parenchyma consists of living cells with or without intercellular spaces. The cells are mostly thin-walled with different shapes and with different physiological functions. They are present in the cortex, pericycle and pith of stem and roots, and the midrib and mesophyll tissue of leaves and fleshy part of fruits. These cells also occur in xylem and phloem.

The parenchymatous cells may be polygonal, lobed, armed, round or stellate, etc. The wall is composed of cellulose, pectin or hemicellulose. They play an important role in wound-healing and regeneration. The parenchyma cells containing chloroplasts take part in photosynthesis and are called chlorenchyma. Parenchyma in aquatic plants enclose large air cavities called aerenchyma. The main functions of parenchyma are storage and photosynthesis. The parenchyma also contains various substances like enzymes, tannins, mucilaginous and resinous substances, oils, crystals of various types, etc. The thick-walled parenchyma cells that give mechanical support are called prosenchyma.

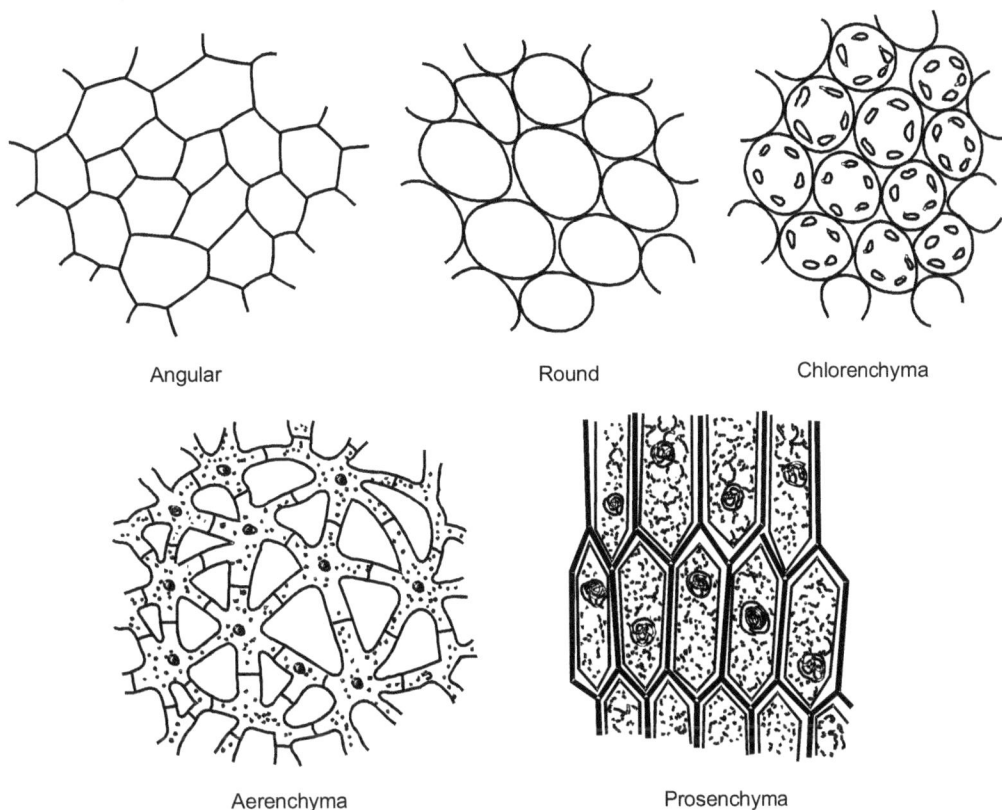

Figure 16.7   Types of parenchyma

Collenchyma

Collenchyma is a living tissue, and the cells are more or less elongated with unevenly thickened walls. In transverse sections, the cells may be rounded, or polygonal in outline. It functions as a supporting tissue in growing young organs such as stem, petiole, floral parts and midrib of leaf. Collenchyma is absent in the stems and leaves of many monocotyledons. Collenchyma always develops below the epidermis to form the hypodermis. The collenchyma may be of plate type, angular type or lacunate type, depending upon the type of wall thickening. In plate type (lamellar type), the thickening material is laid down on the tangential walls of the cells, e.g., stem of sunflower. In *Datura* stem, the thickenings are localized at the angles of the cells to represent angular type, whereas in the stem of *Cucurbita*, the thickening material is deposited on the walls bordering the intercellular spaces and this type is callled lacunate type. The cell walls are composed of a large amount of pectin, hemicelluloses and cellulose.

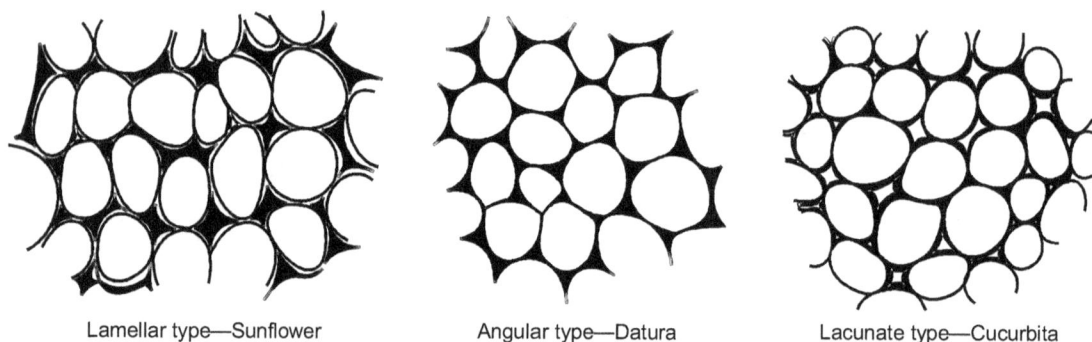

Lamellar type—Sunflower   Angular type—Datura   Lacunate type—Cucurbita

Figure 16.8 Types of collenchyma

Sclerenchyma

The cells with evenly thickened secondary walls are called sclerenchyma. The secondary walls may be lignified or non-lignified and their principal function is to give mechanical support. They occur in all parts of the plant body. These cells exhibit elastic properties. Sclerenchyma cells are of two types: fibres and sclereids.

Fibres   The narrow, elongated, thick-walled cells of sclerenchyma with pointed or tapering ends are called sclerenchymatous fibres. The fibres occur in different parts of the plant body. The cell wall is composed of cellulose and lignin. The mature fibres have well-developed lignified secondary walls with very narrow lumen or sometimes the lumen is obliterated due to heavy thickening of walls.

Fibres are most commonly found among secondary vascular tissues, but in many plants, they are also well-developed in ground tissues. Usually in monocotyledons, fibres occur in an uninterrupted hollow cylinder in the ground tissue and also form sheaths around the vascular bundles. Based on the occurrence of fibres in plants, they are classified into the following types.

- ❋ Xylary fibres
- ❋ Extraxylary fibres

Xylary fibres   The sclerenchymatous fibres associated with xylem are called xylary fibres and develop from the same meristematic tissues as do the other xylem elements. Depending upon wall thickness, type, and amount of pits, xylary fibres may be

- ❋ Libriform fibre
- ❋ Gelatinous fibre
- ❋ Septate fibre
- ❋ Fibre-tracheid

The libriform fibres are the true xylary fibres, having lignified secondary walls with simple pits. Libriform fibres with septa are called septate fibres, e.g., Mahogony. Libriform fibres with gelatinous substance are called gelatinous or mucilaginous fibres, which contain α-cellulose and are poor in lignin, e.g., *Quercus*. The inner wall layer of the gelatinous fibre is called "G-layer". Hence these fibres are called G-fibres. The fibres that are intermediate in structure between fibre and tracheid are called fibre-tracheids. These are narrow, elongated fibres with moderately lignified walls with blunt tip and have both simple and bordered pits.

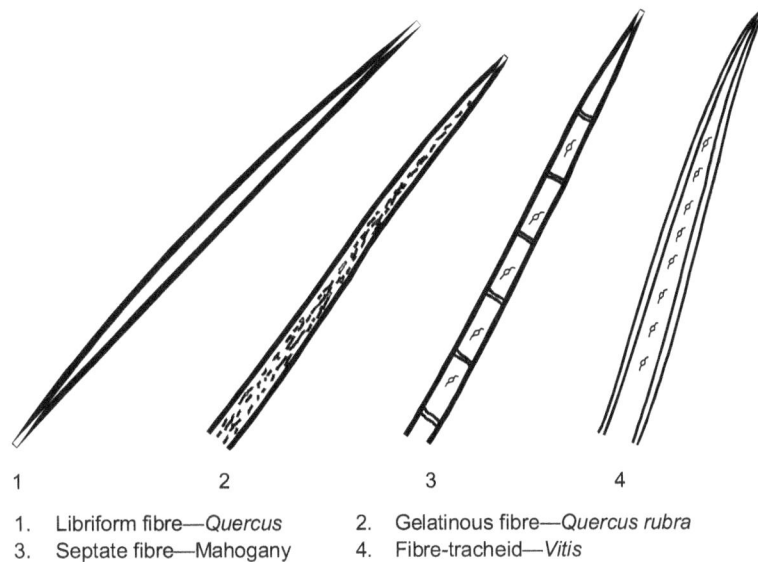

1. Libriform fibre—*Quercus*
2. Gelatinous fibre—*Quercus rubra*
3. Septate fibre—Mahogany
4. Fibre-tracheid—*Vitis*

Figure 16.9 Types of xylary fibres

Extra-xylary fibres These fibres are further classified into different types based on the position in the plant body.

* Cortical fibres
* Pericyclic fibres
* Phloem fibres
* Surface fibres

Cortical fibres The sclerenchymatous fibres associated with cortical cells of the stem are called cortical fibres.

Pericyclic fibres The fibres present in the pericycle region are called pericyclic fibres, e.g., *Cucurbita*.

Phloem fibres The fibres that occur in phloem are called phloem fibres. The phloem parenchyma is converted into phloem fibres. The phloem fibres may be lignified or non-lignified but with secondary wall thickenings, e.g., flax.

Surface fibres The fibres that are found in the testa of various seeds are called surface fibres, e.g., cotton fibres.

Sclereids The extremely thick-walled short cells of sclerenchyma are called sclereids. These cells may be spherical, oval, isodiametrical or irregular in shape. Their walls are lignified, often show well-marked stratification and are traversed by pit-canals, which are usually funnel-shaped or branched. The cell lumen is small and cell contents of diagnostic significance may be present. In many cases the sclereids occur as hard masses of cells, in the region of thin parenchymatous tissue or as individual cells. In many plants, sclereids appear as idioblasts, i.e., which are readily distinguished from the surrounding cells of the tissue by their size, shape and thickness of wall.

Sclereids are present in the hard outer coats of seeds and fruits and in the bark and pericyclic regions of woody stems.

The sclereids are further divided into four main types on the basis of shape and distribution. They are as follows:

Brachy-sclereids (stone cells) These are more or less isodiametric sclereids. They are usually found in the bark, pith, phloem, cortex, hard endocarp and fleshy portions of many fruits, e.g., Pulp of *Pyrus*.

Macrosclereids (rod cells) These sclereids are elongated rod-shaped cells, chiefly found in the outer seed-coat layers of many members of the pea family (Leguminosae).

Ostero-sclereids (prop cells) These are rodlike sclereids with dilated ends and are distributed in the leaf of *Hakea*.

Astrosclereids Stellate type of sclereids are called astrosclereids and these occur in the leaf of *Nymphaea*.

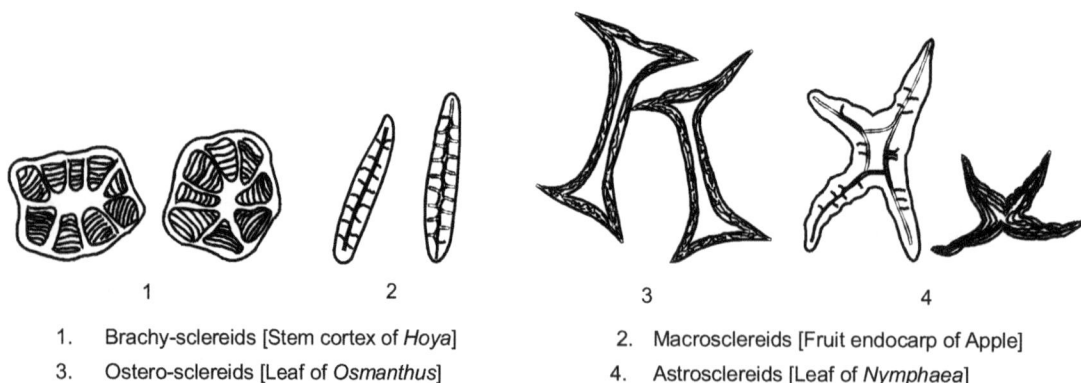

1. Brachy-sclereids [Stem cortex of *Hoya*]
2. Macrosclereids [Fruit endocarp of Apple]
3. Ostero-sclereids [Leaf of *Osmanthus*]
4. Astrosclereids [Leaf of *Nymphaea*]

Figure 16.10 Types of sclereids

VASCULAR TISSUE SYSTEM

Vascular tissue system consists of complex tissues such as xylem and phloem. These two tissues may be separated by a strip of meristematic tissue called cambium which is responsible for secondary thickenings. It occurs in the stems of dicotyledons and gymnosperms and in their roots cambium is differentiated at the time of secondary growth. Cambium is absent in monocotyledons, hence, there is no secondary thickening in these plants. Vascular tissue system conducts food and water. Phloem conducts food material from leaves to the different parts of the plant, whereas xylem tissue conducts water from roots to the leaves and other parts of the plant.

Xylem Tissue

Xylem forms a part of the vascular system and consists of several types of cells. The xylem produced by the procambium in the primary plant body is called the primary xylem. In the primary xylem, the element that differentiates earlier is the protoxylem, and the elements that develop later, constitutes metaxylem. The xylem produced as a result of the activity of the vascular cambium is called the secondary xylem.

It consists of tracheary elements such as tracheids and vessels, xylem fibres or wood fibres, xylem parenchyma or wood parenchyma and xylem rays. The xylem elements of some plants also contain laticifers.

Tracheids   Tracheids are elongated tubes with thick and lignified walls that lack protoplast. Tracheids have a large cell cavity and their end walls are usually tapering or oblique in nature. Their walls are provided with bordered pits. The wall thickenings may be annular, spiral, scalariform or pitted types. Tracheids occur in pteridophytes, gymnosperms and in primitive angiosperms. Tracheids give strength to the plant body but their main function is conduction of water from the root to the leaf.

Vessels   Vessels are elongated to short drum-shaped cells, placed one above the other, i.e., end to end. Their walls are thick and lignified, and the thickening may be annular, spiral, scalariform, reticulate and pitted types. The vessel member lacks protoplast and the cell lumen is very wide and thus helps in transport of more water. The transverse walls with perforations are called perforation plates. They may be simple perforation plates having one perforation or multiple perforation plates with many perforations. The shape, size and distribution of vessel members in plants help in the correct identification of species. The secondary wall thickening is composed of lignocellulose.

Wood fibres   The sclerenchymatous fibres that are associated with wood or secondary xylem are called wood fibres. They occur abundantly in woody dicotyledons and add to the mechanical strength of the xylem and the plant body as a whole due to the deposition of lignin. The wood fibres mostly consist of libriform type of fibres and in addition, septate, gelatinous and fibre-tracheids are also present in different species.

Xylem parenchyma The parenchyma cells that are associated with xylem tissue are called xylem parenchyma. The cells contain living protoplasts and their walls are primary in nature with cellulose and they help in the storage of food. Xylem parenchyma cells in secondary xylem are classified into wood parenchyma or parenchyma axial and ray parenchyma. Wood parenchyma cells that are associated with vessels are called paratracheal. If they are not in contact with vessel elements in wood they are called apotracheal.

Xylem ray The parenchyma cells that form the secondary medullary ray of the wood region are called xylem rays or ray parenchyma and these cells help in radial conduction of food materials. These cells also contain protoplasts and also help in storage of food.

Protoxylem can be classified depending upon its position and the number of protoxylem groups present in root or stem. Depending upon the position of the protoxylem, there are three types of primary xylem—endarch, exarch and mesarch. In endarch, the protoxylem elements are situated close to the centre of the axis and the metaxylem develops towards the periphery, e.g., stem. In exarch, protoxylem elements occur away from the centre of the axis, e.g., root. In mesarch, the protoxylem elements are situated in the centre surrounded by metaxylem elements, e.g., dicot leaf.

Based on the number of protoxylem groups present in the primary xylem, they are classified into monarch wherein, one group of protoxylem is present; diarch wherein, two groups of protoxylem are present; triarch wherein, three groups of protoxylem extend outward; tetrarch with four groups of protoxylem, e.g., roots of dicotyledons, and polyarch wherein, more than four groups of protoxylem extend outward near the pericycle, e.g., roots of monocotyledons.

Phloem Tissue

Phloem is also a part of the vascular system. It consists of components such as sieve elements (sieve tubes and sieve cells), companion cells, phloem parenchyma, bast fibres and phloem rays. It helps in conduction of food from the leaf to the storage organs and the growing regions. The primary phloem is similar to the primary xylem in origin, therefore, it also develops from the procambium. The primary phloem is divided into protophloem and metaphloem. Sometimes laticifers and resin ducts may also be found in the phloem tissue.

Sieve tubes Sieve tubes are the advanced type of sieve elements that are slender, tubular structures and composed of elongated cells, placed end to end. Their walls are thin, made of cellulose and lack nucleus in their protoplast. They help in the transport of food from leaves to other regions of the plant. They have oblique cross walls with sieve areas called sieve plates. The sieve plate may be a simple sieve plate, if it contains one sieve area, or compound sieve plate, if it contains many sieve areas. Sieve tubes are always associated with companion cells. Sieve tubes with sieve plates on their end walls are present in angiosperms, whereas in gymnosperms and pteridophytes, the sieve elements have sieve plates on their lateral walls, and called sieve cells, and there is no distinct arrangement of the sieve cells in linear rows.

Companion cells   The parenchyma cells associated with sieve tubes are called companion cells. They are smaller in size as compared to the sieve-tube with which they are associated. The companion cells are living and possess cytoplasmic content with a large nucleus. Companion cells are present in angiosperms. They are absent in pteridophytes and gymnosperms. This cell walls are thin and made up of cellulose.

Phloem parenchyma   These are parenchymatous cells that are present in the phloem tissue of most of the dicots and pteridophytes. These cells are absent in monocotyledons. These cells are thin-walled with cellulose and they possess cytoplasm and nucleus and stored food materials in the form of starch and fats. Phloem parenchyma cells in the secondary phloem are of two types: 1. Those elongated in vertical direction called axial parenchyma, and 2. those elongated in radial direction constitute secondary phloem ray cells, and contain lot of stored food in the form of starch, protein and fats, and both types are responsible for conduction of food.

Phloem fibres   The phloem parenchyma cells are transformed into phloem fibres by lignifications of their walls. They are absent or fewer in primary phloem, but are abundant in secondary phloem and are of great economic value. In the primary phloem, their walls have cellulose thickenings, whereas in the secondary phloem, the wall is always thickened with lignin. These fibres lack living protoplasts and their cell-cavity is narrow. The chief function of the phloem fibres is to give strength and rigidity to the organ. The phloem fibres vary in length as well as in the shape of their apices.

Cambium

Cambium is a lateral meristem, present between the primary phloem and xylem of the vascular bundles of dicot stems and gymnosperm stems. It is absent in young roots, but appears on maturity in plants between the radially placed phloem and xylem, of the roots to produce secondary xylem on the inner side and secondary phloem on the outside. The cells of the cambium are highly vacuolar and uninucleate. They have thin walls made up of cellulose and showing numerous pits.

Types of vascular bundles   Based on the presence or absence of cambium, the vascular bundles can be classified as follows:

Open vascular bundle The vascular bundles containing cambium in between phloem and xylem is called open bundle. It is a characteristic feature of the stems of dicots and gymnosperms.

Closed vascular bundle A vascular bundle without cambium is called a closed bundle. These bundles are found in the stems of monocotyledons.

Depending upon the arrangement of the xylem and phloem in the bundle, the vascular bundles are classified as follows:

Radial vascular bundles In roots, the xylem and phloem occur as separate strands on different radii alternating with each other. In between them a parenchymatous conjunctive tissue may be present. The xylem is always exarch in roots.

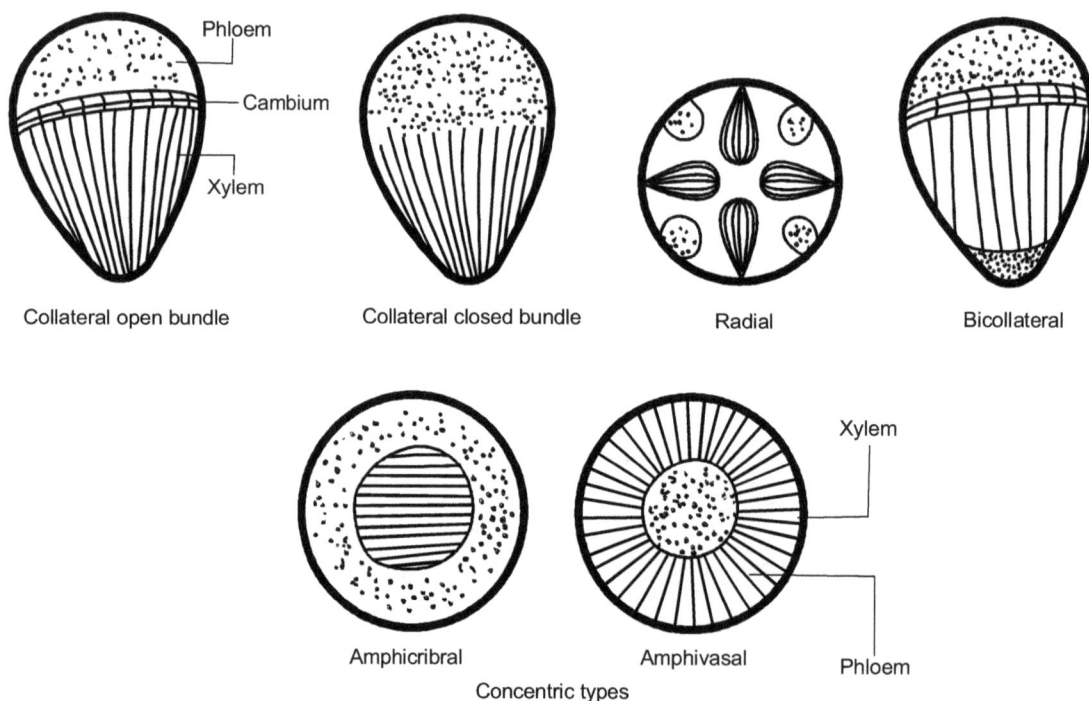

Collateral open bundle   Collateral closed bundle   Radial   Bicollateral

Amphicribral   Amphivasal

Concentric types

Figure 16.11 Types of vascular bundles

Conjoint vascular bundles The xylem and phloem occur one above the other on the same radius to form a bundle. It is further classified into the following types:

❋ *Collateral type* Here, the phloem occurs outside and xylem occurs on the inner side on the same radius of the bundle, with or without cambium in between. Collateral vascular bundles are very common in stems and leaves. It may be open (e.g., *Helianthus*) or closed type. (e.g., Maize) .

- *Bicollateral type* Here, the phloem occurs on both sides of xylem with cambium (i.e., the sequence is outer phloem, outer cambium, central xylem, inner cambium and inner phloem) on the same radius. It is always of the open type, e.g., *Cucurbita* stem.
- *Concentric type* Here, one kind of vascular tissue completely surrounds the other tissue, to form a bundle. It is always of the closed type. It is further classified as amphicribral with phloem surrounding xylem e.g., *Polypodium* and amphivasal type with xylem surrounding phloem, e.g., *Dracaena*.

In dicot stems, the vascular bundles are arranged in the form of a ring, enclosing a large pith, hence, the stele is called eustele whereas, in the stems of monocotyledons, the bundles are found to be scattered, hence the stele is called atactostele.

SECRETORY TISSUES

There are a large number of plants which have special cells or groups of cells that secrete products from the plant body. These secretory cells may occur at certain localized areas or may be distributed throughout the plant body. The secretions may be either useful or of no use to the plant. The former is called excretion, whereas, the latter is called secretion. Hormones and enzymes are useful to the plants. Most of the secretions that are not of any use to the plants are of great economic value to man, e.g., alkaloids, resins, latex, gums, oils, etc.

Classification of Secretory Tissues

Secretory tissues can be classified into two types—laticiferous tissue and glandular tissue.

Laticiferous tissue These tissues are generally long, thin-walled, profusely branched and multinucleate tubelike structures that contain a milky juice with varying refractive index known as latex. Latex is secreted by the cytoplasm.

Latex may occur either in a series of cells or in single long cells. Both these specialized structures are termed as laticifers. Laticifer cells help in the storage of food in the form of starch, and also act as translocatory tissues. The latex of rubber plant is of great industrial importance. The latex of *Carica papaya* is rich in proteolytic enzyme papain. In *Argemone*, the latex is poisonous and serves to protect the plant against grazing animals. Opium (alkaloid) is present in the latex of *Papaver somniferum*.

The cell wall of laticifers is entirely primary and may be as thick as or thicker than the neighbouring parenchyma cells. The thick walls contain cellulose and a high proportion of pectic substances and hemicelluloses. Laticiferous tissue is found in a large number of plants belonging to the families Euphorbiaceae, Asclepiadaceae, Papaveraceae, Apocynaceae, Musaceae, etc.

On the basis of the structure of the laticifer, it is further divided into two types:

Latex cells These are also called non-articulated latex ducts. They develop from a single cell which greatly elongates with the growth of the plant. Latex cells may be branched or unbranched. Such laticifers are also termed as laticiferous cells. Latex cells differ from the latex vessels in that, they are not formed due to cell fusions. Without fusing together they form a network. The latex cells are found in *Catharanthus, Ficus, Euphorbia*, etc. The non-articulated laticifers usually do not anastomose. Non-articulated laticiferous cells are characteristic of various members of the families of Apocynaceae, Asclepiadaceae, Euphorbiaceae, Moraceae and Urticaceae.

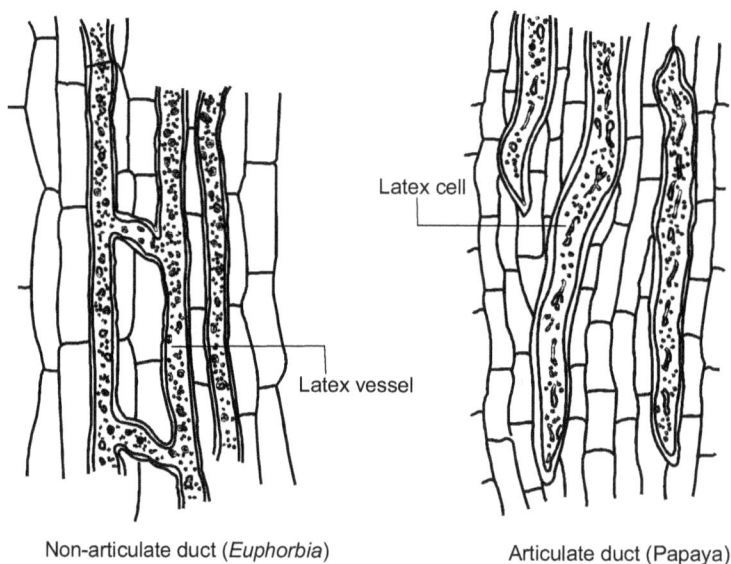

Non-articulate duct (*Euphorbia*)   Articulate duct (Papaya)

Figure 16.12   Laticiferous tissue

Latex vessels These are also called articulated laticifers or compound laticifers. They are composed of a large number of cells, placed end to end with their transverse walls dissolved so as to form long vessels. Such laticifers are also termed as laticiferous vessels. A characteristic feature of the latex vessels is that, they may be single, or two or more vessels anastomose with each other and their branches fuse with each other forming a network. The latex vessels are characteristic of different members of the Compositae (Asteraceae), Convolvulaceae, Papaveraceae, Euphorbiaceae, Caricaceae, Sapotaceae, Liliaceae and Musaceae families. In poppy, the latex contains alkaloids and is a source of morphine and opium. The latex of rubber plant yields rubber of commerce. Latex in *Argemone mexicana* is yellow and poisonous. In *Musa*, the latex is watery.

Glandular tissue As the name indicates, this tissue is present in the form of glands in various parts of the plant.

A gland is a specialized group of cells that are endowed with the capacity to secrete or excrete products. The products are secreted by the cytoplasm of the cell and may be stored in the vacuoles of the cells or in a central cavity, which may be schizogenous, or lysigenous. The cells composing these tissues are generally living, thin-walled and made up of cellulose. Their secretions include gums, resins, honey, etc. Glandular tissue may consist of two types of glands namely external and internal.

External glands They generally occur on the epidermis, in the form of an outgrowth. They are of the following types:

Hydathodes These are also called water stomata, through which water exudes in the form of drops. They are found in the grass leaves, mostly at the tips of marginal veins of plants growing in humid places.

Glandular hairs These hairs are present in the epidermal layers of the leaves and are of various kinds. They may be unicellular or multicellular with thin or thick walls. The hairs may consist of unicellular or multicellular stalk and globular head composed of one or more secretory cells, e.g., *Geranium* and *Citrus*. The oils secreted by these glands are usually odoriferous and commercially much useful.

Stinging hairs These are present on the surface of the leaf of *Urtica dioica*. Their walls are silicified or calcified and brittle. They secrete poisonous substances that cause irritation of skin.

Nectaries These are special structures that secrete nectar and are present on various organs of flowers, e.g., petals (*Ranunculus*), stamens (*Brassica*), involucres (*Euphorbia*), and also on the leaves (*Passiflora*). The nectaries may be extrafloral or floral nectaries, based on their position of the organs.

Digestive glands These are otherwise called enzyme-secreting glands. A few plants like insectivorous plants possess the power of digesting proteins from the bodies of insects. Therefore, they secrete certain digestive enzymes by means of glands or glandular hairs, to obtain nitrogen from foreign bodies, e.g., *Drosera*.

Internal glands There are many plants that possess glands embedded in the various tissues of the plant body. They are of the following types:

Oil glands These glands are found in the mesophyll of the leaves, cortex of stems, and petals of flowers. Oil glands are formed lysigenously, e.g., leaves of *Eucalyptus* and *Citrus*.

Mucilage-secreting ducts   Mucilage-secreting ducts are found in the leaves of Malvaceae. In *Cycas* also, there are ducts that secrete mucilage, which is used for preparing sago.

Resin ducts   Resin ducts are a common feature of conifers and are present in the primary body. In the secondary body they are more liable to be influenced by external factors. The resin ducts are formed schizogenously. They appear like tubes and are lined by small parenchymatous cells, e.g., leaves and stems of *Pinus*.

Gum ducts   The phenomenon termed gummosis occurs primarily as a result of metamorphosis of the organized cell wall materials to unorganized amorphous material, such as gums or resins. In extreme cases, gummosis leads to the formation of lysigenous gum or resin cavities or ducts, in which gum or resin is secreted, e.g., *Acacia*.

GROSS ANATOMY OF SELECTED PLANTS

1. SENNA LEAFLET

Botanical name   *Cassia angustifolia* Vahl.

Common name   Indian Senna

i.   Morphology of the Leaflet

Type	Leaflets of paripinnately compound leaf
Lamina	Isobilateral and entire
Petiole	Short petiole
Size	2.5–6 cm × 7–8 mm (Length × Width)
Shape	Lanceolate
Base	Less asymmetrical
Margin	Entire
Apex	Less acute with a sharp tip
Venation	Pinnately reticulate
Surface	Pubescent with scattered trichomes on both sides of the leaflet
Colour	Yellowish green
Texture	Thin and flexible

ii. Histology

A senna leaflet has an isobilateral structure. In a transverse section the lamina and the midrib show the following anatomical features:

Lamina

a. *Upper and lower epidermis* Both the layers show similar features.

* ❈ They consist of a single layer of polygonal cells and these cells are covered with a thick cuticle on the outer side.
* ❈ The cells have straight anticlinal walls and some epidermal cells contain mucilage.
* ❈ Both surfaces contain scattered, non-glandular, unicellular, short, thick and non-lignified, warty hairs and are curved near the base.
* ❈ The epidermal layers are interrupted with stomata of paracytic type and the number of stomata is equal on both sides.

b. *Mesophyll* It is differentiated into upper and lower palisade layers and middle spongy tissue. It contains clusters of crystals.

Upper palisade The upper palisade is single-layered and compact with elongated, narrow, columnar cells, continuing over the midrib region.

Lower palisade Unlike the upper palisade, the lower palisade cells are smaller and restricted only to the wing region. The cells have wavy walls and are loosely packed.

Spongy parenchyma It is present in between the upper and lower palisade tissues. It is parenchymatous, thin-walled, narrow and loosely arranged. It contains vascular strands and sphaeraphides. The vascular strand consists of a radiating xylem and phloem with an arc of fibres.

Midrib

* ❈ The midrib is biconvex.
* ❈ A zone of collenchyma is present below the upper and lower epidermal layers at the midrib.
* ❈ The epidermal layers are continued over the midrib region.
* ❈ The cells of the lower epidermis are smaller with thick cuticle.
* ❈ The upper palisade cells are relatively smaller at the midrib.
* ❈ The central portion of the midrib is occupied by a large collateral vascular bundle with upper xylem and lower phloem.
* ❈ The vascular bundle is covered on both sides by groups of sclerenchymatous fibres and these fibres are ensheathed in a layer of parenchyma.
* ❈ Calcium oxalate crystals are present in the parenchymatous cells.

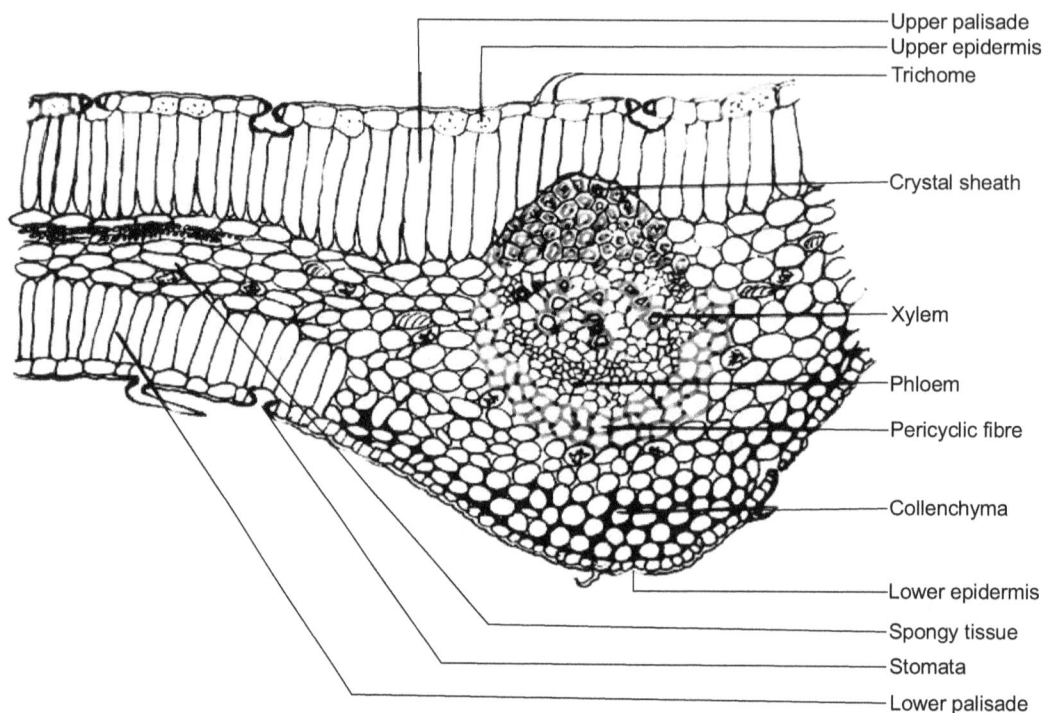

Figure 16.13   T.S. of Indian Senna leaflet (×100)

2. DATURA LEAF

Botanical name   *Datura stramonium* Linn.

Common name   Stramonium

i.   Morphology of the leaf

Type	Simple
Lamina	Dorsiventral, twisted, shrunken and show adnation
Petiole	Petiolate
Size	(Length × Width) 8–25 cm × 7–5 cm
Shape	Cuneate
Base	Sub-cordate
Margin	Sinuate to dentate
Apex	Acuminate
Venation	Palmately reticulate

Surface	Pubescent
Colour	Dark grayish green
Texture	Chartaceous with hairs on both sides

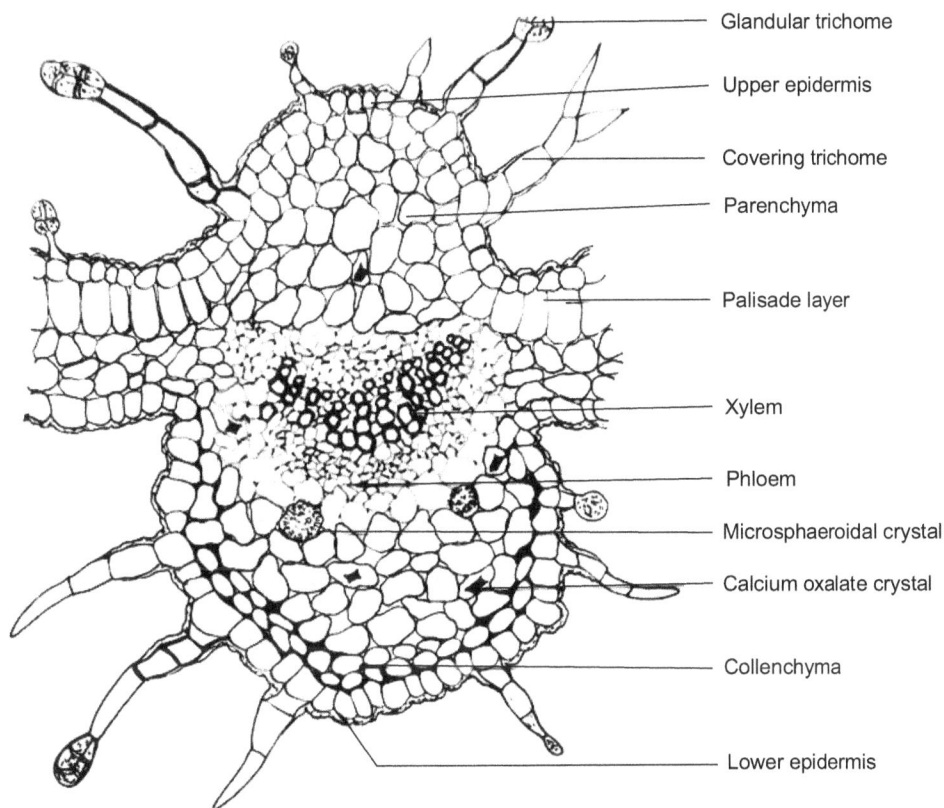

Figure 16.14   T.S. of Datura leaf (×100)

ii. Histology

A transverse section of a leaf of *Datura* shows a bifacial structure. The following are the important characters of the lamina and midrib.

Lamina

a. *Upper and lower epidermis*   Both the layers are single-layered and the cells are rectangular.

* The epidermal layers are covered with a smooth cuticle.
* The epidermal trichomes are uniseriate, multicellular, warty and blunt at the apex.

* Small glandular hairs with one- or two-celled stalk and oval head of 2–7 cells are also present.

* The leaf is amphistomatic with anisocytic type of stomata that are more on the lower side than on the upper side.

* The cells of the lower epidermis have wavy walls.

Mesophyll   The mesophyll is differentiated into upper palisade and lower spongy tissue.

* Palisade is single-layered and compact, and the cells are radially elongated.

* Spongy parenchyma is multilayered and loosely arranged with intercellular spaces.

* The spongy tissue consists of vascular bundles, microspheroidal crystals and sphaeraphides in the upper region.

Midrib   The upper region of the midrib is ridged with epidermal covering hairs. The upper and lower hypodermis consist of 2–3 layers of collenchymatous tissue. A collateral vascular bundle is present at the centre of the midrib. It consists of adaxial xylem and abaxial phloem tissue. The xylem elements are arranged in short radial rows with protoxylem on the adaxial side of the leaf. The ground tissue is formed of cortical parenchyma and it contains prisms of calcium oxalate and microsphaeroidal crystals. Both the covering layers are continuous in the midrib region of the leaf.

3. CINCHONA BARK

Botanical name   *Cinchona succirubra* Pav.

Common name   Cinchona

i. **Morphology of the Bark**

Form	In quills or small curved pieces
Size	(Length and thickness 30 cm and 2–6 mm)
Colour and outer surface	Reddish brown, yellowish brown or grey/rough due to longitudinal and transverse ridges, fissures and wrinkles; grayish lichenous patches may be present.
Colour and inner surface	Dark reddish brown or pale yellowish brown with longitudinal striations.
Fractures	Short fractures in the external layers and fibrous in the inner layers.
Odour	Light and characteristic
Taste	Bitter and astringent

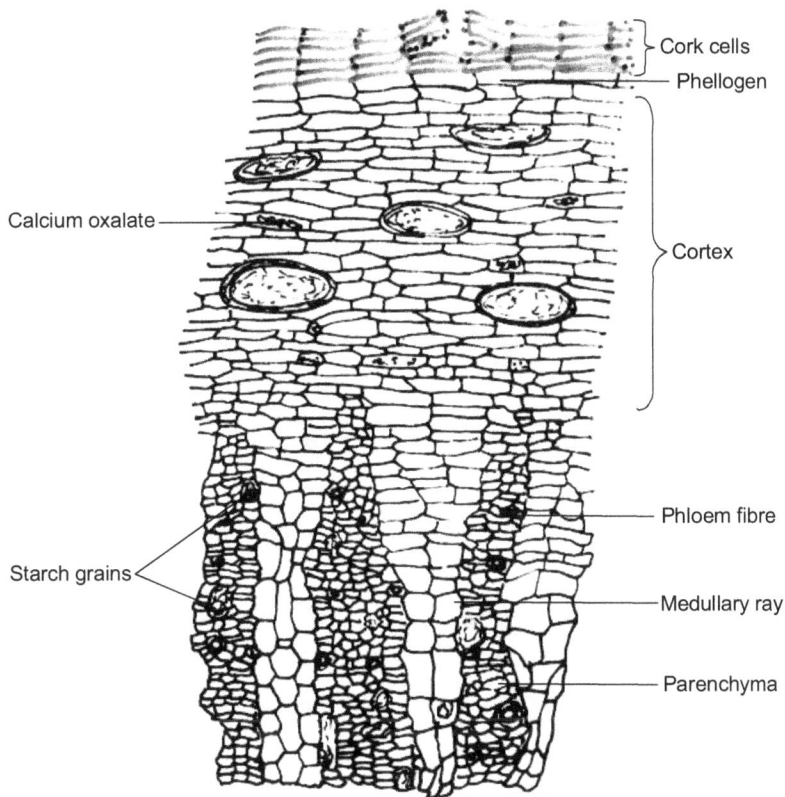

Figure 16.15 T.S. of cinchona bark (×100)

ii. Histology

A transverse section of the bark of *Cinchona* shows the following microscopical features.

Periderm Anatomically, the periderm consists of outer cork cells (Phellum), middle phellogen (cork cambium), and inner phelloderm (secondary cortex).

Cork The cork consists of multilayered thin-walled cork cells that are arranged in regular radial rows. In surface view, these cork cells appear polygonal in shape and are dark reddish in colour due to impregnation of suberin.

Phellogen It is otherwise called cork cambium and occurs inner to the cork cells as 2–3 layers of thin-walled rectangular cells, which are arranged in regular rows. These cells contain dense cytoplasmic content, and are meristematic in nature. These cells cut off the outer cork cells and the inner secondary cortical cells.

Secondary cortex It is otherwise called phelloderm and occurs inner to the phellogen as multi-layered, tangentially elongated, thin-walled cells. The cells contain amorphous reddish-brown substance or small starch grains or even microspheroidal crystals. Idioblasts containing microcrystals of calcium oxalate and secreting cells are also found scattered in the secondary cortical cells.

Bast The secondary phloem which is present inner to the secondary cortex is called bast. It consists of phloem parenchyma, phloem fibres, narrow sieve tubes, companion cells, and phloem medullary rays. The phloem fibres are larger and spindle-shaped with thick conspicuously striated walls traversed by funnel-shaped pits. The fibres are present along with phloem parenchyma and in between the phloem rays. Phloem fibres are numerous, mostly isolated, rounded to oval in shape and yellow in colour. Fibres are thick- walled, heavily lignified with a narrow lumen. The distribution of phloem fibres differ in various species. The phloem medullary rays are 2 to 3 cells wide, thin-walled and the cells are somewhat radially elongated, and contain starch grains.

4. CINNAMON BARK

Botanical Name *Cinnamomum zeylanicum* Bl.

Common name Cinnamon

i. Morphology of the Bark

Form	Single or compound double quills of about 30 or more pieces
Size	(Length, diameter and thickness) Up to 1 m; 6–10 mm; 0.5 mm
Outer surface	Yellowish brown, with wavy lines of pericyclic fibres, scars and holes
Inner surface	Dark brown, with longitudinal striations
Fracture	Short and splintery
Odour	Aromatic
Taste	Warm, sweet and aromatic

ii. Histology

Cinnamon bark is an example of inner bark; the cork cells and cortex are absent. However, occasionally, patches of cork and underlying parenchyma are present. A transverse section of the cinnamon bark shows the following features.

Pericyclic fibres   The outermost region consists of 3 to 4 layers of lignified sclerenchymatous cells that occur as continuous bands called pericyclic sclerenchyma.

On the outer margin there are small groups of about 6 to 15 lignified thick-walled cells that occur at intervals and called pericyclic fibres and these fibres have walls with stratification and pit canals.

Secondary phloem   It consists of phloem parenchyma, phloem fibres, sieve tubes, companion cells, and medullary rays.

Phloem parenchyma   It consists of thin-walled cells, some of the cells containing starch grains and calcium oxalate crystals. The parenchyma cells are sub-rectangular in shape. Some of the phloem parenchyma cells contain tannins. The secretory cells contain volatile oil or mucilage.

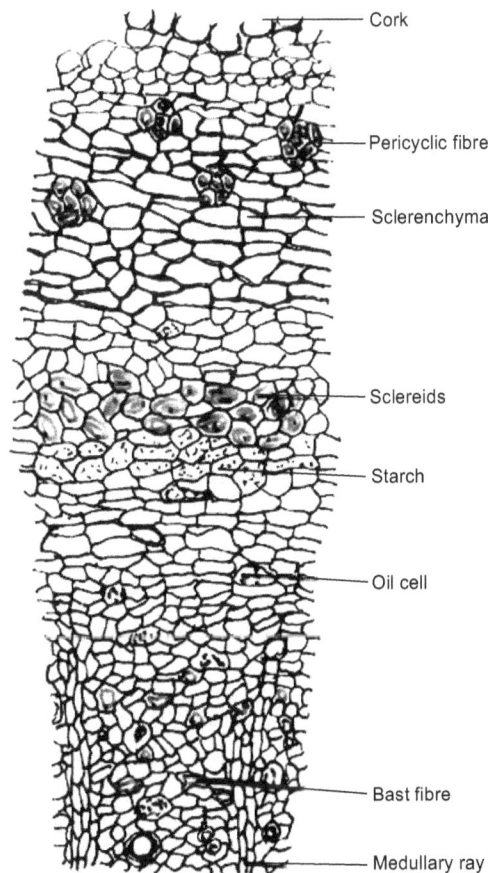

Figure 16.16   T.S. of cinnamon bark (×100)

Phloem fibres These fibres occur as isolated or in tangential rows and are abundantly present towards the inner part of the bark. Phloem fibres are lignified.

Sieve tubes and companion cells These are arranged in tangential bands and are completely collapsed in the outer layers.

Phloem rays These are uni-biseriate near the cambial zone but remain broader towards the outside. The rays are 7–14 cells in height and are radially elongated and thin-walled with yellow-brown cell contents and calcium oxalate crystals.

5. FENNEL FRUIT

Botanical name *Foeniculum vulgare* Gaertn.

Common name Fennel

i. Morphology of the Fruit

Type	Cremocarp with two one-seeded mericarps
Shape	Oval or elliptical; straight or slightly curved, laterally compressed
Size	(Length × breadth) 5–10 mm × 2–4 mm
Surface	Each mericarp has two surfaces: a dorsal surface and a commissural surface. The dorsal surface is glabrous; each mericarp shows five straight, prominent, straw-coloured primary ridges. A bifid stylopod is present at the apex. The commissural surface shows the carpophores which hold together the two mericarps.
Colour	Greenish or yellowish brown
Odour and taste	Strongly aromatic

ii. Histology

Transverse section of a single fennel mericarp shows two prominent surfaces—the commissural surface and the dorsal surface. The commissural surface is flat, containing two prominent ridges with the carpophore in the middle. The dorsal surface has five ridges. Mericarp is divided into pericarp, testa and bulky endosperm. The pericarp is differentiated into outer epicarp, middle mesocarp and inner endocarp.

Pericarp

The percarp consists of an outer epicarp, middle mesocarp and the innermost endocarp.

Epicarp (Exocarp)   It is the outermost layer and encircles the entire mericarp. It consists of a layer of polygonal and tangentially elongated cells with smooth cuticle.

Mesocarp   It lies between the epicarp and endocarp. The bulk of mesocarp is made of parenchyma and bicollateral vascular bundles below the primary ridges. The vascular bundles are surrounded by reticulate and lignified parenchyma. Schizogenous oil ducts are present in the cortex called vittae, which are yellowish brown and elliptical. Four vittae are present between the ridges on the dorsal surface and two on the commissural surface.

Endocarp   It occurs as a single layer between the mesocarp and testa. The cells of endocarp show parquetry arrangement (groups of cells arranged in different directions), which is a typical character of the family Apiaceae (Umbelliferae).

Testa   It is single-layered and is yellow in colour.

Endosperm   It consists of a wide, polygonal and thick-walled parenchyma, which is colourless. The cells contain oil globules, aleurone grains and minute calcium oxalate crystals. It also contains raphae in the middle of the commissural surface in front of the carpophore. A crescent-shaped embryo is observed in sections passing through the apical region of the mesocarp.

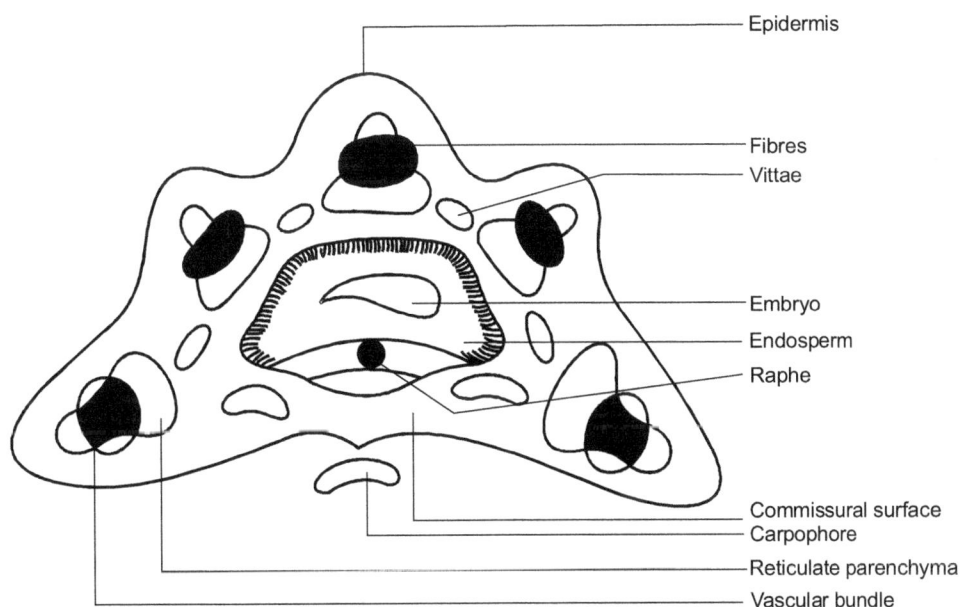

Figure 16.17   T.S. of mericarp of fennel fruit (× 50)

6. CLOVE FLOWER BUD

Botanical name *Syzygium aromaticum* Linn.

Common name Clove

i. Morphology of the Flower Bud

Each flower bud consists of

* ❋ a lower solid stalklike portion called hypanthium
* ❋ A dome-shaped upper portion called head or crown

Hypanthium	It is subcylindrical, four-sided, slightly flattened and tapering below. It is 10 to 13 mm long, 4 mm wide and 2 mm thick. It contains numerous schizogenous oil glands in the hypodermis. The upper part contains inferior bilocular ovary.
Crown	It consists of calyx, corolla, stamens and style. Calyx consists of 4 thick sepals. Corolla contains four pale yellow bowl-shaped petals. It encloses many stamens and forms the head of the bud. Gynoecium contains inferior bilocular ovary, with numerous ovules on axile placentation. Style is single and erect.
Odour	Strong, spicy and aromatic.
Taste	Pungent and aromatic.

ii. Histology

A transverse section of clove hypanthium shows the following structures:

Epidermis It consists of a single layer of small cells with straight walls. The cells are heavily cuticularized, and ranunculaceous type of stomata are present in the epidermis.

Cortex A zone of parenchymatous cells called cortex occurs below the epidermis and is differentiated into 3 different regions. The peripheral region is formed of 2–3 layers of radially elongated parenchymatous cells, in which numerous schizolysigenous oil glands are embedded. The oil glands are ellipsoidal in shape and the long axis of the gland is placed radially and consists of an epithelium with 2–3 layers of flattened cells.

The middle region of the cortex contains one or two rings of bicollateral vascular bundles, which are associated with a few pericyclic fibres. The xylem tissue consists of 3–5 vessels, which are lignified.

The inner cortex consists of a zone of aerenchyma enclosing a number of air spaces that are separated by lamellae with one-cell thickness.

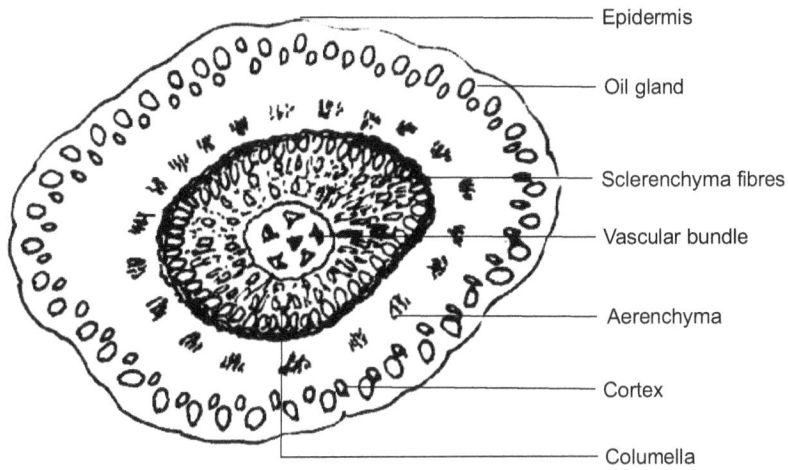

(a)

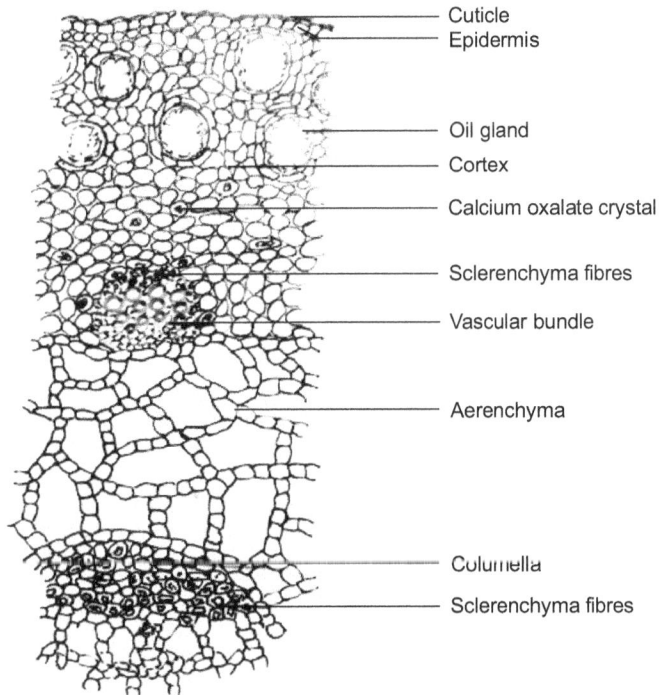

(b)

Figure 16.18 Clove flower bud (a) T.S. of Hypanthium (G.P.) (b) T.S. of of flower bud with a portion enlarged (× 50)

Columella The central part of the hypanthium forms the columella and is composed of thick-walled parenchyma. It is rich in clusters of calcium oxalate crystals. The outer region of the columella contains a ring of about 17–18 vascular bundles.

The hypanthium, in the region of the ovary, shows epidermis, a layer of oil glands and a ring of bicollateral bundles. The basal part of the ovary is parenchymatous tissue. The placentas contain vascular bundles and are rich in clusters of crystals.

7. GINGER RHIZOME

Botanical name *Zingiber officinale* Rosc.

Common name Ginger

i. Morphology of the Rhizome

Form	Thick, sympodially branched, laterally compressed with horizontal rhizome
Size	(Length, width and thickness) 5–15 cm; 3–4 cm; and 1–1.5 cm
Surface	Rhizome is longitudinally striated with rarely projecting fibres. A depressed scar is present at the apex. Nodes, internodes and scales are differentiated in fresh samples, but not clear in dried materials.
Fracture	Short and fibrous
Odour	Agreeable and aromatic
Taste	Pungent

ii. Histology

A transverse section of unpeeled ginger rhizome shows the following structures:

Cork It consists of an outer zone of parenchymatous cells of a few layers, which are dark brown in colour and are irregularly arranged. The inner zone of cells are regularly arranged in radial rows of parenchymatous tissue.

Cortex The cortex contains thin parenchymatous cells with intercellular spaces. These cells contain abundant starch grains, that may be simple, ovoid or sack-shaped grains with a distinct eccentric hilum. Suberized cells with yellowish brown oleoresin are also found in the cortex. The cortex also contains scattered vascular bundles, that are surrounded by sclerenchymatous sheath.

Endodermis It consists of a single-layered, thick-walled cells that lack starch grains.

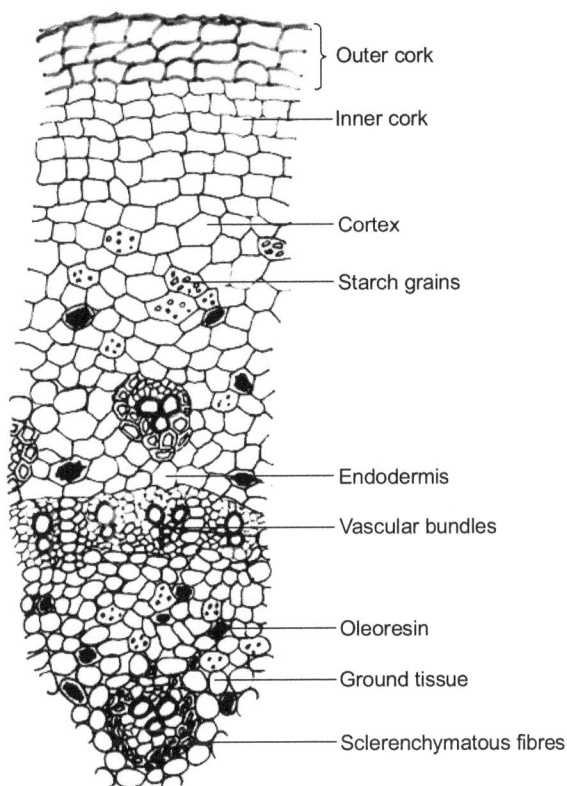

Figure 16.19 T.S. of rhizome of ginger (×100)

Ground tissue It contains large parenchymatous cells with cellulosic walls. The cells are rounded to polygonal with abundant starch grains, oleoresin and vascular bundles. The vascular bundles are collateral, conjoint, and surrounded by a sheath of septate and non-lignified fibres. Each vascular bundle consists of the outer phloem with distinct sieve tubes and companion cells, and xylem composed of annular, spiral or reticulate vessels. The outer ring of vascular bundles of ground tissue is not covered by sclerenchymatous fibrous sheath.

8. IPECAC ROOT

Botanical name   *Cephaelis ipecacuanha* (Brot.) A. Rich.

Common name   Ipecac

i. Morphology of the Root

Shape	Roughly cylindrical and tortuous
Size	(Length and thickness) 5–15 cm and 4–6 mm

Surface	Closely annulated externally; the annulations are broad and rounded, and completely encircle the root.
Fracture	Short at the bark and splintery at the wood
Fractured surface	Wide grayish bark on the outer side and small dense wood on the inner side
Colour	Reddish brown to dark brown
Odour	Slight
Taste	Bitter

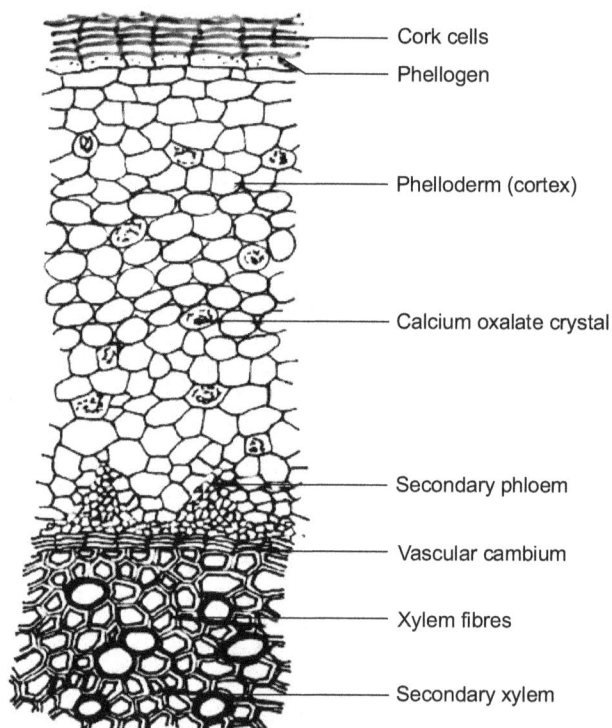

Figure 16.20   T.S. of root of Ipecac (×100)

ii. Histology

A transverse section of the root of ipecac shows periderm, cortex and vascular tissue.

Periderm   The outermost part of the periderm consists of 3 to 5 layers of tangentially elongated, thin cork cells that are brown in colour. This region is followed by 2 layers of phellogen, which occur below the cork. The cells of phellogen are shining and arranged tangentially.

Cortex It consists of multi-layered, thin-walled parenchymatous tissue, with small intercellular spaces. The cells are polyhedral in shape. The cortical cells contain muller-shaped starch grains and acicular raphides of calcium oxalate crystals either in bundles or are scattered.

Vascular tissue The central part of the vascular tissue consists of primary xylem, which is surrounded by a dense wood of secondary xylem. The compact secondary xylem is composed of small vessels, fibres and xylem parenchyma. The xylem is traversed by rows of radially elongated medullary rays, which contain starch grains.

The xylem is externally surrounded by a small zone of vascular cambium and the secondary phloem lies outer to the cambium as thin patches of sieve tubes, companion cells and parenchyma.

9. NUX-VOMICA SEED

Botanical name *Strychnos Nux-vomica* Linn.

Common name Nux-vomica

i. Morphology of the Seed

Shape	Disc-shaped or somewhat flat, depressed on one side and arched on the other or sometimes irregularly bent and concavo-convex.
Size	(Diameter and thickness) 10–30 mm and 3–5 mm
Outer surface	The outer surface of the seed is hard and grey to greenish grey, in colour. The testa of the seed is covered with numerous silky hairs that are unicellular and lignified. These hairs are closely appressed and radiating from the centre of the seed. Hilum is present at the centre of the flat surface. Micropyle occurs as a small projection in the margin. Hilum and micropyle are connected by a ridge.
L.S. of seed	The L.S. of the seed shows that the embryo is located at the micropylar end and it has a cylindrical radicle and two cordate cotyledons. Endosperm is translucent and grayish in colour with a white embryo.
Odour	Odourless
Taste	Extremely bitter taste

ii. Histology

A transverse section of the seed of nux-vomica shows the following features:

Epidermis The epidermis is hairy and single-layered. Each epidermal cell forms a thick-walled, bent and lignified trichome. The base of the trichome is bulbous, with a terminal projection. Each projection is narrowly elongated and slightly bent beyond the base. The upper part of the trichome is nearly at right angle to the base and has about ten longitudinal ridgelike thickenings. The epidermal hairs run parallel in one direction giving the testa a silky appearance.

Below the epidermis lies a two-layered collapsed parenchyma, which contains yellowish brown contents. It is further followed by the endosperm.

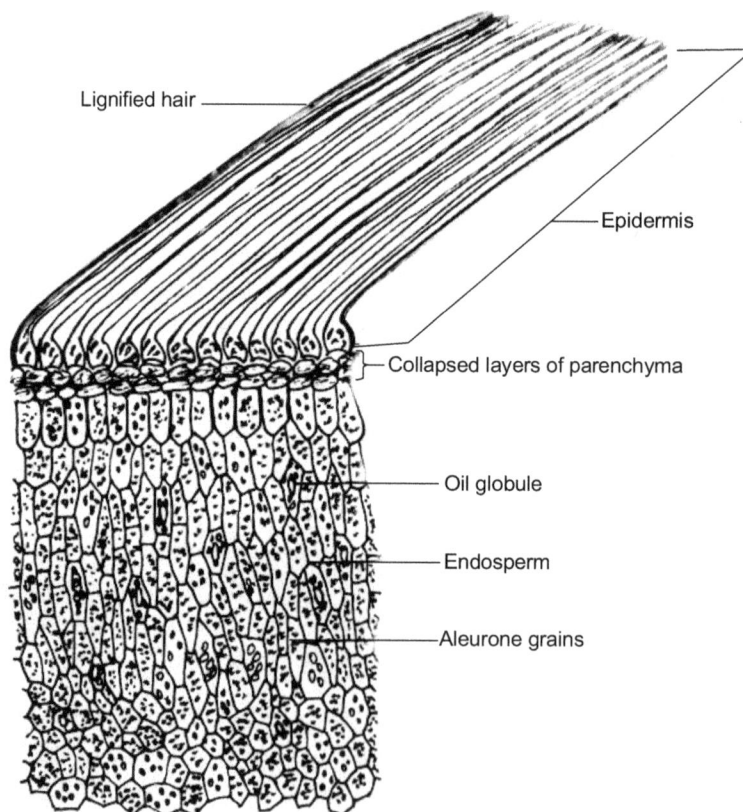

Figure 16.21 T.S. of seed of nux-vomica (×100)

Endosperm It consists of large, thick-walled cells. The cells are non-lignified and composed of carbohydrate. The outer layers of the endosperm consist of palisadelike cells, whereas the inner layers contain polygonal cells and the cells are interconnected by plasmodesmata. The endosperm cells contain aleurone grains and oil droplets. The inner layer of the endosperm contains abundant strychnine and brucine in the outer part of the endosperm.

REVIEW QUESTIONS

State whether the following sentences are true or false

1. Dumb-bell shaped guard cells occur in the members of dicotyledons.
2. The latex cells are distributed in Ficus.
3. The epidermal cells of the seeds of nux-vomica form bent thick-walled and lignified trichomes.
4. Epiblema is the other name for leaf epidermis.
5. The specialized cells of stomata are called guard cells.
6. Mechanical support is provided by thick-walled prosenchyma.
7. G-fibres contain gelatinous substance.
8. *Cucurbita* seeds produce surface fibres.
9. Cambium is absent in monocotyledons.
10. The stele of dicot stem is atactostele.

Choose the correct answer

11. Cremocarp is the fruit of _____
    a. poppy   b. clove   c. fennel
12. Pericyclic fibres occur in the bark of _____
    a. cinnamon   b. sandal   c. *Cephaelis*
13. Multiple epidermis is a characteristic feature of the leaf of _____
    a. sunflower   b. senna   c. *Nerium*
14. Knoll and Ruska discovered _____
    a. electron microscope   b. dialator   c. scanning

15. Caryophyllaceous type of stomata is otherwise referred to as _____
    a. paracytic   b. diacytic   c. anisocytic
16. Vesicles of endoplasmic reticulum secrete _____
    a. oil   b. nectar   c. resin
17. Periderm consists of phelloderm, phellogen and _____
    a. periblem   b. plerome   c. phellum
18. Septate fibres are observed in _____
    a. teak   b. Mahogany   c. mango
19. Seed coat of pulses contains _____
    a. brachy-sclereids   b. astrosclereids   c. macrosclereids
20. Nucleus is absent in the protoplast of _____
    a. parenchyma   b. companion cells   c. sieve tubes

Answers				
1. False	2. True	3. True	4. False	5. True
6. True	7. True	8. False	9. True	10. False
11. c	12. a	13. c	14. a	15. b
16. b	17. c	18. b	19. c	20. c

Answers the following

1. What are epidermal appendages?
2. Define the term gummosis?
3. What is prosenchyma? Mention its significance.
4. Explain the classification of trichomes.
5. What are fibres? Give an example.
6. "Stomatal types play a major role in the identification of plant species"–Discuss.
7. Explain dermal tissue system with examples.
8. Comment on sclereids.
9. Write an essay on vascular tissues.
10. Describe the various components of phloem tissues.
11. Explain the types of vascular bundles that you have studied.
12. Distinguish between amphicribral and amphivasal vascular bundles.

13. Differentiate stomata from hydathodes.

14. With suitable examples, explain the distribution of secretory tissues and their importance in plants.

15. Draw and explain the anatomical structure of Senna lealfet.

16. With suitable diagrams bring out the gross anatomical structure of the bark of *Cinchona*.

17. With the help of anatomical features, how will you identify the fruit of fennel.

18. Write the anatomical features of a clove bud.

19. Draw and label the parts of T.S. of rhizome of ginger.

20. Write down the anatomical characters of the seed of nux-vomica.

SYSTEMATIC EXAMINATION OF SELECTED POWDERED DRUGS

17

INTRODUCTION

Herbal drugs made from roots, stem, leaves, flowers, fruits and seeds of a plant are available in different forms in the market. A common method of dispersal of a drug is mostly in its powdered form. As collection of herbal drugs are done by untrained collectors who have no knowledge about the plants, the powdered form of any crude drug is very easily prone to abuse. Therefore, in order to identify the drug easily and correctly and also to distinguish it from other drugs, the systematic description of the drug in powdered form is very essential because of the following reasons:

* Use of whole drugs, where identification is more reliable, and preparation of the powder from these whole drugs only prior to its incorporation in a formulation, is optimal, but manufacturers are tempted to store or even buy drugs in the powdered form for ease of handling and storage. When quality-control officials come across powdered drugs, samples are drawn for analyses, and are required to be tested as per the pharmacopoeia thus making the description of the drug necessary.

* Chances for adulteration, spoilage, insect infestation or misidentification increase for a powdered drug.

* Knowledge of the diagnostic features of a powdered drug also helps in the identification of ingredients in a compound formulation when such powders are added *in situ* to them.

Drugs can be classified as organized or unorganized depending upon whether they have a definite structure or not.

EXAMINATION OF ORGANIZED DRUGS

Organized drugs are those which have a definite structure, e.g., bark, leaves, flowers, fruits, seeds, etc.

About four 200 mg samples of a selected drug are taken.

✻ One sample is mixed thoroughly with water in a watchglass.

✻ The second is warmed, cleared in chloral hydrate and washed thoroughly in water.

✻ The third is stained with phloroglucinol and conc. hydrochloric acid. It is then washed thoroughly in water.

✻ The fourth is stained with dilute iodine solution and again washed thoroughly with water.

✻ A drop from each of the above is mounted on slides in glycerin. It is observed under a microscope and the salient characters are noted. The microscopical characters of selected organized drugs are given in Table 17.1.

Table 17.1 Microscopical characteristics of different organized drugs

Organ	Microscopic features
Wood	Vessel elements, fibres, wood parenchyma and medullary rays are seen. Usually, all the cellular structures are lignified (except parenchyma). The length and description of the fibres are necessary when making a diagnostic test. The length and width of medullary rays as seen in radial and tangential sections has to be noted; tyloses, cell inclusions like starch, crystal, resin, secretory cells are also important.
Bark	Shows the presence of sieve tubes, cork cells, secondary parenchyma, stone cells, pericycle fibres and phloem fibres, cell inclusions like crystals and starch grains, but rarely secretory cells. Structures like epidermal cells, palisade cells, xylem vessels, tracheids and aleurone grains are absent.
Leaves	Epidermis, stomata, cellulose parenchyma, trichomes, palisade cells and crystals are present. A detailed description of the trichomes and stomata is important.
Seeds	Endospermic cells, aleurone grains, oil globules, starch grains, hemicellulose and a small vascular tissue are present.
Fruits	Contains all the histological features of the seed and also prominent vascular elements, pericarp (containing lignified elements), epidermal cells and endocarp (sclerenchymatous in nature).
Flowers	Contains pollen grains, anthers, surface papillose, sepal and petal epidermis and trichomes.
Herbs	Shows the presence of all the histological features of flowers, leaves, wood and bark.

EXAMINATION OF UNORGANIZED DRUGS

Unorganized drugs, as the name indicates, are drugs that show no definite structure, e.g., latex, resin, gums, mucilage, etc. Unorganized drugs are fairly homogeneous and may be solids, semi-solids or liquids. For the examination of an unorganized drug, the following macroscopical data can be recorded:

* Form tears, lumps, semi-solid masses, etc.
* Size, colour, odour and taste.
* Surface appearance as such and freshly broken—shiny, smooth, rough, dusty, dull.
* Fracture brittle, granular, conchoidal, clean, or hard to fracture, waxy, etc.
* Any distinctive feature of fractured surface: any other peculiarity.
* Reaction towards water, acids and solvents.
* Fluorescence as such and after preparing solutions, or after chemical reaction with certain reagents.

Unorganized drugs may be sorted by observing the solubility towards alcohol and, then applying other tests.

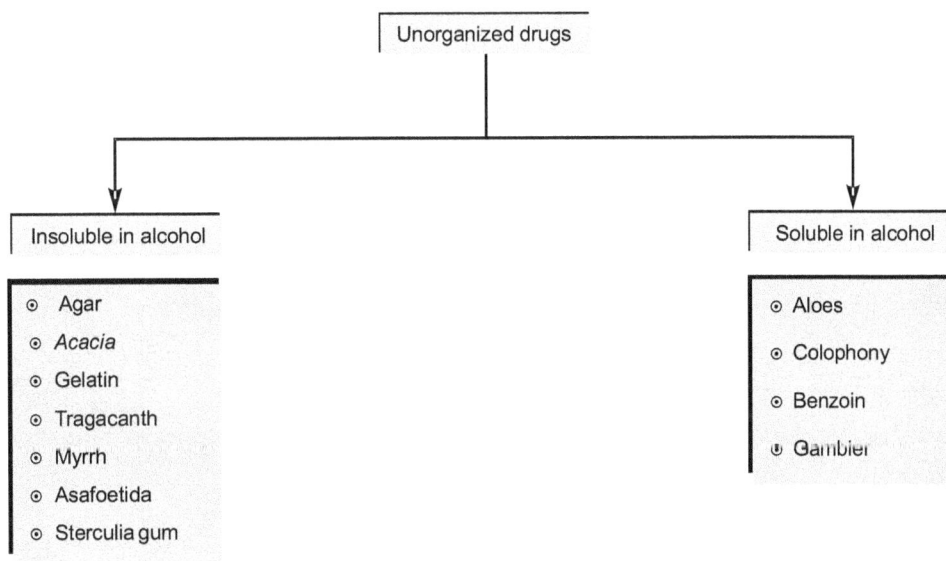

```
                    ┌────────────────────┐
                    │  Unorganized drugs │
                    └────────────────────┘
         ┌────────────────────────────────────────┐
         ▼                                          ▼
┌────────────────────┐                  ┌────────────────────┐
│ Insoluble in alcohol│                 │  Soluble in alcohol │
└────────────────────┘                  └────────────────────┘
  ⊙ Agar                                   ⊙ Aloes
  ⊙ Acacia                                 ⊙ Colophony
  ⊙ Gelatin                                ⊙ Benzoin
  ⊙ Tragacanth                             ⊙ Gambier
  ⊙ Myrrh
  ⊙ Asafoetida
  ⊙ Sterculia gum
```

Preliminary Tests for Drugs Insoluble in Alcohol

i. Triturate the sample with water; if it gives the characteristic smell of asafoetida or myrrh, it may contain these compounds.

ii. Heat the sample with soda lime in a dry test tube; observation of ammoniacal vapours may indicate the presence of gelatin or albumin.

iii. One part of the selected sample is boiled with 100 parts of water; if formation of a stiff jelly-like mass is observed on cooling, it may be agar; if there is coagulation of liquid, it may be albumin.

iv. The sample is warmed in the presence of 5% caustic soda solution. Formation of a canary yellow colour indicates the presence of agar or tragacanth.

v. A small quantity of the sample is mounted in alcohol and moistened with water. The behaviour of the particle of the sample is observed. If the particles of the sample dissolve without swelling, it may be acacia. If the particles undergo swelling and ultimately dissolve or become diffused, it may be tragacanth, sterculia gum, gelatin or agar.

vi. Mount the sample in ruthenium red. Formation of pink colour may indicate the presence of agar or sterculia gum.

vii. The sample is mounted in N/50 iodine solution; formation of olive green colour with blue specks may indicate the presence of tragacanth; formation of crimson or brown colour may indicate the presence of agar; if there is no colour reaction, it may be acacia.

viii. Presence of an acetous odour may indicate the presence of sterculia gum.

Preliminary Tests for Drugs Soluble in Alcohol

i. Test for hydroxymethyl anthraquinone derivatives: Observation of rose-pink colour indicates the presence of aloes (exception: Zanzibar aloes).

ii. Treat the sample with potassium permanganate solution. Observation of benzaldehyde odour (bitter almond smell) may indicate the presence of benzoin.

iii. Treat the sample with alcoholic ferric chloride solution. Observation of green colour indicates the presence of gambier.

iv. If an alcoholic solution of the drug is acidic to litmus, it indicates the presence of colophony.

After the preliminary examinations, the unorganized drugs need to be identified by specific chemical tests.

Chemical Tests for Different Unorganized Drugs

1. Agar   The following tests can be performed on the sample to detect the presence of agar.

Test	Observation	Inference
Boil 1 g of the sample with 10 ml of water until the solution is completely boiled. Cool to room temperature.	A jelly mass is formed.	Agar is present.
To a 10 ml hot solution of 0.2% drug, add 1ml of hot solution of 10% tannic acid.	No ppt. is formed.	Agar is present (distinction from gelatin).
Mount a small quantity of sample powder in a solution of ruthenium red and examine microscopically.	Particles acquire red colour.	Agar is present.
Add 1 drop of N/10 solution of iodine to 10 ml decoction of the sample and rapidly cool under tap water to room temperature.	Pale yellow colour is produced.	Agar is present.
Add 0.5 ml of concentrated hydrochloric acid to 4 ml of 0.5% solution of the sample. Heat it on a water bath for 30 minutes. Cool at room temperature, and divide into two portions:		
a) Add 3 ml of 10% NaOH solution and Fehling's A and B solutions. Heat over a water bath.	Red precipitate of cuprous oxide is obtained.	Reducing sugar is present.
b) Add 10% barium chloride solution.	Slight white preciptate is obtained.	Presence of sulphate in agar (Distinction from acacia and tragacanth).

2. Aloe   Boil 0.5 g of the sample with 50 ml of water until it is nearly dissolved. Cool and add 0.5 g of kieselguhr. Filter and use the filtrate for the following tests.

Test	Observation	Inference
Borax test for anthranol (Schonten's reaction)		
To 5 ml of filtrate add 0.2 g of borax. Heat until dissolved. Add few drops of this solution into a test tube filled with water.	A green fluorescence is observed.	Aloe is present.
Mix 2 ml of filtrate with 2 ml of bromine water.	A pale yellow ppt. of tetrabromaloin is obtained (due to isobarbaloin).	Aloe is present.
Klunge's isobarbaloin test (cupraloin test) Mix the filtrate with an equal volume of water. To this, add few drops of saturated copper sulphate solution, 1 g of sodium chloride and 10 ml of 90% alcohol.	1. Wine red colour persists.	Indicates Curacao aloe.
	2. Wine red colour changes to yellow.	Indicates Cape aloe.
	3. No colour change.	Presence of other two types of aloe.
Mix 5 ml of filtrate with 2 ml of HNO_3:	1. Yellow-brown colour is produced.	Indicates Cape aloe.
	2. Vivid green colour is produced after storing the mixture for some time.	Indicates Curacao aloe.
	3. A deep brownish red/ wine-like colour is obtained.	Indicates Socotrine aloe.
Nitrous acid test (for isobarbaloin)		
To 5 ml of filtrate, add 5 ml of sodium nitrite solution and a few drops of dilute acetic acid.	Pink/purple colour is formed.	Indicates Curacao or Cape aloe.
Modified anthraquinone glycoside test (modified Borntrager's test)		
Shake 0.1 g of the powdered sample with 5 ml of ferric chloride solution (5%). To this add 5 ml of dilute hydrochloric acid and immerse in a water bath and then heat for 5 minutes. Cool to room temperature. Shake gently with 5–10 ml of benzene. Separate the benzene layer and shake with dilute ammonia solution.	Rose–pink to cherry red colour is produced in the aqueous layer.	Aloe is confirmed.

3. Asafoetida/Devil's Dung   Asafoetida can be detected by the following tests.

Test	Observation	Inference
Triturate the sample powder with water.	Yellowish orange emulsion is observed.	Asafoetida is present.
To the fractured surface, add sufficient sulphuric acid.	Appearance of red or reddish brown colouration. If washed with water, the colour changes to violet.	Asafoetida is present.
Over the fractured surface of the powder, add 50% nitric acid.	Appearance of green colour.	Asafoetida is present.
Combined umbelliferone test Boil 0.5 g of asafoetida with 8 ml of hydrochloric acid and filter. To the filtrate, add equal volume of alcohol and excess of liquor ammonia.	Blue fluorescence (ferrulic acid, umbellic acid, umbelliferone) is produced.	Asafoetida is confirmed.

4. Balsam of Tolu   The following tests can be used to detect the presence of balsam of Tolu.

Test	Observation	Inference
Prepare an alcoholic solution of the sample and dip red litmus in it.	Litmus turns blue colour.	Shows acidic nature. It could be balsam. The following tests have to be done to confirm it.
Prepare an alcoholic solution of the drug and $FeCl_3$.	Green colour (Tolu resinotannol) is formed.	Tolu balsam is present.
Add 1 g of the drug to 5 ml of water and boil. Filter, and heat the filtrate with 30 mg of potassium permanganate.	Benzaldehyde odour (smell of bitter almonds) is released	Tolu balsam is confirmed.

5. Beeswax   Beeswax is sometimes adulterated with other fatty substances or resins. The purity of beeswax can be verified by performing the saponification cloud test.

Saponification Cloud Test

Boil 0.5 g of the sample of beeswax for 10 minutes in 10 ml of sodium hydroxide solution, to make up the original volume, and filter through the glass wool, and then acidify with hydrochloric acid. If fatty substances, fatty acids or resins are present, they will be precipitated. Fats may be saponified by boiling with aqueous sodium hydroxide, but waxes are unaffected by aqueous alkali though they can be saponified by strong alcoholic potassium hydroxide.

6. Benzoin   The presence of benzoin can be detected by performing the following tests.

Test	Observation	Inference
Heat 0.5 g of powder slowly in a dry test tube.	Melts and white fumes are evolved which condense on the walls of the test tube to form crystalline sublimate.	Benzoin is present.
Warm 1 g of powder with 5 ml of potassium permanganate.	Distinct odour of benzaldehyde (bitter almonds) is evolved.	Benzoin is present.
Triturate 0.1 g of powder with 95% alcohol (5 ml) and filter. To the filtrate, add 5 ml of 5% w/v ferric chloride solution in alcohol.	Absence of green colour.	Sumatra benzoin is present.
Digest benzoin with 5 ml of ether for 5 minutes. Pour the ethereal solution in a china dish containing 2–3 drops of concentrated sulphuric acid and rotate the dish.	Reddish brown colour is produced.	Benzoin is confirmed.

7. Black Catechu   Black catechu is completely soluble in hot water and alcohol. It can be identified by the following tests.

Test	Observation	Inference
Test for tannins		
i. Prepare 10% solution of powder in water by heating. Add 5% $FeCl_3$ solution to it.	Green colour is produced.	Catechu (tannins) is present.
ii. Add NaOH solution to the above mixture.	Colour changes to purple.	Catechu (tannins) is present.
The powdered substance is dissolved in water and heated. Add 1 drop of this solution to lime water.	Brown colour is produced. After standing, a red ppt. is formed.	Black catechu is present.
Add vanillin hydrochloric acid to original substance.	Pink or red colour (due to phloroglucinol) is produced.	Black catechu is present.

8. Castor oil   The following tests can be used to detect the presence of castor oil.

Test	Observation	Inference
Add 5 ml of light petroleum (40–60%) to 10 ml of the test sample oil.	A clear solution results and on increasing the petroleum to about 15 ml, turbid mixture will be obtained.	Castor oil is present.
Add oil and equal volume of alcohol and cool to 0°C for 3 hours.	A clear liquid is obtained.	Castor oil is present.

9. Colophony   The following tests can be performed on the drug to confirm the presence of colophony.

Test	Observation	Inference
Dissolve 0.1 g of the drug in 10 ml of acetic anhydride in a dry test tube. Heat gently, cool and add a few drops of sulphuric acid.	Bright purplish red colour is produced which rapidly changes to violet.	Colophony is present.
Dissolve the drug in light petroleum and filter. To the filtrate, add about 2–3 volumes of copper acetate solution.	Petroleum layer shows light green colour (due to copper salt of acetic acid).	Colophony is present.
Dissolve the drug in alcohol.	Blue litmus turns red.	Colophony is present.

10. Gelatin   The presence of gelatin can be confirmed on heating the powder with soda lime, ammonia gas is evolved. Gelatin can also be identified by performing a number of tests on a test solution. The test solution is obtained by dissolving 0.5 g of the substance in 100 ml of water and heating.

Test	Observation	Inference
Add Millon's reagent to the test solution.	White ppt. forms which becomes red on heating.	Gelatin is present.
1 ml of 10% sulphuric acid is added to 1 ml of the test solution. Boil gently for 30 seconds. Add 2 drops of 1% sodium nitrite solution.	Red precipitate is produced.	Gelatin is present.
Biuret test Add 1 ml of 5% sodium hydroxide to 3 ml of the test solution to make it strongly alkaline. Add 3 drops of 1% copper sulphate solution.	Violet or pink colour is obtained.	Gelatin is present.
A few drops of 10% tannic acid is added to the test solution.	Violet or white colour is formed.	Gelatin is confirmed.

11. *Gambier/Pale catechu* The following tests can be performed to confirm the presence of pale catechu.

Test	Observation	Inference
Warm 0.3 g of powder with 2 ml of 90% alcohol. Cool and filter. To the filtrate, add 2 ml of light petroleum, shake and allow to separate.	Upper layer shows greenish fluorescence (due to gambier fluorescin).	Pale catechu is present.
A small amount of powder is mounted on a slide with lactophenol and examined under a microscope.	Minute acicular crystals (catechin) are formed.	Pale catechu is present.
Dip a matchstick into the extract. Dry the stick and dip in concentrated hydrochloric acid. Warm near the flame.	Matchstick develops stain (due to phloroglucinol formed with catechin).	Pale catechu is confirmed.

12. Honey The presence of honey can be confirmed by adding 4 ml of alcohol to 1 ml of the substance to form a slight turbidity. Honey is often adulterated with reducing sugars and artificial invert sugar. Fiehe's test is used to detect the presence of artificial invert sugar.

Test	Observation	Inference
Test for reducing sugars		
0.5 g of the substance is dissolved in 10 ml of water and boiled with Fehlings solution A and B.	Brick-red ppt. of cuprous oxide is produced.	Reducing sugars are present.
Fiehe's test		
Add 5 ml of ether to 10 ml of honey and shake thoroughly. Allow the two layers to separate. The upper layer is collected and evaporated to dryness. To the residue, add 1% of resorcinol hydrochloric acid (1 drop only).	a) Transient red colour is obtained.	Honey is present.
	b) Permanent red colour is formed.	Artificial invert sugar in honey is detected.

13. *Indian gum (Gum acacia)* Prepare a solution of the sample by adding 10 ml of water to 0.5 g of the sample, stirring well. Use this solution for the following tests:

Test	Observation	Inference
To a 2 ml solution, add 50 mg of borax.	Stiff translucent mass is formed.	Acacia is present.
Dilute 1 ml of the solution by adding 4 ml of water and add a few drops of dilute lead acetate solution.	Bulky gelatinous white mass/ppt. is formed.	Acacia is present.
To a 3 ml solution, add 0.5 ml of solution of hydrogen peroxide and 0.5 ml of benzidine in ethyl alcohol (10%).	Blue colour is produced.	Presence of acacia.
To 1 ml of solution add 4 ml water and few drops of concentrated hydrochloric acid. Boil over a water bath for a few minutes and add few drops of NaOH solution to neutralize the acid. Then add Fehling's solution A and B and boil over a water bath.	Red precipitate is obtained.	Reducing sugars are present.

14. Lanolin   The presence of lanolin can be detected by checking the presence of cholesterol.

Dissolve 0.5 g of the substance in 5 ml of chloroform. Add 1 ml of acetic acid and 2 drops of concentrated sulphuric acid. A deep green colour confirms the presence of lanolin.

15. Myrrh   The following tests can be used to detect the presence of myrrh.

Test	Observation	Inference
Triturate powdered drug with water.	Yellowish emulsion is formed.	Myrrh is present.
Digest the powder with ether. Dry the ethereal extract and		
i. Treat it with bromine vapours.	Extract becomes reddish brown	Myrrh is present.
ii. Add one drop of HNO_3 to the extract.	Purple colour is produced.	Myrrh is present.

16. Sesame oil   Oil can be identified as sesame oil by shaking 1 ml of the oil sample with a solution of 0.5 g sucrose in 10 ml of hydrochloric acid for 30 minutes. The acid layer becomes bright red, finally changing to dark red. This confirms the presence of sesame oil.

17. Shark-liver oil   The presence of shark-liver oil can be confirmed by performing the following tests.

Test	Observation	Inference
Dissolve 1 drop of the sample oil in 1 ml of chloroform and add 1 drop of concentrated sulphuric acid, and shake well.	A pale violet colour is produced, changing to purple, brown or blue (due to Vit. A).	Shark-liver oil is present.
Add 10 ml of chloroform and a saturated solution of antimony trichloride to 1 ml of the oil.	Blue colour is produced.	Shark-liver oil is present.

18. Sterculia gum/Indian tragacanth The same tests used to confirm the presence of gum tragacanth (explained later) can be used to detect the presence of sterculia gum. Besides these, the following tests can be used.

Test	Observation	Inference
Mount a small quantity of the powder in ruthenium red and examine under a microscope.	Particles acquire pink colour.	Mucilage is present. The drug could have sterculia.
Warm the powder with 5% KOH solution	Brownish colour is noted.	Sterculia is confirmed.

19. Storax The following tests can be used to confirm the presence of storax.

Test	Observation	Inference
Warm 0.5 g of the drug in chloral hydrate.	Clear solution is produced.	Storax is present.
Add 2 ml of potassium chromate and 1 ml of sulphuric acid to 1 g of the drug and warm it.	Odour of benzaldehyde (bitter almonds) is produced.	Storax is present.
4 ml of 2% potassium permanganate solution is added to 1 g of storax.	Odour of benzaldehyde is evolved.	Storax is present.

20. Tragacanth   The presence of tragacanth can be confirmed by the following tests.

Test	Observation	Inference
To 4 g of 0.5% w/v solution, add 0.5 ml of hydrochloric acid, and heat for 30 minutes on a water bath. Divide the hydrolysed product into two parts.		
i. Part 1 Add 1.5 ml of NaOH solution to neutralize, and then add Fehling's solution A and B and heat on a water bath.	Red ppt. (of cuprous oxide) is produced	Reducing sugars present. It could be agar, acacia or tragacanth.
ii. Part 2 Add barium chloride solution.	No ppt.	Distinction from agar.
To a few ml of 5% w/v solution, add a few drops of 20% w/v solution of lead acetate.	Heavy white flocculent ppt. is formed.	Distinction from acacia.
Mount a small quantity of powder in ruthenium red and examine microscopically.	Particles do not acquire pink colour.	Distinction from Indian tragacanth and agar.
To 0.1 g of powder, add N/50 iodine.	The mixture acquires greenish colour.	Distinction from acacia and agar.
Powder is warmed with 5% alcoholic KOH solution.	Canary yellow colour is produced.	Tragacanth is confirmed.

REVIEW QUESTIONS

State whether the following sentences are true or false

1.   Gambier gives green colour with alcoholic ferric chloride.

2.   Beeswax is adulterated with resins.

3.   Appearance of acicular crystals with lactophenol indicates the presence of powdered drug of pale catechu.

4.   Colophony in alcohol turns blue litmus to green colour.

5.   Formation of violet or white colour with tannic acid confirms the presence of gelatin.

Choose the correct answer

6. _____ is/are an example of organized drug.

   a. mucilage   b. leaves   c. resins

7. The terms brittle, granular, conchoidal refer to _____

   a. fracture   b. size   c. colour

8. Drugs without definite structure are called _____

   a. adulterated   b. unorganized   c. organized

9. Appearance of pale yellow precipitate of tetrabromaloin with bromine water indicates the presence of _____

   a. honey   b. sesame oil   c. aloe

10. Formation of green colour with nitric acid [50%] over the fractured surface of the drug shows the presence of _____

    a. myrrh   b. asafoetida   c. wax

Answers				
1. True	2. True	3. True	4. False	5. True
6. b	7. a	8. b	9. c	10. b

Answer the following

1. Why is systematic evaluation of drug necessary?
2. Write down the procedure for systematic analysis of powdered drugs.
3. Describe the various microscopical characters used for identification of organized drugs.
4. What are the macroscopical features used for the identification of unorganized drugs.
5. Enumerate the chemical tests to detect the following drugs:

   i. Agar   ii. Aloe   iii. Balsam of Tolu

   iv. Benzoin   v. Gambier   vi. Gelatin

   vii. Storax   viii.Tragacanth

6. What are the tests carried out in the lab to identify colophony, myrrh, honey and beeswax.

FIBRES, SUTURES AND SURGICAL DRESSINGS

18

FIBRES

Fibres can be derived from natural sources (plants and animals) or prepared artificially. The natural fibres are either carbohydrates or proteins and made up of long-chain molecules (cotton and wool) whereas, the synthetic fibres are prepared from polymers (nylon and terylene).

TYPES OF FIBRES

Based on the nature of source and chemical component, fibres are classified as natural fibres and artificial fibres.

1. Natural Fibres

These are further divided into three types. They are as follows:

i. *Vegetable fibres* These are elongated thick-walled sclerenchymatous cells with tapering ends and narrow lumen. Cellulose and lignin are the components of cell walls, e.g., cotton, flax, hemp, jute, etc.

ii. *Animal fibres* These are proteinaceous compounds containing peptide linkage, e.g., silk and wool.

iii. *Mineral fibres* Mineral fibres consist of sand (silica) mixed with oxides of aluminium, calcium, boron and magnesium, e.g., glass and asbestos.

2. Artificial Fibres

These fibres are categorized into two types:

a. *Regenerated fibres* These fibres are prepared from naturally occurring polysaccharides and composed of either cellulose or protein, e.g., cellulose acetate (acetate rayon), viscose rayon, oxidized cellulose, nitrocellulose, etc.

b. *Synthetic fibres* These fibres are more stronger than natural fibres. They are prepared from organic molecules by polycondensation process. Nylon, terylene, orlon, polyethylene, etc. are the polymers used as pharmaceutical aids.

USES OF FIBRES

The important uses of fibres in pharmacy are as follows:

* For preparing surgical dressings, e.g., absorbent cotton and nylon.
* For absorbing blood, mucus and pus, e.g., absorbent cotton.
* For filtration and straining, e.g., wool.
* For the manufacture of yarns, ropes, coarse bags, medicated tows, rugs, domett, cellophone, enzyme, crepe bandage, flannel, packing materials and also in textiles, e.g., jute, wool, hemp and flax.
* As absorbable haemostatic dressings, and also in neurosurgery, endaural and dental surgery, e.g., alginate fibres and oxidized cellulose.
* For stabilizing lotions, suspensions, pastes, ophthalmic preparations and some ointments, e.g., methyl cellulose.

IDENTIFICATION TESTS FOR FIBRES

Vegetable and Regenerated Fibres

* When heated with aqueous picric acid solution, they are not stained permanently.
* When treated with a mixture of iodine and sulphuric acid, they produce a blue colour.
* With Molisch's reagent, they produce a violet colour.
* When ignited or boiled with soda lime, they do not produce foul smell.
* With Millon's reagent, they do not produce red colour.
* Vegetable fibres dissolve in copper oxide ammonia (cuoxam) to form a blue colour.

Animal Fibres

✤ Stained permanently with picric acid.

✤ Positive result to Millon's test.

✤ Dissolve in 5% aqueous potassium hydroxide solution.

✤ Produce disagreeable odour on ignition.

Synthetic and Mineral Fibres

✤ On heating, glass fibres melt to form beads.

✤ Heating has no effect on Asbestos fibres.

✤ Tests for vegetable and animal fibres give negative results to tests for synthetic and mineral fibres.

Table 18.1   Tests for identification of vegetable and regenerated fibres

Test	Absorbent cotton	Carbohydrate fibres		Calcium alginate
		Flax	Viscose rayon	
Ignition				
Advance slowly towards flame or heat in crucible.	Rapid burning with flame. No foul odour or fumes. White ash. No bead	Same as cotton	Same as cotton	Smoulders in flame and chars. Goes out when removed from flame.
Solubility				
i. 80% w/w sulphuric acid	Soluble	Slowly soluble	Soluble	Soluble with swelling.
ii. 60% w/w sulphuric acid	Insoluble	Insoluble	Soluble	Insoluble.
iii. Cuoxam	Raw: soluble with ballooning		Soluble with uniform swelling	Soluble with uniform swelling.
iv. 5% potassium hydroxide	Insoluble	Insoluble	Insoluble but acquires yellow tint	Insoluble but acquires yellow tint.
v. Concentrated hydrochloric acid	Insoluble	Insoluble	Insoluble	Insoluble.
vi. Cresol or 90% phenol	–	–	–	–
vii. 5% sodium citrate solution	–	–	–	Soluble
Moisten with iodine (N/50) followed by a drop of 8% w/w sulphuric acid.	Blue colour	Blue colour	Blue colour	No blue colour, but brownish red colour.
Warm/boil in a test tube with picric acid solution. Then rinse well.	No permanent yellow stain	No permanent yellow stain	No permanent yellow stain	No permanent yellow stain.
Warm with Millon's reagent.	No red stain	No red stain	No red stain	No red stain
Warm with 5% potassium hydroxide solution. Then add lead acetate solution.	–	–	–	–
Iodine (N/50)+ 8% sulphuric acid.	Blue colour	Blue colour	Blue colour	Brownish red colour.

Table 18.2 Tests for protein and synthetic fibres

Test	Protein fibre		Synthetic fibre	
	Wool	Silk	Nylon	Polyester
Ignition				
Advance slowly towards flame or heat in crucible	Brown gases alkaline to litmus. Foul odour (like burnt hair or flesh). It burns slowly giving bead followed by white ash.	Same as in wool	Melts leaving hard bead	Melts leaving hard bead.
Solubility				
i. 80% w/w sulphuric acid	Insoluble	Soluble	Soluble	Insoluble
ii. 60% w/w sulphuric acid	Insoluble	Soluble	Soluble	Insoluble
iii. Cuoxam	Insoluble but swells and scales separate	Partially soluble	Insoluble	Insoluble
iv. 50% potassium hydroxide	Soluble	Soluble	Insoluble	Insoluble
v. Concentrated hydrochloric acid	Soluble	Soluble rapidly	Soluble and disintegrates on boiling in dilute HCl	Insoluble
vi. Cresol or 9% Phenol	–	–	Soluble	Insoluble
vii. 5% sodium citrate solution	–	–	–	–
Moisten with iodine (N/50) followed by a drop of 8% w/w sulphuric acid.	Yellow	Yellow	No blue colour	No blue colour
Warm/boil in a test tube with picric acid solution. Then rinse well.	Permanent yellow stain	Permanent yellow stain	No permanent yellow stain	No permanent yellow stain
Warm with Millon's reagent.	Red stain	Red stain	No red stain	No red stain
Warm with 5% potassium hydroxide solution. Then add lead acetate solution.	Black precipitate	White precipitate	–	–

VEGETABLE FIBRES

1. ABSORBENT COTTON

Biological source　It is the purified form of epidermal hairs of seed coat cells of the species of *Gossypium herbaceum* Linn., *G. hirsutum* Linn., *G. arboreum* Linn., and *G. barbadense* Linn. of the family Malvaceae.

The common names include purified cotton; surgical cotton; absorbent wool; panchu.

Geographical distribution　The plants are shrubs reaching a height of 1–1.3 m. The plants are commonly cultivated in India, Egypt, USA, South America, China, South Africa and Pakistan.

Preparation of cotton

❖ Cotton plant produces fruits called capsules, the bolls. Each boll is a spherical or ovoid capsule consisting of 3 to 5 locules or chambers. The content of each division is called a ' lock,' within which 6 to 9 seeds are present.

❖ The surface of each seed is covered with epidermal hairs. The hairs may be long hairs (lint, floss or staple) or short hairs (fuzz or linters).

❖ The capsule cracks at maturity and a white fluffy mass is pushed outside the capsule wall.

❖ At maturity the fibres undergo twisting and this peculiarity greatly helps in the spinning operation.

❖ A mature fibre looks like a translucent, flattened, twisted more or less tubular structure with a broad base and an untwisted tapering apical end.

❖ The outer surface of the fibre is covered by a protective waxlike covering.

❖ Raw cotton discarded by the textile industry is used as combers waste.

❖ The cotton is subjected to combing process so as to separate short fibres which are utilized for making absorbent cotton, while the long fibres are spun and woven into cloth.

❖ The combers waste consisting mainly of short fibres is boiled with dilute caustic soda and soda ash solution for 10–15 hours at a pressure of 1–3 atmosphere.

❖ The treatment will remove fatty cuticle of the trichomes making the wall absorbent. It is then thoroughly washed with water.

❖ Bleaching is done by treatment with sodium hypochlorite solution and dilute hydrochloric acid. It is washed with water and dried.

❖ The washed cotton is then carded into flat sheets. The carding machine makes thin continuous films of cotton wool. Several such thin films are placed one above the other and packed into packages, which are finally sterilized.

Characteristics of cotton fibre Absorbent cotton is a very pure form of cellulose. It is a unicellular hair, which is a microscopic structure and is 2–4 cm in length and 25–40 μ in diameter. It is white, soft, odourless and tasteless. This cotton should be stored in a cool place to retain its absorbent property. If cotton is kept in moist and warm conditions, it is attacked quickly by bacteria and fungi, and thus loses its absorbing property.

Chemical constituents of absorbent cotton Cotton entirely consists of cellulose with 6 to 7% moisture. The cellulose molecule is built up of glucose residues united by 1, 4-β-glucosidic links. The unit is the "cellobiose unit", many of which are united in the polysaccharide molecule of cellulose.

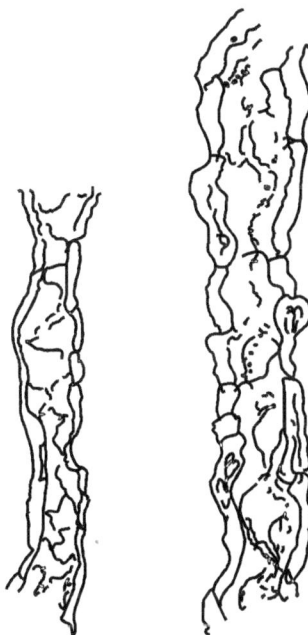

Figure 18.1 Absorbent cotton

Chemical tests

❖ Soak cotton fibres in iodine H_2O and dry. Add a few ml. of 80% sulphuric acid. Cotton turns purplish blue or bluish green in colour.

❖ Absorbent cotton dissolves in ammoniacal copper oxide solution with uniform swelling.

✤ Cotton is insoluble in dilute sodium hydroxide and HCl.

✤ Cotton is soluble in 66% sulphuric acid.

Uses

✤ Cotton is used as a filtering medium, in surgical dressings and as an insulating material.

✤ It is used to absorb blood, mucus and pus, and to protect the wounds from microbial infection.

Absorbent gauze, microcrystalline cellulose, purified rayon, ethyl cellulose, oxidized cellulose, etc. are used in pharmacy.

2. HEMP

Biological source   The pericyclic fibres of the stem of *Cannabis sativa* Linn. of the family Cannabinaceae constitute hemp.

The common names include Indian hemp; Ganja; Indian cannabis.

Geographical distribution   Hemp is a tall, hollow-stemmed annual herb. It is mostly cultivated in India, Africa, America, Russia, Italy, France and Japan.

Preparation

✤ The best grade of hemp fibre is obtained from the male plants. The plants are collected when the male flowers begin to shed their pollen grains.

✤ The plants are cut with a hemp knife, about 2–3 cm above ground level. The chopped stems are spread on the ground for drying.

✤ The dried stems are put into stagnant water in the form of bundles for about 10–20 days for retting. In order to protect the bundles from sunlight, coconut leaves or straw are spread over the bundles. If needed, the surface of the bundles are heavily loaded with logs of wood or stones.

✤ During this period, the soft tissues like parenchyma and phloem get disintegrated and separated from the bark.

✤ The fibres are stripped off, cleaned and washed thoroughly with water and dried in the sun after bleaching.

✤ The dried hemp fibres are graded, and the fibres are put into bales of 175–200 kg and sent to market for sale.

Characteristics of hemp fibre

* Primary and secondary pericyclic fibres are used.

* The number of primary fibre bundles varies from about a dozen to as many as 40, each consisting of 10–35 individual fibre cells.

* The length of the individual fibre is about 25 to 40 mm with a diameter of 22 μ.

* Hemp fibres have blunt ends; some have forked ends.

* The lumen of the fibre is large, with thick walls having cross lines.

* The strands of the fibre are usually yellowish or greenish, or grey or dark brown in colour and are quite variable due to their length, strength and durability.

* Fibres are not affected by water and cannot be bleached easily.

* Due to lack of flexibility and elasticity, these fibres cannot be spun into yarn.

Chemical constituents   It consists of 67% cellulose and 16% hemicellulose and less amount of lignin.

Chemical tests

* Hemp fibres show slightly red colour with phloroglucinol.

* With iodine and sulphuric acid, the inner wall of the fibre gives blue colour and middle lamella yellow.

* Treatment with chloro-zinc iodine gives purple to yellow colour.

Uses   Hemp fibres are used in the manufacture of sail-cloth, twine, ropes, artificial sponges, etc.

3. FLAX

Biological source   Flax consists of pericyclic fibres of the stem obtained from *Linum usitatissimum* L. of the family Linaceae.

The common name is flax.

Geographical distribution   Flax is an annual herb and is commonly found in countries such as Argentina, Canada, Russia, Scotland, India, Holland, Belgium and USA.

Preparation

* The plants are harvested when the lower part of the stem becomes yellow in colour and the leaves have fallen off.

❖ The plants are pulled out from the field to prevent the deterioration of the cut end of the straw. The harvested plants are allowed to dry in the field for 2 to 3 days.

❖ The extraction of flax fibre involves four steps: rippling, retting, scutching and hackling.

❖ All the leaves and fruits are stripped off from the dried stems without damaging, a process called rippling.

❖ The denuded stems in small bunches are then subjected to retting by immersing them in tanks or ponds for several days to give superior fibres.

❖ After retting, the straw is dried thoroughly either by artificial means or by spreading it in the field for a few days to prevent further fermentation.

❖ The fibres are separated from the woody "core" or "boon" as it is called, by "scutching" that involves two distinct operations, namely breaking and crushing the boon and stripping which separates the fibres from the inner tissues of the stem.

  ❖ In the first step, the woody portion is reduced to short pieces either by beating with wooden mallets or by passing it through fluted rollers.

  ❖ The fibres are then removed from the woody pieces with the help of scutching machines.

  ❖ Finally the short and tangled fibres are separated from the long fibres. The long unbroken fibres are called line fibres.

  ❖ The fine grades of flax are woven into fabrics such as cambrics, damasks and sheeting lace for apparel and linen thread.

Characteristics of flax fibre

❖ The fibres occur in discrete groups or aggregates of many cells in the pericycle, each bundle consisting of 10 to 40 individual fibres.

❖ The fibres are non-lignified with sharply pointed ends.

❖ The length of the fibre is in the range of 25–50 mm and the diameter in the range of 12–25 μ. The lumen is very narrow or obliterated due to thickening.

❖ The fibres are pale yellow or cream in colour and lustrous due to wax.

❖ Flax has good lustre and has much greater tensile strength than cotton.

❖ The flax fibres are known for their fineness, durability, flexibility, heat conductivity, moisture absorbency and great strength, when wet.

Chemical constituents Pecto-cellulose is the main constituent of flax with 64% cellulose, 17% hemicellulose and 2% lignin.

Chemical tests

* Fibres turn slightly pink in colour with phloroglucinol.

* On treatment with iodine and sulphuric acid, fibres become blue or violet in colour.

* When treated with chloro-zinc iodine, it gives purple to yellow colour.

Uses

* Flax fibres are used in the manufacture of yarn into linen fabrics.

* They are used in the preparation of rugs, lacelawn and filtering medium.

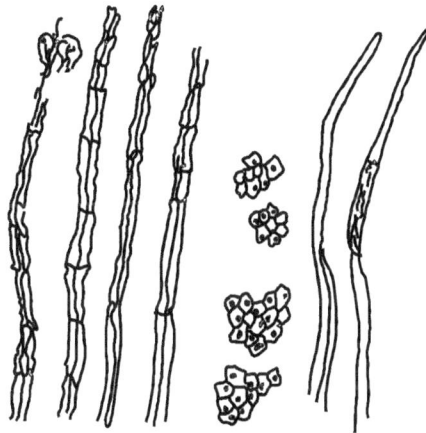

Figure 18.2 Flax fibres

4. JUTE

Biological source Jute consists of phloem fibres of the stem of the cultivated species of *Corchorus olitorius* L. (tossa jute), *C. capsularis* L. (white jute), and other species of *Corchorus* of the family Tiliaceae.

The common names include gunny and chanal.

Geographical distribution Jute is an annual plant that grows to a height of 3–4 m. It is mostly cultivated in Orissa, Assam and West Bengal.

Preparation and processing

* Harvesting is done in the month of July,when 50 per cent of the plants are in fruiting stage. At this stage, both the yield and quality of the fibres are good.

* The plants are chopped close to the ground with a sickle or pulled out by hand. The cut stems are bundled and left in the fields for 2 to 3 days to dry.

* During drying, the leaves and fruits wilt and finally fall off. If needed, the leaves and fruits are removed using a comblike device.

* The dried stems are bundled and immersed flat in a pool or water tank or ditch for retting. Another layer of bundles is placed on top at right angle to the first.

* The bundles are immersed in water for about 15–20 days for retting and the surface of the bundles are loaded with stones or weight for proper retting.

* During retting, disintegration of soft tissues takes place and the phloem and parenchymatous tissues of the bark get separated.

* The fibres are removed by beating and stripping and cleaned and washed thoroughly with water.

* The cleaned fibres are then bleached and dried in the sun. The dried jute fibres are graded according to colour, length and glossiness.

* The prepared fibres are made into bales of 175–200 kg and sent to the market.

Characteristics of jute fibre

* Jute is pale yellow or yellowish white in colour.

* The commercial fibres are 1–3 m in length and 10–25 μ in diameter and possess a silklike lustre.

* The apex is bluntly pointed and rounded, wall is without markings and the lumen is varying in size.

* It is difficult to bleach.

* Jute gives deep red colour with phloroglucinol and yellow with iodine and sulphuric acid.

Chemical constituents   Jute consists of 63% cellulose, 20% hemicellulose, 10% lignin, 13% moisture and 11–12% fats. It also contains waxes, gums and minerals.

Chemical tests

* Jute fibre becomes deep red in colour with phloroglucinol and hydrochloric acid.

* It gives yellow colour with iodine and sulphuric acid and also with chloro-zinc iodine.

Uses   Jute is used in the manufacture of yarns and ropes, and medicated tows, sags and gunny bags, rugs, blankets, carpets, twine, wrapping textiles, padding splints, and as filtering and straining medium.

Figure 18.3   Jute fibres

ANIMAL FIBRES

1. SILK

Biological source   Silk fibres are obtained from the cocoons of the mulberry silkworm, *Bombyx mori* Linn. and other species of Bombyx of the family Bombycidae and also from species of *Antheraea* such as *A. mylitta* and *A. asama* of the family Saturniidae.

The common name is *pattu knool*; silk.

Geographical distribution   Silk fibres are commonly manufactured in India, China, Italy, Asia Minor, Japan, France and other countries.

Preparation of silk Silkworm produces larvae, a stage in its life cycle, which is very important for the production of silk of commerce. The larva or caterpillar then passes to the next stage called pupa or chrysalis stage. Before it passes from the larval to pupal stage, it secretes around itself an oval structure called cocoon, of about 2–5 cm length and consists of a continuous thread up to 1200 m length. This is because, at this stage larvae produce fibres called silk fibroin from the glands of their mouth. The fibroin unites together with a help of a gumlike secretion known as sericin and forms the cocoon. As cocoons are important to get silk fibres, they are steam-heated to 60–80°C for a short period. The heated cocoons are then put into hot water to dissolve the gum secretion and to separate the silk fibres.

Figure 18.4 Cocoon of silkworm on mulberry leaves

If the pupa is allowed to mature, the insect will escape as butterfly damaging the cocoon. Therefore, the cocoons are collected at the right stage, that is at chrysalis (pupa) stage to get the silk fibres.

The process of cultivation of silkworm from the stage of egg up to the completion of cocoon for the production of silk and also the production of mulberry trees to feed the silkworm is called sericulture.

Silk containing sericin is called raw silk. Sericin is removed by treating with hot soap solution and this process is called stripping or degumming. The stripping method gives a lustre, semi-transparency and a smooth surface to silk. It is sometimes treated with metallic salt, a finishing substance, to improve its quality, density and weight.

Characteristics of silk

* Silk is a continuous filament with remarkable tensile strength.

* Silk fibres are about 1200 m in length and 5 to 25 μ in diameter.

* The fibres are very fine, soft, solid and smooth.

* The fibres can be stretched to about 20% beyond their original length before breaking.

* Silk fibres are usually yellow in colour.

* The fibres are insoluble in water, alcohol, ether and dilute alkali.

Chemical constituents　Silk fibres are made up of protein called fibroin, and the main amino acids constituting fibroin are glycine and alanine.

Chemical tests

* Silk fibres dissolve in Cuoxam, 66 per cent sulphuric acid and concentrated hydrochloric acid.

* Solution of silk in potash does not give black colour when treated with lead acetate solution.

Uses　Silk is used for making sieves, ligatures and special types of sutures.

2. WOOL

Biological source　Wool is obtained from the fleece or protective covering of the sheep, *Ovis aries* Linn. of the family Bovidae and Order Ungulata.

The common names include sheep's wool and animal wool.

Geographical source　Large quantities of wool are produced in Australia, Argentina, Russia, USA and India.

Preparation of wool

* Wool is removed from the body of the sheep at shearing time using a wool sorter, and wool of different sizes and qualities are separated. Simultaneously it is beaten over the netting to remove dust and dirt.

* The burrs and straw pieces are picked up.

* The wool is washed in tanks containing warm, soft, soapy water to remove wool grease.

- The clean and defatted wool is subjected to bleaching, washed again and dried. It is then squeezed between rollers and dried, and the fibres are mechanically loosened.
- The loosened fibres are carded and spun into yarns.
- "Wool grease" is removed by mechanical means or by using organic solvents.
- Purified "wool grease" is called wool fat or anhydrous lanolin.
- Wool fat is used in the preparation of cosmetics and ointments.

Characteristics of animal wool

- Wool consists of elastic, lustrous and smooth hairs.
- The hairs are loosely fitted, hygroscopic and slippery to touch.
- Anatomically wool contains cuticle, cortex and medulla. The outermost surface is the cuticle, which consists of imbricated, flattened, translucent epithelial scales.
- Wool hairs have the tendency to cling together and they have excellent affinity for dyestuff.
- The wool hair may retain about 17% of moisture of its weight.
- It has good resistance to dry-cleaning solvents, strong alkalies and high temperature.

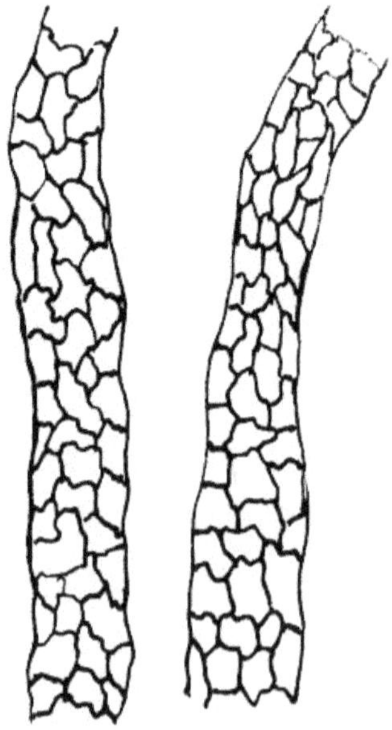

Figure 18.5 Wool fibres

Chemical constituents

* Pure wool contains the sulphur-containing protein known as keratin, which is responsible for the elastic nature of the fibres.

* Keratin is rich in the amino acid cystine, which contains sulphur in it.

* The unstable form of keratin is known as β-keratin, and the stable form is called α-keratin.

Chemical tests

* Wool is insoluble in warm hydrochloric acid and in cold concentrated sulphuric acid and cuoxam.

* A solution of wool in caustic soda when added to lead acetate gives a black precipitate due to high sulphur content.

Uses   Wool is used as a medium for filtration and straining and also in the manufacture of dressing like domett and crepe bandages and flannel.

REGENERATED FIBRES

These fibres are prepared from naturally occurring polysaccharides. These compounds are modified to yield regenerated fibres, e.g., viscose rayon and alginate yarn.

1. VISCOSE RAYON

Common names are regenerated cellulose; man-made regenerated fibre; rayon.

Preparation of viscose rayon

* The coniferous wood pulp or cotton linters are processed to give a white pulp cellulose (80–90%) and some hemicellulose.

* Hemicellulose is removed by treating the product with sodium hydroxide. The remaining alkali-cellulose or sodium cellulosate is then treated with a mixture of carbon disulphide and sodium hydroxide solution to get a viscous solution of sodium cellulose xanthate.

* As this solution is viscous in nature, the rayon obtained from it is called viscous rayon.

* The viscous solution is allowed to ripen, filtered and the filtrate is forced through a spinnert equipped with fine nozzles.

* It is then immersed in a bath containing dilute sulphuric acid and sodium sulphate to give continuous filaments of regenerated cellulose.

❋ These filaments are drawn together to form yarns.

❋ The washed yarns are combined, twisted to strengthen, and then treated with sodium sulphide to remove free sulphur, and then bleached, washed and dried, and 10% moisture content is adjusted and used for preparing surgical dressings.

❋ The viscose solution is allowed to pass through long narrow slits into a regenerating bath to get sheets of viscose. These sheets are washed, bleached and treated with a glycerin solution and dried to produce cellophane.

Characteristics of viscose rayon

❋ Viscose rayon used in surgical dressing has a diameter similar to that of cotton.

❋ Viscose rayon is a white, highly lustrous, and pure form of cellulose with tensile strength and the tensile strength is greater than cotton.

❋ The molecules contain 450 glucose residue units as compared to 9000 glucose units in wood cellulose.

❋ Viscose fibres are solid, transparent, 15–20 μm in diameter, slightly twisted and contain grooves.

❋ The fibre ends are abrupt and peculiar.

❋ They can be delustred by addition of the white pigment, titanium oxide, to the solution before preparation of the yarns.

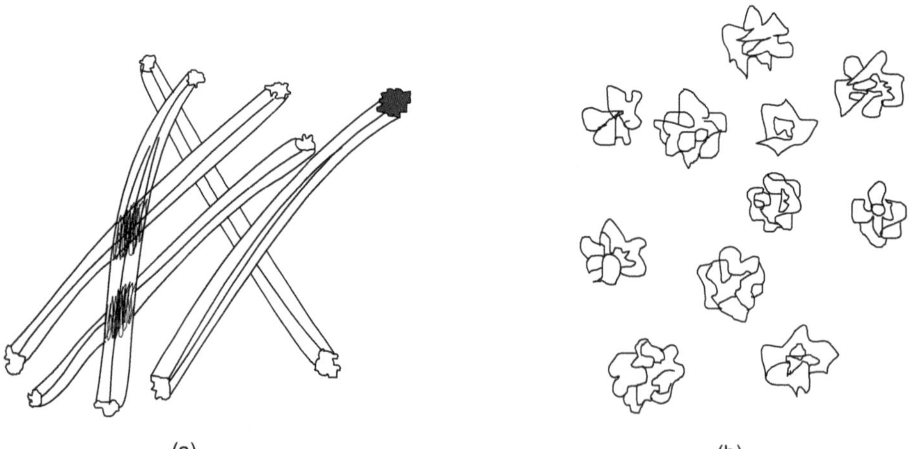

(a)   (b)

Figure 18.6 (a) Threads of viscose rayon   (b) T.S. of viscose fibres

Chemical constituents It is a true form of cellulose and contains 10% of moisture.

Chemical tests

❖ On ignition, it gives a trace of ash which contains sulphur.

❖ Viscose rayon follows the general tests of vegetable fibres.

Uses

❖ It is used for making fabrics and in the preparation of surgical dressings, enzymes, cellophane, etc.

❖ Gauzes and dressings made from viscose rayons are superior to cotton dressings, in that, they show no loss of absorbents on storage.

❖ Cellophane is a heat-sealable packing material and is also used as a dialysing membrane, as a protective dressing and as a substituent of oiled silk.

2. ALGINATE YARN/FIBRE

It is a regenerated fibre consisting of calcium alginate. It is prepared from algenic acid, which is a polysaccharide that forms the main cell wall component of the brown algae.

Geographical distribution   Alginate yarn is manufactured in USA, Canada, Japan and Scotland.

Preparation

❖ A large quantity of brown algae is collected from the sea coasts and the collected sea weeds are washed, dried and milled and then macerated with dilute sodium carbonate solution.

❖ The resultant is diluted and the clear liquid is decanted.

❖ An aqueous solution of sodium alginate is pumped through a spinnerret, which is immersed in a bath containing calcium chloride solution, acidified with dilute hydrochloric acid.

❖ In the bath, sodium cations are substituted with calcium cations and the insoluble calcium alginate is precipitated as continuous filaments.

❖ These filaments are collected, washed and dried and used as alginate fibres for surgical purposes.

❖ The filaments are cut to give stable form of length 1 to 8 inches for preparing calcium alginate wool or fabrics.

❖ A small amount of substances are added to the calcium alginate to inhibit microbial growth.

Characteristics of alginate fibre

* Alginate fibres are pale cream coloured and fairly lustrous filaments.
* The fibres may be converted to calcium alginate wool or a gauze or into absorbable haemostatic dressings.
* Alginic acid absorbs many times its own weight of water and swells up.

Chemical constituents

* Alginate fibres are composed of calcium salts of long-chain molecules of alginic acid (polysaccharide).

Chemical tests

* Alginic acid is insoluble in cold water and slightly soluble in hot water.
* It is soluble in 5% sodium citrate solution
* It gives results similar to vegetable fibres for general tests.

Uses

* Alginate fibres are used as absorbable haemostatic dressings.
* They are also used in neurosurgery, endaural and dental surgery.
* They are used internally to arrest bleeding and form protective dressing for burns or sites from which skin grafts have been taken.
* Alginate fibres are non-toxic, non-irritant and compatible and possess antibiotic properties.
* Calcium alginate wool is used as a swab for pathological work or bacterial study.

SYNTHETIC FIBRES

Synthetic fibres are prepared by polycondensation of organic molecules which are more stronger than the natural fibres. Nylon, terylene, orlon and polyethylene are the polymers used as pharmaceutical aid.

1. NYLON

Common names are perlon; caprolan; grilon.

Preparation of nylon

* It is a synthetically manufactured fibre and the fibre-forming substance is a long-chain synthetic polyamide, which is a polymer of adipic acid and hexamethylene diamine.

* Nylon is prepared by condensing adipic acid with hexamethylene diamine.

* The molten polymer is pumped through a spinnerret to produce filaments.

Characteristics of nylon filaments

* The nylon filaments are smooth, solid cylinders of uniform diameter.

* These fibres soften at 210°C and melt at 223°C. Moisture regaining is about 4%.

* Nylon fibres show different sizes of cross sections.

* They may be slightly lustrous to dull, white or coloured and are very strong.

* Swelling capacity of this fibre is low.

* They are immune to microbiological attack, and resistant to most organic chemicals.

Chemical constituent   It is a polymer of adipic acid and hexamethylene diamine. Chemically, nylon is represented as: H—$[NH(CH_2)_6 CO]_n$ OH; n = approx. 200.

Chemical tests

* Nylon fibre dissolves in phenol, cresol and strong acids.

* On ignition, the fibres melt and form a hard bead.

* Nylon fibres are resistant to abrasions and are not affected by alkali.

* They are soluble in 5 M hydrochloric acid, 90% formic acid, and 90% phenol, and insoluble in acetone.

Uses

* Nylon fibres are used for making fabrics, and have twice the strength and natural resilience of silk.

* They are also used to prepare filter cloth, sieves, non-absorbable sutures, nylon syringes, film, monofilament, tire cord and fishing lines, tow ropes, etc.

2. ORLON

Common names are polyacrylonitrile, fibre A.

Preparation   Orlon is obtained by polymerizing acrylonitrile. It is denoted as $[CH_2 CN(CN)]_n$.

Characteristics of orlon fibre

* It is a white fibre.

* It sticks at 235°C; ironing temperature above 160°C may cause yellowing.

* Its specific gravity is 1.17.

* It resists attack by moulds, mildew and insects.

Chemical tests

* It can be degraded by strong alkalies.

* It has very good resistance to mineral acids and excellent resistance to common solvents, oils, greases, neutral salts and sunlight.

* Its flammability is similar to that of rayon and cotton.

Uses   Orlon fibre is used for furnishing (tents, furniture, awnings), in anode bags in electroplating, in knitwear, rugs and dressings, etc.

3. POLYETHYLENE

The common names are polythene, ethene homopolymer.

Preparation

* Polyethylene is prepared by polymerization of liquid ethylene at high temperature under pressure.

* The polymer is a plastic solid of milky transparency, tough and flexible at room temperature.

* Melting point of polyethylene is 85–110°C.

* The polymer $(CH_2-CH_2)_n$ is transformed into filaments by melt spinning and heat-sealable packing film by a process similar to that adopted in nylon.

Characteristics of polyethylene fibre

* It is a good electrical insulator.

* It is stable to water, non-oxidizing acids and alkalies, alcohols, ethers, ketones and esters at ordinary temperature.

Chemical tests

* It burns, but hardly supports combustion.

* It is attacked by oxidizing acids like nitric acid and perchloric acid, benzene, gasoline, lubricating oils, and aromatic and chlorinated hydrocarbons.

* It has flexibility over a wide range of temperature.

* An important property of polyethylene is that, it is used both as filament and resin, in its low specific gravity (0.92).

* A low softening point (110°C) limits it application in wearing apparel uses.

Uses

* It is used as a laboratory tubing, in making protheses, electrical insulation, packing materials, kitchenware, tank and pipe linings, paper coatings and textile.

* It is also used in filtering fabrics.

SUTURES AND LIGATURES

The sterile thread or string specially prepared for sewing or stiching together tissues, muscles and tendons with the help of a needle in surgery is called suture.

If these threads or fibres are used to tie severed blood vessels to stop bleeding, without the use of needles, they are called ligatures.

PROPERTIES OF SUTURES

Sutures must possess certain qualities and are as follows:

* They must be sterile and non-irritant.

* They should have finest possible gauge and adequate mechanical strength.

* If absorbable, their time of absorption must be known.

* They are intended to be used for one occasion, with minimum time of absorption.

PREPARATION

Sutures are prepared from intestinal tissues and tendons of animals and birds, vegetable fibres, camel hair, human hair, synthetic threads or metallic wires.

METHODS OF STERILIZATION OF SUTURES

Chemical sterilization and irradiation are the two methods used for sterilizing sutures. Sterilization should not affect the properties of sutures or their utility.

Heat sterilization Depending upon heat resistance, sutures are classified into boilable and non-boilable types.

Irradiation method In this method sterilization is carried out by electron particles or by gamma rays from cobalt 60. The dose of gamma rays used for treatment is 2.5 megarad.

TYPES OF SUTURES

Based on the nature of absorption, i.e., digestion in the tissues of the body, sutures are categorized into absorbable and non-absorbable types and haemostatics.

Absorbable Sutures

The sutures that are broken down and digested in animal tissues are called absorbable sutures. It is of the following types: Surgical catgut (small intestine of ox, sheep and deer), and synthetic polyesters.

Surgical catgut   Surgical catgut is the classical example of absorbable suture. Catgut is a sterilized fibre or strand prepared from collagen of connective tissues obtained from healthy cattle and sheep.

The preparation of surgical catgut is as follows:

* The submucosal layer of the small intestine of a freshly killed sheep is normally used for the preparation of catgut.

* The intestine is removed and cleaned and split longitudinally into ribbons.

* The outer muscularis and serosal layers and the innermost mucosa are removed mechanically or with the help of a machine leaving behind the submucosa.

* One to five such ribbons are stretched, spun or twisted and dried under tension to form a uniform strand (fibres).

* These strands are polished to get smooth strings, and gauzed for their diameter and cut into suitable lengths.

* These strands are then kept in boiling liquids like toluene or xylene and sterilized using autoclave.

* Sterilization is also carried out by irradiating the suture by gamma rays (cobalt-60) or by electron particles.

* Chromicized surgical catguts are prepared by immersing the ribbons in solutions of chromium salts for tanning the tissues and these fibres are resistant to proteolytic enzymes in the body and their absorption in the body is a slow process.

* Kangaroo tendons are used for bone repairs and hernia.

* These fibres are prepared from the tails of kangaroo and the preparation is similar to catgut.

* The Cargile membrane is a proteinaceous suture obtained as a thin sheet of pliable tissue from the appendix of the deer, ox and fascia lata.

* Fascia lata is obtained from ox fascia, and is a strong suture, generally used in hernia repair.

Synthetic polyesters These are otherwise called synthetic absorbable sutures.

* The polymers obtained by condensation of cyclic derivatives of glycolic acid (glycolide) with lacticide, which is a cyclic derivative of lactic acid, are used to prepare synthetic polyesters.

* These sutures have high tensile strength.

* Synthetic polyesters are degraded by hydrolysis and completely absorbed in the tissue.

Non-absorbable Sutures

The sutures that are not absorbed and that remain in the body as such for a long period are called non-absorbable sutures. They are removed after healing of wounds. Silk and cotton are the examples of natural non-absorbable sutures, and nylon and metallic sutures are examples of synthetic non-absorbable sutures.

Natural non-absorbable sutures

i. *Cotton sutures* These sutures have uniform size and have reproducible strength and low order of tissue reactivity. Therefore, they are recommended in critical parts where strength of the suture is required for a long time.

ii. *Silk sutures* These sutures are prepared by spinning or twisting the silk fibres into a single strand with varying diameters. The silk sutures are smooth and strong and used in their braided form combining several twisted yarns into a compact mass. The strands are sterilized and boiled with water to make them soft.

Synthetic non-absorbable sutures

i. *Nylon suture* It is a synthetic polyamide available in the form of microfilaments of nylon braided into strands of required diameter. Nylon sutures are strong and water-resistant and mainly used in skin and plastic surgery.

ii. *Metallic suture* These are wires of metals like silver or stainless steel and available as twists and braids. It is used to assist surgical repairs.

Haemostatics

It consists of oxidized cellulose and absorbable gelatin sponge. They are as follows:

Oxidized cellulose It is a cellulose of varied carboxyl content retaining the fibrous structure. It is prepared by oxidizing cotton wool or gauze with nitrogen dioxide until the number of carboxylic groups formed by the oxidation of the primary alcohol groups of the glucose moieties of the cellulose molecules reaches 16–22 per cent. After reaction, the cellulose molecule contains glucuronic acid residue units and some glucose residue units. It is identical with normal cotton in appearance. It is used as a local absorbable haemostatic in surgery and in chromatography. But it delays bone repair and cannot be sterilized by heat.

Absorbable gelatin sponge Absorbable gelatin sponge is a sterile, white, tough and finely porous spongy material, which is water-insoluble and absorbable.

It is prepared from a warm solution of gelatin, which is whisked to a foam of uniform porosity and then dried, cut into pieces of specific size and finally, sterilized at 150°C for one hour.

It is absorbed in the body fluids. About 9 g of absorbable gelatin sponge takes up to 45 times its weight of well-agitated oxalated whole blood.

It is used as a haemostatic. It is moistened with sterile sodium chloride solution and put within a surgical incision where it gets absorbed in 4–6 weeks. It is also used as a local coagulant and haemostatic.

SURGICAL DRESSINGS

The materials used for dressing of wounds are called surgical dressings.

PURPOSE OF DRESSING

Dressings of the injured site serve various functions.

* They remove wound exudates from the injured site and prevent infection.
* They give physical protection to the healing wound and mechanical support to the supporting tissues.

Various fibres are used in the preparation of surgical dressings. A good quality of dressing should be durable, easy to handle, properly sterilized, formed from loose threads and fibres and they should not attack the granulating surface.

REQUIREMENTS OF SURGICAL DRESSINGS AS MENTIONED IN PHARMACOPOEIA

* They should be sterilized before use.
* They should be stored in dry, well-ventilated place at a temperature not exceeding 25°C.
* Permitted antiseptics in prescribed concentrations are to be used.
* Adhesive products should not be allowed to freeze.
* There should not be any loose threads and fibre-ends in dressings.

TYPES OF SURGICAL DRESSINGS AS RECOGNIZED BY BRITISH PHARMACOPOEIA

The following materials are allowed to be used as surgical dressings:

* Fibres and related materials, i.e., viscose, animal, wood cellulose wadding.
* Carded products, i.e., absorbent cotton, viscose wadding, etc.
* Non-extensible adhesive woven products, i.e., belladonna, adhesive plaster, zinc oxide surgical adhesive tape.
* Non-extensible, non-adhesive woven products, i.e., absorbent muslin, domett bandage, etc.
* Extensible non-adhesive woven products, i.e., cotton bandage, elastic web bandage, etc.
* Extensible adhesive woven products, i.e., elastic adhesive bandage, extension plaster, etc.
* Non-woven products, i.e., impermeable plastic surgical adhesive tape, permeable plastic surgical adhesive tape, etc.

CLASSIFICATION OF SURGICAL DRESSINGS

Surgical dressings are classified as:

* Primary wound dressings
* Absorbents
* Bandages
* Adhesive tapes
* Protectants

Primary Wound Dressings

The surgical dressings that are placed over the wound surface to absorb wound secretions such as pus, mucus and blood and minimize maceration are called primary wound dressings, e.g., gauzes of suitable mesh. Some dressings may adhere to the wound surface and cause pain on removing them. To overcome this problem, nowadays non-adherent dressings are available. They include petrolatum-impregnated gauze, viscose gauze impregnated with a bland, hydrophilic oil-in-water emulsion or an absorbent pad faced with a soft plastic film having openings, etc.

Absorbents

A material that is used to absorb wound secretions is called absorbent. Cotton is commonly used as a surgical absorbent material to absorb wound secretions. The absorbent cotton is prepared from raw cotton fibre by different processes and it is further shaped into cotton balls of various sizes. Larger cotton balls are used in obstetrical practices while the medium-sized ball is used for applying antiseptics or medication locally or for cleansing wounds. The small cotton ball is used for skin cleansing before giving injections. Composite absorbent dressings have also been developed for certain specific cases. In addition to cotton, other materials such as rayon wool, gauze pads, laparotomy sponges, sanitary napkins, disposable cleaners, eye pads, nursing pads, cotton tip applications, etc. are also used as absorbent materials. They are used either in the shape of balls or pads.

Bandages

A bandage is a material that holds dressing at the required site by giving support, and checks haemorrhage. The bandages may be elastic or non-elastic in nature. Common gauze roller bandage and muslin bandage rolls are commonly employed. Elastic bandages are of different types such as woven elastic bandage, crepe bandage and conforming bandage.

Adhesive Tapes

Based on forms, strength and adhesion, the surgical adhesive tapes are categorized into rubber-based adhesive or an acrylate adhesive. Rubber adhesive tapes are very cheep, superior and provide strength of backing while acrylate adhesive tapes are used in case of operative or post-operative procedures to reduce skin trauma.

Protectants

The materials that are used to cover wet dressings and hot or cold compresses are called protectants or protectives. They are also used as a covering for poultices and for retention of heat. They prevent the escape of moisture from the dressing. Some of the common examples of protectives are plastic sheeting, rubber sheeting, waxed or oil-coated papers and plastic-coated papers.

REVIEW QUESTIONS

State whether the following sentences are true or false

1. Defatted wool is the purified wool grease.
2. Hemp fibres cannot be spun into yarn due to lack of flexibility and elasticity.
3. Synthetic fibres are prepared from natural polysaccharides.
4. Alginate fibres are used as non-absorbable haemostatic dressings.
5. Sutures should be non-irritant and sterile.
6. Ligatures are used for dressing of wounds.
7. The classical example of absorbable suture is nylon suture.
8. Jute fibres turn deep red in colour with phloroglucinol and HCl.
9. Vegetable and regenerated fibres produce blue colour with iodine and H_2SO_4.
10. The chemical component of alginate yarn is calcium alginate.

Choose the correct answer

11. The natural fibres obtained from vegetable source is _____
    a. silk   b. wool   c. hemp
12. Absorbent cotton is formed of _____
    a. cellulose   b. pectin   c. lignin
13. The botanical source of flax is _____
    a. *Linum usitatissimum*   b. *Gossypium*   c. *Corchorus olitorius*
14. Silk is a/an _____
    a. mineral fibre   b. vegetable fibre   c. animal fibre
15. The synthetic fibres are _____
    a. nylon and orlon   b. terylene and jute   c. flax and polyethylene

16. The raw cotton discarded by the textile industry is called _____

    a. combers waste   b. linters   c. staple

17. The removal of all leaves and fruits from the dried stems is called _____

    a. nibbling   b. rippling   c. scotching

18. The morphology of jute fibre is _____

    a. cortical fibre   b. xylem fibre   c. phloem fibre

19. Wool is obtained from _____

    a. *Ovis aries*   b. *Bombyx mori*   c. fascia lata

20. Tendons used for bone repairs and hernia are derived from _____

    a. ox   b. zebra   c. kangaroo

Answers				
1. False	2. True	3. False	4. False	5. True
6. False	7. False	8. True	9. True	10. True
11. c	12. a	13. a	14. c	15. a
16. a	17. b	18. c	19. a	20. c

Answer the following

1. What are linters?
2. Define retting.
3. Differentiate synthetic fibres from natural fibres.
4. Explain the types of fibres with suitable examples.
5. How are fibres used in pharmacy?
6. How will you identify the different fibres used in pharmaceutical industry?
7. Write the biological source, preparation, characters and importance of any two vegetable fibres.
8. Explain the preparation of silk and wool for the pharmaceutical industry.
9. Comment on the importance of wool.
10. What are regenerated fibres? Explain the method of preparation, characters and importance of viscose rayon and alginate fibre.
11. Discuss the preparation, characters and pharmaceutical use of any two synthetic fibres studied by you.

12. What are the chemical tests used to identify the following fibres?

    i. Nylon

    ii. Absorbent cotton

    iii. Flax

    iv. Silk

    v. Alginate fibre

13. Give the biological source, preparation, chemical constituents and use of absorbent cotton.

14. Explain how hemp fibre is obtained and processed to be used in pharmacy.

15. Distinguish between regenerated fibres and synthetic fibres.

16. What are sutures?

17. Explain the types and the importance of sutures.

18. What are ligatures?

19. What are surgical dressings?

20. Why are surgical dressings needed?

21. Explain the classification of surgical dressings?

22. Comment on properties of sutures.

23. How and why are sutures sterilized in the laboratory?

24. Define the following:

    i. Dressing

    ii. Sutures

    iii. Bandage

25. Describe the method of collection and preparation of silk.

26. Enumerate the types of fibres used for surgical dressings with suitable examples.

19

TECHNICAL PRODUCTS AND PHARMACEUTICAL AIDS DERIVED FROM PLANTS AND ANIMALS

TECHNICAL PRODUCTS

A number of natural products used in medicine also have applications in other fields as beverages, condiments, flavouring agents, spices, in confectionaries, in paint and varnish, textiles and cosmetics. These products are termed as technical products. Some of the natural substances used as technical products are given in Table 19.1.

Table 19.1   Natural substances used as technical products

Commercial use	Natural substances as technical products
1. Cosmetic industry	
i. in perfumes	i. Lavender, sandalwood, *Citronella*, balsam of Peru, balsam of Tolu.
ii. in incense	ii. Myrrh, *quillaia*.
iii. in soaps	iii. Fatty acids of castor oil, cottonseed oil, peanut oil.
iv. in shampoos	iv. *Quillaia*.
v. in creams	v. Wool fat, spermaceti, white wax.
vi. in lotions	vi. Benzoin.
vii. in hair dressings	vii. Coconut oil, castor oil, henna
2. Food industry	
i. in confectionary and bakery products	i. *Acacia*, agar, alginates, starch, sterculia gum.
ii. in beverages [soft drinks]	ii. Citrus fruits, ginger.
iii. as spices and condiments	iii. *Capsicum*, nutmeg, cardamom, cinnamon, clove, coriander, caraway, dill.
iv. as vegetable oils	iv. Coconut, sesame, cottonseed, peanut, mustard.

(Contd.)

Table 19.1 (Continued)

Commercial use	Natural substances as technical products
3. Tobacco industry in cigarettes, cigars, snuffs	*Glycyrrhiza, Vanilla.*
4. Brewing industry in the manufacture of beer	Hops, yeast, malt.
5. Paint and varnish industry	Colophony, copaiba, fixed oils—castor and linseed oils.
6. Textile industry	*Acacia*, agar, alginates, catechu, cotton, gambier, rosin, starch, sterculia gum.

PHARMACEUTICAL AIDS DERIVED FROM PLANTS AND ANIMALS

The main objective of the pharmaceutical industry is to produce quality products that are acceptable to be used as drugs in the diagnosis and treatment of diseases. For the production of such drugs various techniques such as purification, filtration, adsorption, solubilization, emulsification, clarification, etc. are employed. A number of natural products are used in these techniques. Agents like flavour, colour, and perfumes are also added to make the drugs acceptable for human consumption. These agents are of very little or no therapeutic value but are essentially used in the manufacture of various pharmaceutical products. These agents are called pharmaceutical aids. Pharmaceutical aids may be of plant, animal, mineral or synthetic origin as shown in Table 19.2.

Table 19.2 Pharmaceutical aids of plant and animal origin

Pharmaceutical aids	Pharmaceutical use
Plant products	
Colouring agents	
Saffron, chlorophyll	As natural dyes
Gums	
Tragacanth, acacia	As binding, suspending and emulsifying agents.
Sterculia, tragacanth	As laxatives
Guar gum	As thickening and disintegrating agent
Mucilages	
Isapgol, linseed	As demulcent and soothing agent

(Contd.)

Table 19.2 (Continued)

Pharmaceutical aids	Pharmaceutical use
Other carbohydrates	
Starch	As disintegrating agent and also as dusting powders
Sodium alginate	As stabilizing, thickening, emulsifying, deflocculating, gelling and filming agent
Glucose, sucrose and honey	As sweetening and coating agents
Agar	As emulsifying agents and in culture media
Saponins	
Quillaia	In the preparation of coal tar emulsion
Fixed oils and fats	
Olive, sesame, cottonseed, almond, and castor oils, theobroma oil.	As emollients and vehicles for drugs
Wax	
Carnauba wax	As ointment bases.
Volatile oils	
Caraway, cardamom, cinnamon, clove, coriander, fennel, lemon, orange	As flavouring agent
Miscellaneous	
Pyrethrum	As insecticide
Fibres	As filtering media and surgical dressings
Animal products	
Animal proteins	
Gelatin	As suspending agent and for preparing capsules
Fixed oils and fats	
Wool fat, beeswax, lard, spermaceti	As ointment bases
Colouring agent	
Cochineal	As natural dyes

1. SPERMACETI

Biological source It is a waxy substance obtained from the head and blubber of spermwhale, *Physeter macrocephalus* of the family Physteridae.

The common names include spermwax, cetaceum.

Geographical distribution Pacific, Atlantic and Indian Oceans.

Collection and preparation   Whales attain a length of 20 m and have a circumference of 6–9 m. In the head of the animal, there is a special cavity, which is filled with sperm oil. The animal is killed with torpedo harpoons, the cranial cavity is opened and the oil is removed either by pumping or by buckets. A single whale can yield 10 to 15 barrels of oil. The collected oil on cooling, deposits as a crystalline mass known as spermaceti wax. The crude material is pressed, melted and strained and further bleached with caustic alkalies. The purified spermaceti is made into cakes after cooling. Spermaceti is further refined by boiling with alcohol for pharmaceutical use.

Characteristics   Spermaceti is a white, somewhat transparent mass with crystalline fracture and pearly lustre. It is hard, odourless and tasteless. It is insoluble in water and cold alcohol, but dissolves in chloroform and ether. It melts between 45–50°C.

Chemical constituents   Spermaceti contains 85% of the total esters chiefly of a mixture of cetyl esters, cetyl palmitate, cetyl myristate, acetyl laurate and cetyl stearate. In addition it also contains free cetyl alcohol and esters of higher alcohols.

Uses   It is used as a pharmaceutical aid for cold creams, as a base for ointments, cerates, etc. It is also used in the manufacture of soaps, cosmetics, laundry wax and candles.

2. LARD

Biological source   Lard is the purified internal fat obtained from the abdomen of hog, *Sus scrofa* Linn. of the family Suidae.

The common name is Adeps.

Preparation   Lard is prepared from the internal abdominal fat obtained as a by-product of the meat industry. The prepared fat is called "flare", which is subjected to a temperature of about 50–55°C to melt the lard. It is separated by passing through a muslin cloth. After cooling, it is dried, till it becomes anhydrous. The development of rancidity in the lard is retarded by adding 1 per cent Siam benzoin.

Characteristics   It is a soft, white, homogeneous fat, with slight odour and bland taste. It is insoluble in alcohol and soluble in benzene, ether, carbon disulphide and chloroform. The melting point is 34–41°C.

Chemical constituents   Lard contains about 60% mixed liquid glycerides such as olein and 40% stearin and palmitin.

Uses   Benzoinated lard is used as an emollient and protective. It is also commonly used as an ointment base.

3. COCHINEAL

Biological source   Cochineal consists of dried female insects of *Coccus cacti* Linn. of the family Coccidae enclosing the young larvae.

The common name is Coccus.

Geographical distribution   Mexico and Peru.

Collection and preparation   The insects usually feed on a cactus called *Opuntia coccinellifera.* (L.) Mill. The insects are collected from the cactus by means of special brushes or brooms. They are killed in boiling water or on exposure to charcoal fumes.

Chemical constituents   Cochineal contains 10% carminic acid, which is a C-glycoside anthraquinone derivative. Carmine is a dye obtained from cochineal, which contains 50% of carminic acid.

Use   It is used as a colouring agent.

4. GELATIN

Biological source   Gelatin is a heterogeneous mixture of animal proteins, obtained from the collagenous tissues like skin, ligaments, bones and tendons of specific animals of the family Bovidae.

Some common names of the product are gel foam; gelatinum; puragel.

Preparation   Commercial preparation of gelatin may consist of the following steps:

i.   **Preliminary treatment**   The skin trimmings are obtained as discard by-products of slaughtered cattle, sheep, ox and hogs or from leather industry. These products are treated with soda lime for 10–30 days to remove the fats attached to the skin and washed with water several times. The bones are defatted and decalcified with hydrochloric acid to collagen materials.

ii.   **Extraction**   The material obtained above is heated with water at 85°C to dissolve collagen to get gelatin.

iii.   **Clarification**   An electrolyte is added to the solution to remove the impurities and is done by sedimentation method.

iv.   **Decolorization**   To decolorize the crude gelatin solution, it is passed through charcoal or sand.

v.   **Bleaching**   The colour of the material is bleached by treatment with sulphurous acid, or any oxidizing agent.

vi.   **Evaporation**   The bleached gelatin solution is then concentrated under reduced pressure to a specific gelatin content and allowed to set in shallow trays. Such moulded gelatin is dried rapidly in a drying room at temperature between 30° and 60°C to eliminate moisture.

Characteristics   Gelatin is colourless or pale yellow in colour, transparent and brittle. It is odourless and tasteless. The protein products are available in thin sheets or occur in flakes or as coarse to fine powder. In dry condition, it is stable in air, but in moist condition or in solution it is degraded due to microbial attack. It is insoluble in cold water but softens and swells, but is soluble in hot water and sets into jelly on cooling. It is also soluble in glycerol and acetic acid but insoluble in organic solvents. It is amphoteric.

Chemical constituents   Gelatin is mainly composed of the protein glutin. The jelly formation is due to the presence of the nitrogenous substance. Gelatin contains eighteen different amino acids and on hydrolysis, it gives a mixture of amino acids. The adhesive nature of gelatin is due to the presence of glutin.

The quality of gelatin is expressed as "bloom strength". It is the weight in gramme. A jelly strength of not less than 150 bloom is recommended by British pharmacopoeia for pharmaceutical uses such as coating of pills, as vehicles for suppositories and as an emulsifying agent. But for gelatin capsule manufacture and microbial culture media, higher jelly strengths are used. The jelly strength is designated by the bloom gelometer number.

Chemical tests   For confirming gelatin, the following chemical tests are applied:

❁ Gelatin dissolves in water on heating, and on cooling forms a jelly. The gelatinizing power of gelatin is reduced by boiling and the quality of gelatin is judged from its "jelly strength".

❁ A yellow precipitate is formed with picric acid or tannic acid (10%).

❁ It answers positive to the following tests.

   ❁ Biuret test

   ❁ Xanthoproteic test

   ❁ Millon's test

   ❁ Ninhydrin test

❁ On heating gelatin (1 g) with soda lime, fumes and smell of ammonia are produced, indicating it to be a protein. This test distinguishes it from agar, which does not respond to this test.

Uses   Gelatin is used for preparing pastes, suppositories, pastilles, pill-coatings, gelatin sponge, as suspending agent, tablet binder, as stabilizer, for manufacturing rubber substitutes, adhesives, cements, lithographic and printing inks, plastic compounds, artificial silk, photographic plates and films, matches, light filters for mercury lamps, clarifying agent, in sizing paper and textiles, for inhibiting crystallization in bacteriological work for preparing cultures and as a nutrient. It is used to make hard and soft gelatin capsules.

It is also used in the cosmetic industry and in confectionary. There are two grades of gelatin, grade A and grade B. Grade A gelatin has isoelectric point 7–9 and is not compatible with anionic substances like agar, tragacanth and acacia. Grade B gelatin with isoelectric point 4.7–5.0 is compatible with these pharmaceutical aids.

5. CARNAUBA WAX

Biological source   It is an exudate obtained from the leaves of Brazilian wax palm tree *Copernicia cerifera* Mart. and *C. prunifera* of the family Palmae (Arecaceae).

The common names of this product are Palar wax and Brazil wax.

Geographical distribution   It is found in North Brazil to Argentina and South Africa.

Preparation   The leaves are collected from the palm and dried. By brushing and beating, the wax is separated. It is then melted, processed further to purify and poured into the moulds.

Characteristics   It is a hard, greenish, solid and yellow wax with crystalline fracture. It is odourless with a bland taste. It is soluble in fat solvents. It is commercially available in the form of flakes or powder. The melting point of wax is 82–85°C.

Chemical constituents   It contains 80% of esters of hydroxylated fatty acids, i.e., carnaubic and cerotic acids and myricyl cerotate and 10% of monohydric alcohols and traces of resin.

Uses   It is used as an agent for tablet coating, in the preparation of cosmetic products, depilatories and deodorant sticks. It is also used in the manufacture of high-quality shoe polishes and automobile waxes.

6. THEOBROMA OIL

Biological source   It is derived from the seeds of *Theobroma cacao* L. of the family Sterculiaceae.

The common names are cocoa butter and cacao butter.

Geographical distribution   The plant is native to tropical America (Mexico), Ecuador, Curacao, Central America, Brazil, Srilanka, Philippines, Nigeria, Ghana and India.

Collection and processing   The plant is a small tree attaining a height of about 12 m. Each fruit has ten longitudinal furrows and five rows of seeds, 10–12 in each row. The seeds are separated from pods and packed in boxes and allowed to ferment at 42°C for 3 to 6 days. During fermentation, the colour of the seeds are changed from white to dark reddish brown due to enzymatic reaction. After sun-drying the seeds are roasted at 140°C to remove water and to develop a distinct odour and taste. The roasted seeds are then cooled immediately and passed through a "nibbling" machine to crack and remove the seed coats. The kernels are separated from the cracked seed coats and fed to hot rollers to yield oil.

Characteristics   Theobroma oil is a yellowish-white solid that has a pleasant chocolate odour and chocolatelike taste. It melts at 30–35°C. The oil is soluble in boiling absolute alcohol, ether and chloroform, but insoluble in water.

Chemical constituents   The fixed oil consists chiefly of glycerides of oleic (37%), stearic (34%), palmitic acid (26%) and small amount of arachidic acids.

Uses   It is widely used for making suppositories and as a vehicle in the manufacture of some creams and cosmetics and used as a base for ointment. It is also used as a lubricant in massage.

7. HONEY

Source   Honey is a sugary liquid deposited in the honeycomb by the honeybees (hive bees), *Apis mellifera* Linn. and other species of *Apis.* of the family Apidae.

The common names are madhu; me; theynu.

Geographical source   Honey is produced in India (Himachal Pradesh), England, California, Canada, West Indies, Chile and parts of New Zealand, Africa and Australia.

Collection and preparation   The worker bees collect the nectar from flowers of different plants with the help of their proboscis. The nectar mainly contains sucrose. The sucked nectar passes through the oesophagus into the honey-sac. The nectar as it passes through the oesophagus into the honey-sac is mixed with an enzyme invertase present in the salivary secretion of the bees. The enzyme invertase hydrolyses the sucrose of the nectar into invert sugar. Worker bees on reaching the hive, deposit the invert sugar of the honey sac into the cells of the honeycomb. The cells of the honeycomb filled with invert sugar are sealed by wax and the stored invert sugar is called honey.

The honeycomb is smoked to remove bees. After removing the bees, the honeycomb is detached and the sealed wax that covers the hive is cut with a knife and the honey is collected either by expression or by drainage or filtered through a cloth. Before sending to the market, the honey is heated to 80°C, so as to avoid fermentation and to remove the impurities floating on the surface and then cooled rapidly to prevent colour change on storage.

Characteristics   Fresh honey is a thick, clear and translucent, syrupy liquid. The colour is pale yellow to reddish brown and the odour is pleasant with sweet taste. It becomes opaque and granular on storage, due to crystallization of glucose. It is soluble in water and insoluble in alcohol and other organic solvents.

Chemical constituents   Honey consists mainly of glucose (40%), fructose (40–50%) and small quantities of sucrose (0.1–10%), dextrin and formic acid. In addition, it also contains volatile oil, traces of enzymes, vitamins, proteins and amino acids.

Uses   It is used as a nutrient for all age groups. It is an ingredient of linctus and cough mixtures. It is used as a vehicle for the administration of a variety of drugs. It is used as a sweetening agent, as a demulcent, mild laxative and antiseptic. It is also applied to burns and wounds and used in creams, lotions, soft drinks and candies.

Adulterants and test for identity   Honey is adulterated with artificial invert sugar, which contains furfural and gives red colour with resorcinol and HCl acid. Commercial liquid glucose and sucrose are also used as adulterants. Adulteration in honey is determined by the following tests:

1. *Fiehe's test for artificial invert sugar*

    * 5 ml of honey is mixed with 2.5 ml of petroleum or solvent ether for 5–10 min.

    * The ethereal layer formed on the upper side is separated and evaporated using a china dish.

    * On adding 1% resorcinol in HCl (1 ml), a transient red colour is produced by natural honey while in artificial honey, the colour persists for some time.

2. *Limit tests*   The limit tests of sulphate, chloride and ash are compared with the pharmacopoeial specifications.

3. *Reduction of Fehling's solution*   To 1 ml of an aqueous solution of honey, 1 ml of Fehling's solution 1 and 2 are added and the mixture is heated on a water bath for 5–10 min.

❋ The colour of the Fehling's solution is reduced.

❋ On further heating, a brick-red colour is obtained due to the presence of reducing sugars.

REVIEW QUESTIONS

State whether the following sentences are true or false

1. Honey is adulterated with invert sugar.
2. Cream is an example of pharmaceutical aid.
3. *Theobroma* oil is a yellowish white solid with chocolate odour.
4. Ninhydrin test is negative to gelatin.
5. Coccus cacti feeds on *Opuntia cochinellifera*.

Choose the correct answer

6. The technical products are _____
   a. saffron, gum, perfumes, beverage
   b. wax, shampoos, starch, fixed oils
   c. soap, spice, cream, beer

7. Perfumes are obtained from _____
   a. acacia   b. lavender   c. hops

8. Cotton is used in _____
   a. varnish industry   b. brewing industry   c. textile industry

9. Sperm oil is obtained from the _____
   a. head of whale   b. tail of whale   c. abdomen of hog

10. Carnauba wax is derived from the leaves of _____
    a. *Calamus rotang*   b. *Copernicia cerifera*   c. *Borassus flabellifer*

Answers				
1. True	2. False	3. True	4. False	5. True
6. c	7. b	8. c	9. a	10. b

Answer the following

1. What are technical products? Give an example.

2. What are pharmaceutical aids? Cite an example.

3. List out the natural products used as technical products.

4. Critically comment on pharmaceutical aids of plant origin.

5. Write the biological source, geographical distribution, collection and preparation, characteristics and use of the following:

    i. Lard

    ii. Cochineal

    iii. Theobroma oil

6. Write a short note on spermaceti.

7. Give the biological source, preparation, characteristics, chemical constituents and importance of gelatin.

8. Comment on the collection, method of preparation and pharmaceutical use of honey.

9. How will you test the adulterants of honey and gelatin?

GLOSSARY

Aborticide A drug that induces abortion/ premature birth.

Abortifacient A drug that causes pregnancy to end prematurely and causes an abortion.

Abortion The separation and expulsion of the contents of the pregnant uterus before the 28th week of pregnancy or termination of pregnancy before the completion of 7th month of an infant.

Abscess A localized collection of pus in a wound or boil in the body.

Ache A continuous pain felt in the body.

Acidity A sign of indigestion due to digestive disorder.

Acne A kind of inflammatory skin eruption that occurs on the face, back and chest usually during puberty and adolescence.

Acrid A sharp pungent taste.

Addiction Physical and psychological dependence on psychloactive substances like narcotics.

Addison's disease A disease characterized by progressive anaemia, extreme weakness, low blood pressure, dyspepsia and bronze discoloration of the skin.

Adenoids The development of spongy tissue internally between nose and throat.

Administer Dispense or apply a drug.

Aflatoxin A specific mycotoxin secreted by a fungus *Aspergillus* which induces tumour.

Agraphia A disorder of the brain collapsing the power to write.

Ague A type of fever coming in periodical fits, accompanied by shivering, and cold.

Ail Illness or physical feeling of pain in the body.

Albuminuria A condition in which albumin is present in the urine.

Alexia A damage to the brain which causes inability to recognize or read written words or letters.

Alexiteric Resisting poison, or the effects of venom.

Allergen Substance(s), which causes an allergic reaction.

Allergy A special reaction of the body cells to specific substances such as antigens and allergens.

Alleviative Moderating pain by making it easier to bear; mitigative.

Alterative A drug that makes a change in the vital functions and re-establishes function of an organ.

Amenorrhoea The state of abnormal suppression or absence of menstruation in women.

Amnesia A state of loss of memory of a person either due to aging or over-dosage of a particular drug.

Amoeba A very minute microscopic protozoan unicellular organism.

Amylolytic The conversion of starch into sugar by the action of acids or enzymes like amylase.

Anaemia The paleness of the body due to deficiency of red corpuscles or haemoglobin due to various iron and B_{12} deficiencies.

Anaesthesia The induction of the loss of tactile sensibility (numbness) or unconsciousness of the body due to a drug, for example chloroform.

Anaesthesis A state of unconsciousness; the condition of insensibility.

Anaesthetic A drug which produces partial or total loss of the sense of pain, touch, etc.

Analeptic A medicine tending to restore a person's health or strength.

Analgesic A drug that acts to relieve pain without loss of consciousness.

Angina A heart disease signalled by pain in chest, shoulder, etc.

Anodynes Curative measures which soothe pain; a medicine that allays pain.

Anorexia A state of loss of appetite.

Anthelmintic A medicine which is used for destroying or expelling intestinal parasitic worms.

Antiallergic A drug that counters allergy or an allergic state.

Anti-amoebic A drug that helps to cure amoebiasis.

Antiaphrodisiac A medicine which decreases libido and represses sexual desire in humans.

Anti-asthmatic A drug that cures asthmatic complaints.

Antiatherosclerotic A substance which prevents fatty degeneration of the middle coat of the arterial wall.

Antibiotic Any drug that kills pathogenic microbes that cause disease, especially bacteria.

Anticancer drug A drug that prevents the division of tumour cells and cures cancer.

Anticoagulant An agent which prevents blood from clotting.

Antidiabetic A substance that helps a person with diabetes control their level of glucose (sugar) in the blood.

Anti-diarrhoeals A drug that controls looseness of bowels.

Antidote An agent that counteracts the action or effect of poisons.

Antifertility drug A drug that inhibits the formation of sperm or ova, or interferes with the process of fertilization.

Antifibrillic An agent which acts against fibrillation—the rapid contraction of muscles of the heart.

Antigenic Any substance that stimulates the formation of antibodies.

Antigout A drug that reduces the inflammation of the smaller joints caused by gout.

Antihypertensive A drug that reduces high blood pressure.

Anti-infective An agent that counteracts any infection.

Anti-inflammatory A drug that suppresses any inflammation in the body.

Anti-leprotic A substance that cures leprosy.

Anti-malarial Drug that cures malarial fever, e.g., Quinine .

Anti-menorrhegic A drug that reduces excessive menstrual discharge.

Anti-neoplastic A drug that inhibits the development of neoplasms (tumour).

Antiperiodic An agent which prevents periodic recurrence of disease as in malaria.

Antiprotozoal A drug that destroys protozoa or inhibits their growth and ability to reproduce.

Antipruritic Preventing or relieving itching.

Antipyretic Drugs used to reduce body temperature in fever.

Anti-rheumatic A medicine that prevents or relieves the pain of joints caused by rheumatism.

Antiscorbutic A drug that cures scurvy.

Antiseptic Antimicrobial substances that are applied to living tissue/skin to reduce the possibility of infection, sepsis, or putrefaction.

Antisomniac A medicine which is used to treat insomnia or the inability to sleep.

Anti-spasmodic A drug which relieves painful muscular contraction and relieves colic.

Antitussive A drug that relieves or suppresses coughing.

Antiviral A drug that is used to treat viral infection.

Aperient A substance which is employed to evacuate the bowels; a laxative.

Aphthae A boil or small ulcer in the mucous membrane in the mouth.

Appendicitis An inflammation of the vermiform appendix of the ceacum.

Appetite A state of craving for food needed to maintain the body.

Appetizer An agent which stimulates and increases appetite.

Aromatic Any substance having pleasant smell or flavour.

Arrhythmia A condition in which the heart beats with an irregular or abnormal rhythm.

Arteriosclerosis A condition in which fatty material collects along the walls of arteries. This fatty material progressively thickens, hardens (forms calcium deposits), and may eventually block the arteries.

Arthritis Arthritis is inflammation of one or more joints, which results in pain, swelling, stiffness, and limited movement.

Asthma A lung disease characterized by wheezing, difficult breathing and feeling of suffocation.

Astigmatism An optical defect in which vision is blurred due to the inability of the optics of the eye to focus a point object into a sharp focused image on the retina.

Astringent A chemical compound that tends to shrink or constrict body tissues, usually locally after topical medicinal application.

Atherosclerosis A type of arteriosclerosis.

Autophonia The hyperperception of one's own voice and breathing, as a consequence of rapid weight loss.

Bacteriocidal The action of a substance which kills bacteria.

Beriberi A disease causing inflammation of the nerves and heart failure ascribed to a deficiency of vitamin B_1 (Thiamine).

Bile Dark green to yellowish brown fluid produced by the liver of most vertebrates, which aids in digestion.

Biliousness General condition of indigestion, constipation and headache concerned with disorder of the liver or bile, etc. caused by derangement of the bile.

Biopsy An examination of tissue removed from a living body to discover the presence, cause, or extent of a disease.

Blister A vesicle formed on the skin due to friction, burning, etc., and filled with serum.

Blood pressure The pressure caused on the walls of blood vessels due to the flow and pumping of the blood by heart.

Blood purifier Any agent that removes impurities from blood.

Boil A small area of inflammation starting in the roots of hairs and due to microbial infections.

Bronchitis An inflammation of the mucous membrane of the bronchial tubes by viral or bacterial infection.

Bronchodilator A substance that dilates the bronchi and bronchioles, decreasing resistance in the respiratory airway and increasing airflow to the lungs.

Cachexia Weakness and wasting of the body due to severe chronic illness.

Calculus A stonelike concretion which forms in certain parts of the body like bladder, kidney, gall bladder, etc.

Cancer Any malignant growth or tumour formed by uncontrolled proliferation of certain cells in any part of the body.

Carcinogen A substance which induces cancer.

Carcinoma Cancer that begins in the skin or in tissues that line or cover body organs.

Cardiac A term relating to heart.

Cardialgia Burn of the heart.

Cardiotonics Substances which strengthen the heart.

Caries Bacterial infection of the enamel and dentine of the tooth leading to decalcification and cavitation (decay).

Carminative An agent or drug that promotes expulsion of gas from the intestines and stomach.

Cataract A medical condition in which the lens of the eye becomes progressively opaque, resulting in blurred vision.

Catarrh Excessive discharge or build-up of mucus in the nose or throat, associated with inflammation of the mucous membrane.

Cathartic A substance used to evacuate bowels; a powerful purgative, especially a laxative.

Cerate A thick ointment composed of oils mixed with wax, resin, etc.

Cerebral haemorrhage Escape of blood from a ruptured blood vessel that occurs within brain tissue.

Cerebral thrombosis A stroke resulting from a blood clot (thrombus) formed in an artery that supplies blood to the brain.

Chancroid A sexually transmitted infection characterized by painful sores on the genitalia.

Cholagogues An agent that stimulates gallbladder contraction to promote bile flow.

Cholecystectomy The removal of gall bladder by surgery.

Cholera An acute infectious disease, caused by *Vibrio cholerae* resulting in diarrhoea and vomiting and suppression of urine.

Choleretic Substances that increase the volume of secretion of bile from the liver as well as the amount of solids secreted.

Cholinergic Any drug that functions to enhance the effects mediated by acetylcholine in the central nervous system, the peripheral nervous system, or both.

Cinchonism A disturbed condition of the body characterized by dizziness, ear ringing, temporary deafness and headache due to overdose of quinine.

CNS depressant A drug that suppresses the activity of the central nervous system.

CNS stimulant A drug which activates the central nervous system.

Colic Severe pain in the abdomen caused by wind or obstruction in the intestines and suffered especially by babies.

Colitis An inflammation of the colon.

Colostomy An incision (cut) into the colon (large intestine) to create an artificial opening or "stoma" to the exterior of the abdomen.

Coma A state of unconsciousness lasting more than six hours, in which a person cannot be awakened, fails to respond normally to various stimuli, lacks normal sleep–wake cycle and does not initiate voluntary action.

Constipation A condition in which there is seldom or incomplete, irregular or delayed evacuation of the bowels.

Contagious A disease spread from one person or organism to another by direct or indirect contact.

Convalescent Recovering from an illness or medical treatment.

Convulsions A violent irregular motion of the limb or body due to involuntary contraction of muscles.

Cyanosis A condition when the skin becomes blue due to insufficient oxygen in the blood.

Cyst A closed sac having a distinct membrane, that contains a fluid or semifluid.

Cystitis An inflammation of the urinary bladder.

Cytotoxic Substance toxic to cells.

Dandruff Small pieces of dead skin in a person's hair.

Debility A state of physical weakness, or poor health.

Dehydration The process of losing a large amount of water from the body.

Deliriant A mental disturbance marked by illusions, hallucinations, etc.

Delirium A state of perverted consciousness in which an irregular discharge of nervous energy goes on, causing incoherent talk, delusion and ill-regulated muscular action.

Demulcent A substance having a soothing action on the skin and mucous membranes.

Dengue An infectious fever with acute pain in the joints.

Dental A term relating to teeth.

Dentifrice A toothpowder, paste, or liquid used to help maintain oral hygiene.

Dentistry A branch of medicine involved in the study, diagnosis, prevention and treatment of diseases of the oral cavity.

Denture False or artificial teeth.

Deobstruent A substance that clears the natural ducts of the fluids of the body.

Depressant A drug that reduces functional or nervous activity.

Dermatitis A condition of the skin in which it becomes red, swollen and sore, sometimes with small blisters, resulting from direct irritation of the skin by an external agent or an allergic reaction to it.

Diabetes insipidus A condition characterized by excessive thirst and excretion of large amounts of severely diluted urine, with reduction of fluid intake having no effect on the latter.

Diabetes mellitus A group of metabolic diseases in which a person has high blood sugar, either because the body does not produce enough insulin, or because cells do not respond to the insulin that is produced. This high blood sugar produces the classical symptoms of polyuria (frequent urination), polydipsia (increased thirst) and polyphagia (increased hunger).

Dialysis A clinical process primarily used to provide an artificial replacement for lost kidney function in people with renal failure.

Diaphoretic The state of perspiring profusely, or something that has the power to cause increased perspiration.

Diarrhoea The passing of increased amounts (more than 300 g in 24 hours) of loose stools often caused by a virus or bacteria and can be acute (short term) or chronic (long term).

Diphtheria An upper respiratory tract illness caused by *Corynebacterium diphtheriae*, a facultative anaerobic gram-positive bacterium. It is characterized by sore throat, low fever, and an adherent membrane (a pseudomembrane) on the tonsils, pharynx, and/or nasal cavity.

Disinfectant Substances that are applied to non-living objects to destroy microorganisms that are living on the objects.

Disinfection The process of killing pathogenic organisms or rendering them inert.

Disinfestation Destruction of insects, rodents, or other animal forms present on the person or their clothes or in their surroundings, and which may transmit disease.

Diuretic A drug which increases the secretion and copious excretion of urine by the kidneys.

Dizziness An impairment in spatial perception and stability.

Dropsy An old term for the swelling of soft tissues due to the accumulation of excess water.

Dysentery An inflammatory disorder of the intestine, especially of the colon, that results in severe diarrhoea containing mucus and/or blood in the faeces with fever and abdominal pain.

Dysmenorrhoea A gynaecological medical condition of pain during menstruation that interferes with daily activities.

Dyspepsia Indigestion; difficult to digest.

Dysuria Painful or difficult urination.

Eclampsia A condition in which one or more convulsions occur in a pregnant woman suffering from high blood pressure, often followed by coma and posing a threat to the life of the mother and baby.

Eczema A medical condition in which patches of skin become rough and inflamed with blisters which cause itching and bleeding.

Emaciation Abnormally thin or weak, especially because of illness or lack of food.

Embalming A method of preserving a dead body to prevent decay.

Embolism Obstruction of an artery typically by clot of blood or an air bubble.

Emesis The action or process of vomiting.

Emetic An agent that induces vomiting.

Emmenagogues A substance that stimulates or increases menstrual flow.

Emollient Having the quality of softening or soothening the skin.

Emulsify To form emulsion.

Emulsion A fine dispersion of minute droplets of one liquid in another in which it is not soluble or miscible.

Encapsulation Enclose something in or as if in a capsule.

Encephalitis Inflammation of the brain, caused by infection or an allergic reaction.

Endoscopy A diagnostic method in which an instrument called endoscope which can be introduced into the body to give a view of its internal parts.

Enema A procedure in which liquid or gas is injected into the bowel or rectum to expel its contents or to introduce drugs or permit X-ray imaging.

Enteric fever Another term for typhoid.

Epilepsy A neurological disorder marked by sudden recurrent episodes of sensory disturbance, loss of consciousness or convulsions associated with abnormal electrical activity in the brain.

Episiotomy A surgical cut made at the opening of the vagina during childbirth to aid a difficult delivery and prevent rupture of tissues.

Epistaxis A nasal haemorrhage.

Erysipelas Dermatitis of the hands due to bacterial infection, occurring mainly on handlers of meat and fish products.

Erythema Superficial reddening of the skin, usually in patches, as a result of injury or irritation, causing dilation of the blood capillaries.

Erythrocytes Red blood corpuscles.

Expectorant A medicine which promotes the secretion of sputum by the air passages, used to treat coughs.

Febrifuge Medicine used to reduce fever.

Febrile Having or showing the symptoms of a fever.

Fibrillation Make a quivering movement due to uncoordinated contraction of the individual fibrils.

Fibrositis Inflammation of the fibrous connective tissue, typically affecting the back and causing stiffness and pain.

Fit A sudden attack of convulsions and/or loss of consciousness typical of epilepsy and some other medical condition.

Flatulence Accumulation of gas in the alimentary canal.

Flu Influenza.

Flutter Disturbance of the rhythm of the heart that is less severe than fibrillation.

Furunculosis Simultaneous or repeated occurrence of boils on the skin.

Galactagogue Any substance which is used to stimulate lactation in humans and other mammals.

Gastroenteritis Inflammation of the stomach and intestines, typically resulting from bacterial toxins or viral infection and causing vomiting and diarrhoea.

Germicide A substance or another agent which destroys harmful microorganisms.

Glaucoma A condition of increased pressure within the eyeball, causing gradual loss of sight.

Glycaemia Presence of glucose in the blood.

Glycosuria A condition characterized by an excess of sugar in the urine.

Goitre A swelling of the neck resulting from enlargement of the thyroid gland.

Gonorrhoea A venereal disease in which there is inflammatory discharge of mucous from the urethra or vagina due to infection.

Gout A disease in which defective metabolism of uric acid causes arthritis, especially in the smaller bones of the feet, deposition of chalk-stones and episodes of acute pain.

Haematemesis Vomiting blood.

Haematoma A solid swelling of clotted blood within the tissues.

Haematuria The presence of blood in urine.

Haemolysis The rupture or destruction of red blood cells.

Haemophilia A medical condition in which the ability of the blood to clot is severely reduced, causing the sufferer to bleed severely from even a slight injury.

Haemorrhage An escape of blood from a ruptured blood vessel.

Haemorrhoids A swollen vein or group of veins in the region of the anus. This is also called "piles".

Haemostatics Stopping of a flow of blood.

Hallucination An experience involving the apparent perception of something not present

Hay fever An allergy caused by pollen or dust in which the mucous membranes of the eyes and nose are inflamed, causing running at the nose and watery eyes.

Hemicrania Severe headache confined to one side of the head.

Hemiplegia Paralysis of one side of the body.

Hepatitis A disease characterized by inflammation of the liver.

Hernia A condition in which part of an organ is displaced and protrudes through the wall of the cavity containing it.

Hidrosis The action of sweating.

Hoarseness A harsh, rough or grating voice, typically as a result of sore throat or of shouting.

Hodgkin's disease A malignant, though often curable, disease of the lymphatic tissues, typically causing painless enlargement of the lymph nodes, liver and spleen.

Hydrogogue A drug that induces evacuation of a watery stool.

Hydropathy The treatment of illness through the use of water, either internally or through external means such as steam baths.

Hyperhydrosis A medical condition in which a person sweats excessively and unpredictably.

Hypertension Abnormally high blood pressure.

Hypnotic A sleep-inducing drug.

Hypotension Abnormally low blood pressure.

Hysterectomy A surgical operation to remove all or part of the womb.

Hysteria Exaggerated or uncontrollable emotion or excitement.

Icterus A technical term for jaundice.

Immunity The ability of an organism to resist a particular infection or toxin by the action of specific antibodies or sensitized white blood cells.

Immunization Make immune to infection, typically by inoculation.

Indolent Causing little or no pain.

Infection The process of infecting or the state of being infected.

Inflammation The localized physical condition in which part of the body becomes reddened, swollen, hot and often painful, especially as a reaction to injury or infection.

Influenza A highly contagious viral infection of the respiratory passages causing fever, severe aching, and catarrh, and often occurring in epidemics.

Infusion The slow injection of a substance into a vein or tissue.

Insanity The state of being seriously mentally ill.

Insomnia Habitual sleeplessness; inability to sleep.

Intoxication Cause to lose control of their faculties or behaviour.

Itching An uncomfortable sensation on the skin that causes the desire to scratch.

Labour (Parturition) The process of childbirth from the start of uterine contraction to delivery.

Lactation The secretion of milk by the mammary glands.

Laparotomy A surgical incision into the abdominal cavity, for diagnosis or in preparation for a major sugery.

Laryngitis Inflammation of the larynx, typically resulting in huskiness or loss of the voice, fast breathing and painful cough.

Laxative A drug or medicine tending to stimulate or facilitate evacuation of the bowels.

Leprosy Contagious disease that affects the skin, mucous membranes and nerves, causing discoloration and lumps on the skin and, in severe cases, disfigurement and deformities.

Leucoderma A condition in which pigment is lost from areas of the skin, causing whitish patches often with no clear cause; also termed vitiligo.

Leucorrhoea A thick white or yellowish mucous discharge from the vagina.

Leukaemia A malignant progressive disease in which the bone marrow and other blood-forming organs produce increased numbers of immature or abnormal leucocytes; blood cancer.

Ligature A cord or thread used in surgery especially to tie up a bleeding artery.

Linctus Thick liquid medicine, especially cough mixture.

Liniment An embrocation for rubbing on the body to relieve pain, especially one made with oil.

Lotion A thick, smooth liquid preparation designed to be applied to the skin for medicinal or cosmetic purposes.

Lozenges A small medicinal tablet, originally in the shape of a rhombus or diamond, taken for sore throats and dissolved in the mouth.

Lumbago Pain in the muscles and joints of the lower back.

Malaria An intermittent and remittent fever caused by a protozoan parasite which invades the red blood cells and is transmitted by mosquitoes.

Mania Mental illness marked by periods of great excitement or euphoria, delusions, and overactivity.

Measles An infectious viral disease causing fever and a red rash, typically occurring in childhood.

Medicament A substance used for medical treatment.

Melancholia A feeling of deep sadness.

Meningitis A serious disease in which there is inflammation of the meninges, caused by viral or bacterial infection, and marked by intense headache and fever, sensitivity to light and muscular rigidity.

Menopause The ceasing of menstruation.

Menorrhagia Abnormally heavy bleeding at menstruation.

Micturition Urinate.

Migraine A recurrent throbbing headache that typically affects one side of the head and is often accompanied by nausea and disturbed vision.

Miotic Excessive constriction of the pupil of the eye.

Mucosa A mucous membrane of any part of the body.

Mucous A slimy secretion from the mucous membranes.

Mumps A contagious and infectious viral disease causing swelling of the parotid salivary glands in the face and a risk of sterility in adult males.

Mutilate Inflict violent and disfiguring injury on.

Mydriasis Dilation of the pupil of the eye.

Myope A person with short-sight.

Narcotic An addictive drug which induces drowsiness, stupor, or insensibility, and relieves pain.

Nausea A feeling of sickness with an inclination to vomit.

Nephritis Inflammation of the kidney.

Neuralgia Intense, typically intermittent pain along the course of a nerve, especially in the head or face.

Neuritis Inflammation of a peripheral nerve usually causing pain and loss of function.

Neurosis A relatively mild mental illness that is not caused by organic disease, involving symptoms of stress but not a radical loss of touch with reality.

Noctambulist A person who walks in his/her sleep.

Nocturia The need to wake up and pass urine at night.

Obese The state of being overweight.

Oedema A condition characterized by an excess of watery fluid collecting in the cavities or tissues of the body.

Oestrogen Any of a group of steroid hormones which promote the development and maintenance of female characteristics of the body.

Ointments A smooth oily substance that is rubbed on the skin for medicinal puposes or as a cosmetic.

Ophthalmia Inflammation of the eye.

Oxytocic A drug which is used for hastening childbirth, especially by stimulating contraction of the uterus.

Oxytocin A hormone released by the pituitary gland that causes increased contraction of the womb during labour and stimulates the ejection of milk into the ducts of the breast.

Ozaena A chronic disease of the nose characterized by a foul-smelling nasal discharge and atrophy of nasal structures.

Palpitation A noticeably rapid, strong, or irregular heartbeat due to agitation, exertion, or illness.

Pancreatitis Inflammation of pancreas due to infection.

Paralysis The loss of the ability to move or feel anything in part or most of the body, typically as a result of illness, poison, or injury.

Parkinson's disease A progressive disease of the nervous system marked by tremor, muscular rigidity, and slow, imprecise movement, chiefly affecting middle-aged and elderly people.

Pathogenic The state of producing disease by organisms.

Pediculicide A chemical used to kill lice.

Perspiration The process of sweating.

Pertussis Whooping cough.

Pharmacy The science or practice of the preparation and dispensing of medicinal drugs.

Pimple A small, hard, inflamed spot on the skin.

Pneumonia Inflammation of the lungs caused by bacterial or viral infection in which air sacs fill with pus and may become solid.

Polio A viral disease caused by polio myelitis virus which affects the CNS and can cause temporary or permanent paralysis.

Poultice A soft composition of medicinal preparation of flour, bran, etc. applied to inflamed parts or sores.

Prophylactic A medicine used for preventing or protecting a person from disease.

Psoriasis A chronic inflammatory skin disease marked by reddish, itchy, scaly patches.

Psychopathic Mental illness or disorder.

Psychosis A severe mental disorder in which thought and emotions are so impaired that contact is lost with the external reality.

Pulse A rhythmical throbbing of the arteries as blood is propelled through them typically as felt in the wrist or neck.

Pungent A warm biting sensation or a sharp taste.

Purgative A drug that stimulates peristaltic action and bowel evacuation.

Pyaemia Blood-poisoning caused by the spread in the bloodstream of pus-forming bacteria released from an abscess.

Pyorrhoea A disease marked by purulent discharge from the gums of teeth.

Pyrogen A toxin that causes fever.

Quinine A bitter drug obtained from the bark of the plant cinchona used to cure malarial fever.

Rennin A substance prepared from the stomach of the calf to digest milk.

Repellent Any substance having the effect of driving back.

Rheum A watery discharge from the nose or eyes.

Rheumatism A disease affecting joints and muscles making them stiff, swollen and painful.

Rhinitis The inflammation of the mucous membrane of the nose.

Rickets A vitamin D deficiency disease of the young, affecting the development of bone.

Rickettsiae A group of microorganisms which are intermediate between bacteria and viruses and cause disease.

Rubefacient A substance for external application that produces redness of the skin, e.g., by causing dilation of the capillaries and an increase in blood circulation, and is believed to relieve pain

Sanity The healthy mental state of a person.

Scabies A contagious itching skin disease marked by small raised red spots caused by the itch mite, sarcople scabies.

Scarlatina It is a contagious streptococcal disease characterized by fever, inflammation of throat and scarlet rash.

Sciatica A pain affecting the back, hip, and outer side of the leg caused by compression of the spinal nerve root in the lower back often owing to degeneration of an intervertebral disc.

Scurvy A nutritional deficiency disease due to lack of vitamin C in the diet leading to swollen gums, livid spots, etc.

Sedative A drug taken for its calming or sleep-inducing effect.

Sepsis The presence in tissues of harmful bacteria and their toxins, typically through infection of a wound.

Septic Infected with bacteria.

Serosal The tissue of a serous membrane.

Smallpox An acute contagious viral disease with fever and pustules that usually leave permanent scars.

Somnambulism A state of walking during sleeping.

Somniferous Tending to induce sleep.

Somnolence Sleepiness, drowsiness.

Soporific A drug that induces sleep.

Sores A general term used for variety of morbidity tender tissues, boils, ulcers and wounds.

Spasm A sudden involuntary muscular contraction or convulsive movement.

Spasmolytic A drug that prevents spasms.

Sprain A swelling (but not dislocation) of an ankle, wrist or other joints caused by a wrench or twist in the ligaments.

Stertorous Noisy and laboured breathing.

Stimulant A substance that raises levels of physiological or nervous activity in the body.

Stomachic Promoting the appetite or assissting digestion.

Stupor A state of near inconsciousness or insensibility.

Sudorific A drug that induces sweating.

Suppository A solid medical preparation in a roughly conical or cylindrical shape, designed to be inserted into the rectum, to be dissolved.

Suppressant A drug or other substance which acts to suppress or restrain something.

Suppurate Pus formation.

Suture A stitch or row of stitches holding together the edges of a wound or surgical incision.

Sympathomimetic A drug producing physiological effects characteristic of the sympathetic nervous system by promoting the stimulation of sympathetic nerves.

Syndrome A group of symptoms which consistently occur together, or a condition characterized by a set of associated symptoms.

Syphilis An infectious venereal disease caused by *Treponema pallidium*.

Tetanus A bacterial disease marked by rigidity and spasms of the voluntary muscles.

Thrombosis Local coagulation or clotting of the blood in a part of the circulatory system.

Tinnitus Ringing or buzzing in the ears.

Tonic A medical substance taken to give a feeling of vigour or well-being.

Tonsilectomy A surgical operation to remove the tonsils.

Tonsillitis Inflammation of the tonsils.

Tranquilizer A medicinal drug taken to reduce tension or anxiety.

Trichinosis A disease caused by trichinae, typically from infected meat, characterized by digestive disturbance, fever and muscular rigidity.

Tuberculosis An infactious bacterial disease characterized by the growth of nodules in the tissues especially the lungs.

Tumour A swelling of a part of the body, generally without inflammation, caused by an abnormal growth of tissue, whether benign or malignant.

Typhoid An infectious bacterial fever with an erection of red spots on the chest and abdomen and severe intestinal irritation.

Typhus fever An infectious fever caused by rickettsiae characterized by a purple rash, headache and fever.

Ulcer An open sore on an internal or external surface of the body, caused by a break in the skin or mucous membrane, which fails to heal.

Uraemia A raised level in the blood of urea and other nitrogenous waste compounds that are normally eliminated by the kidneys.

Vaccine An antigenic substance prepared from tha causative agent of a disease or a synthetic substitute used to provide immunity against one or several diseases.

Varicose Referes to a vein especially in the leg,

swollen, twisted and lengthened as a result of poor circulation.

Venereal disease Diseases that are sexually transmitted though sexual intercourse, e.g., syphilis, gonorrhoea, chancroids, etc.

Vermicide A substance that is poisonous to worms.

Vermifuge A drug that when ingested expels intestinal worms.

Vertigo A condition in which the affected person loses balance and feels that the surrounding objects are whirling.

Vesicants Agents that cause tissue blistering.

Vomiting The act of the stomach contents through the mouth.

Vulnerary Drug useful in healing or curing wounds.

Warts A dry hard growth on the skin.

REFERENCES

Atal, C.K. and Kapur, B.M. (1982). Cultivation and Utilization of Medicinal Plants, (vol.1) and Aromatic Plants (vol.11). Jammu-Tawi; R.R.L., C.S.I.R.

Balandrin, M.F. and Klocke, J.A. (1988). "Medicinal, aromatic and industrial materials from plants." In: *Biotechnology in Agriculture and Forestry*. Bajaj, Y.P.S. (Ed.). Springer-Verlag, Berlin, Heidelberg.

Bannerman, R.H. (1982). "Traditional medicine in modern health care." *World Health Forum*. 3(1):8–13.

Bhattacharjee, S.K. and De, L.C. (2005). *Medicinal Herbs and Flowers*. Aavishkar Publishers, Distributors. Jaipur. pp. 434.

Bhattacharjee, S.K. (2004). *Handbook of Medicinal Plants*. Pointer Publishers, Jaipur.

Bianchini, F., Corbetta, F., Dejey, M.A. and Pistoia, M. (1977). *Health Plants of the World*. Newsweek Books, New York.

Blasko, G. and Cordell, G.A. (1990). In: *The Alkaloids*. Vol. **37**. Brossi, A. (Ed.). Academic Press Line, New York. pp.1–76.

Bluden, G., Culling, C. and Jewers, K. (1975). "Steroidal sapogenins: A review of actual and potential plant resources." *Trop. Sci.***17**: 139–154.

Chopra, R.N., Chopra, I.C., Handa, K.L. and Kapur, L.D. (1958). *Chopras' Indigenous Drugs of India*. U.N. Dhur and Sons Pvt. Ltd., Kolkata.

Chopra, R.N. and Chopra, I.C. (1955). *A Review of Work on Indian Medicinal Plants*. Cambridge Printing Works, New Delhi.

Chopra, R.N., Nayar, S.L. and Chopra. I.C. (1956). *Glossary of Indian Medicinal Plants*. CSIR, New Delhi.

Cutter, E.G. (1978). *Plant Anatomy: Part 1: Cells and Tissues*. Edward Arnold, London.

Devassay, A. and Nair, E.V.G. (1983). "Studies on medicinal plants." *Sachitra Ayurved*. **35**:601–604.

Esau, K. (1960). *Anatomy of Seed Plants*. John Wiley and Sons, New York.

Fahn, A. (1967). *Plant Anatomy*. Pergamon Press, Oxford, London, New York, Paris.

Fahn, A. (1981). *Plant Anatomy*, 3rd edn. Pergamon Press, Oxford.

Gamble, J.S. (1957). *Flora of the Presidency of Madras*. Vol.1–3. Botanical Survey of India, Calcutta.

Gershenzon, J. and Mabry, T.J. (1983). "Secondary metabolites and the higher classification of angiosperms." *Nordic J. Bot.* **3**: 5–34.

Gibbs, R. Darneley. (1974). *Chemotaxonomy of Flowering Plants*, **4** Vols. McGill Queen's University Press, Montreal.

Grewal, R.C. (2000). *Medicinal Plants*. Campus Books International, New Delhi. pp. 430.

Guenther, E. (1949–1952). *The Essential Oils*. **1-4**, D. Van Nostrand Co., Inc., New York.

Gupta, R. (1977). "Periwinkle produces anticancer drug." *Indian Farming*. **27**(4): 11–13.

Gurcharan, S. *Plant Systematics. Theory and Practice*. Oxford and IBH Publishing Co.Pvt. Ltd. New Delhi. pp. 359.

Hadley, S.K. and Petry, J.J. (1999). "Medicinal herbs: A primer for primary care." *Hospital Practice*. **34**(6):105–123.

Handa, S.S. and Kapoor, V.K. (1998). *Pharmacognosy*. Vallabh Prakashan, Delhi.

Harborne, J.B. (1964). *Biochemistry of Phenolic Compounds*. Academic Press, New York.

Harborne, J.B. (1973). *Phytochemical Methods*. Chapman, and Hall, International Edition, London.

Henry, T.A. (1960). *The Plant Alkaloids*, 5th edn. McGraw-Hill, New York.

Heywood, V.H. (1976). Plant Taxonomy, 2nd edn. Edward Arnold, London.

Jain, S.K. (1994). *Medicinal Plants*, 5th edn. NBT, New Delhi.

Kapoor, L.D. (1993). "Ayurvedic medicine of India." *J. Herbs-Spices and Medicinal Plants*. 1(4):37–219.

Kay, A.L. (1938). *Microscopical Studies of Drugs*. Bailliere Tindall and Cox., London. pp. 18.

Khanna, N.M. (1999). "Turmeric-national precious gift." *Curr. Sci*. 76:1351.

Kirtikar, K.R. and Basu, B.D. (1935). *Indian Medicinal Plants*. 2nd edn. Lalit Mohan Basu Publications, Allahabad.

Lawrence, G.H.M. (1951). *Taxonomy of Vascular Plants*. Macmillan, New York. pp. 823.

Matthew, K.M. (1983). The Flora of the Tamil Nadu Carnatic. The Rapinat Herbarium, St. Joseph's College, Tiruchirapalli, India, **3** vols.

Metcalfe, C.R. and Chalk, L. (1983). *Anatomy of Dicotyledons*, 2nd edn. Vol. 2. Clarendon Press, Oxford.

Metcalfe, C.R. and Chalk, L. (1950). *Anatomy of the Dicotyledons*, Vol 2. Clarendon Press, Oxford.

Mohammed, A. (1994). *Text Book of Pharmacognosy*. CBS Publishers and Distributors, Delhi.

Morice, J. and Louran, J. (1971). "Study of morphine content in the oil poppy (*Papaver somniferum* L.)." *Annals of Amelior Plants*. 21(4): 465–485.

Naik, V.N. (1984). *Taxonomy of Angiosperms*. Tata McGraw-Hill, New Delhi.

Pagare, P.K. (2007). *Medicinal Plants*. APH Publishing Corporation, New Delhi. pp. 250.

Panda, H. (2005). *Medicinal Plants Cultivation and their Uses*. Asia Pacific Business Press Inc., Delhi.

Pandey, S.N. and Misra, S.P. (2008). *Taxonomy of Angiosperms*. Ane Books India, New Delhi. pp. 620.

Peach, K. and Tracey, M.V. *Modern Methods of Plant Analysis*, Vol. 1–4. Narosa Publishing House, New Delhi.

Prajapati N.D. and Kumar, U. (2005). *Agro's Dictionary of Medicinal Plants*. Agro Bios, India. pp. 398.

Rao, Rama and Gurjar, M.K. (1990). "Drugs from plant resources. An overview." *Pharma Times*. 22(5):19–27.

Raveendra Retnam, K. and Martin, P. (2006). *Ethnomedicinal Plants*. Agrobios, Jodhpur. pp. 285.

Roy, P. (2006). *Plant Anatomy*. New Central Book Agency. Kolkata.

Sambamurty, A.V.S.S. (2005). *Taxonomy of Angiosperms*. I.K. International Pvt. Ltd., New Delhi. pp. 892.

Santapau, H. (1966). *Common Trees*. National Book Trust, New Delhi.

Schellard, E.J. *Exercises in the Evaluation of Drugs and Surgical Dressings*. Pitman Medical Publishing Co. Ltd., London.

Scheuer, P.J. (1973–81). *Marine Natural Products*, 1–2. Academic Press, London.

Shukla, P. and Misra, S.P. (1979). *An Introduction to Taxonomy of Angiosperms*. Vikas Publishing House, New Delhi. pp. 546.

Singh, B. and Pruthi, T.D. (1997). "Biodiversity and herbal medicine." Proc. 6th National Seminar on Biological Diversity and Human Welfare held at Rishikesh, 28–30 September.

Sivarajan, V.V. and Balachandran, I. (1994). *Ayurvedic Drugs and their Plant Sources*. Oxford and IBH Publishing Co. Pvt. Ltd., New Delhi. pp. 570.

Stumpf, P.K. and Conn, E.E. (1980–81). *The Biochemistry of Plants: A Comprehensive Treatise*, 1–8. Academic Press, London.

Swain, T. (1966). *Comparative Phytochemistry*. Academic Press, London.

The Wealth of India, Raw Materials (All volumes), Council of Scientific and Industrial Research, New Delhi.

Trease, G.E. and Evans, W.C. (1983). *Pharmacognosy*, 12th edn. Bailliere Tindall, Eastbourne, U.K.

Trivedi, P.C. (2004). *Medicinal Plants, Utilization and Conservation*. Aavishkar Publishers, Distributors, Jaipur. pp. 431.

Tyler, V.C., Brady, L.R. and Robbers, J.E. (1981). *Pharmacognosy*, 8th edn. Lea and Febiger, Philadelphia.

Varier, P.S.V. (1995). *Indian Medicinal Plants-A Compendium of 500 Species*, Vols. I and II and Vol. III, Orient Longman Ltd., New Delhi.

Wallis, T.E. (1967). *Text Book of Pharmacognosy*, 5th edn. J&A Churchill Ltd., London.

Weiss, T.J. (1983). Food Oils and their Uses, 2nd edn. Ellis Horwood Ltd., England.

Willis J.C. (1973). *A Dictionary of Flowering Plants and Ferns*, 8th edn. Cambridge University Press, London. pp. 1245.

Yasue, T. (1969). "Histochemical identification of calcium oxalate." *Acta. Histochem. Cytochem.* **2**: 83–95.

INDEX

www.ingramcontent.com/pod-product-compliance
Lightning Source LLC
Chambersburg PA
CBHW082128210326
41599CB00031B/5904